Protecting and Exploiting New Technology and Designs

KEITH HODKINSON
Lecturer, Faculty of Law
University of Manchester
Consultant, Marks and Clerk,
Chartered Patent and Trade Mark Agents

London
E. & F.N. SPON
New York

First published in 1987 by
E. & F.N. Spon Ltd
11 New Fetter Lane, London EC4P 4EE
Published in the USA by
E. & F.N. Spon
29 West 35th Street, New York 10001

Printed in Great Britain at the
University Press, Cambridge

ISBN 0 419 13810 2 (Hardback)
ISBN 0 419 13820 X (Paperback)

British Library Cataloguing in Publication Data

Hodkinson, Keith
Protecting and exploiting new technology and designs.
1. Patent laws and legislation — Great Britain
I. Title
344.1064′86 KD1369

ISBN 0-419-13810-2 ISBN 0-419-13820-X Pbk

Library of Congress Cataloging in Publication Data

Hodkinson
Protecting and exploiting new technology and designs.
Bibliography: p.
Includes index.
1. Patent laws and legislation—Great Britain.
I. Title.
KD1369.H63 1987 346.4104′86 86-31321
ISBN 0-419-13810-2 344.106486
ISBN 0-419-13820-X (pbk.)

Contents

Preface

The objective of this book is to provide a general guide to the protection and the exploitation of new technology and design, written from the point of view of the lawyer explaining to the non-lawyer (or the non-specialist lawyer) the essentials of the law and its many implications for business practice in this field.

It is hoped that the work will be useful as a reference book and an occasional 'dip-in' both for the company secretary, or the director of R and D or the licence negotiator, as well as for the non-specialist lawyer wishing to become acquainted with this field for the first time. For that reason the use of detailed legal references, case citations, etc. which are customary in a book directed at a specialist legal audience have been omitted, but the very complexity of the law and the practice in this area requires that some subjects be dealt with in a degree of detail to avoid the severe risks which accompany over-simplification. Unless otherwise stated the law explained is the law of England and Wales.

Inevitably some things have been left out. It is hoped that the selection and allocation of space to material is all in all balanced and useful. As with all law-related books material becomes out of date rapidly and the very recent Government White Paper Cmnd 9712, which is reviewed in Chapter 16, may accelerate the pace of the changes to come, but the essentials will largely remain the same and it is on the essentials that the book has concentrated. Stress has been placed on those areas which experience has shown to be most confusing to the non-lawyer.

Although they are in no way responsible for the content of this book I would like to thank my colleagues at the University of Manchester and at Marks and Clerk, patent and trade mark agents, Manchester, for their helpful comments and answers to queries, as well as to the others in legal practice and in R and D work in the North-West who have knowingly or otherwise contributed to my views on protection and exploitation of new technology and design. Responsibility for all errors remains mine.

Finally, E. & F.N. Spon Ltd, in bringing this book into being, have been patient both with my deadlines and with my typing.

KEITH HODKINSON
Manchester
1 July 1986

1

Introduction

1.1 THE OBJECTIVES OF THIS BOOK

This book is concerned mainly with the protection and exploitation of new technology and designs through the use of intellectual property rights: in other words, through:

(1) Patents
(2) Registered designs
(3) Plant variety rights
(4) Copyright
(5) Trade marks
(6) Know-how and trade secrets

The term intellectual property rights is a general one which covers all of these separate legal rights. Sometimes people will use the term industrial property rights: it means basically the same thing, but it has a rather narrower import, since it excludes those aspects of copyright dealing with artistic and literary works which are generally not thought of as industrial. This book will use the wider term to cover the rights listed above.

The book discusses which types of new technology and design you can protect and how you can protect them using the intellectual property rights both here in the United Kingdom and overseas. It discusses some of the options that are available in exploiting a new piece of technology or design and pitfalls to avoid in managing intellectual property rights. The book also points to the use of intellectual property data as a source of new technical, design and other business information.

The task of the book is to raise your awareness of intellectual property as a means both of protecting and exploiting technology and design; and to give practical advice on the questions which you should be asking your professional advisers, on the information you should be providing them with and on the steps you should take to acquire and exploit an intellectual property right.

The purpose of the book is not to turn you overnight into a do-it-yourself patent agent, lawyer or general business adviser. It could never do that because, in practice, there are no ready-made answers to most of the problems to be discussed.

Instead, it has been written from the point of view of a lawyer or a patent agent explaining to the non-lawyer client (who might be a research worker, or a commercial manager, or even perhaps a personnel manager dealing with employment problems) some of the legal and commercial implications of new technology and designs and how the law can help or hinder the acquisition and exploitation of rights in them according to the various decisions that person has to make. It concentrates on options, when to act, what questions to ask and of whom to ask them.

Above all, the book has a simple message. It is that being aware of intellectual property always means very careful planning ahead and organizing your R and D procedures so as to make the acquisition of intellectual property rights easier. It means having to educate even relatively junior staff about the basic precautions to be taken at a practical level internally to safeguard a company's position. It means ensuring that the sales staff and negotiators dealing with others outside the company are aware of the importance of confidentiality, of the potential of licensing deals and of the need for care in agreeing any licences. It means a constant flow of information between R and D, finance, sales and legal sections or outside advisers: in essence, it means good communication. Therefore the book is concerned with the practical implications of intellectual property for the internal management of a firm and the necessity for a flow of information to maximize its potential for the protection and exploitation of technology and design.

1.2 WHAT IS INTELLECTUAL PROPERTY?

People often only identify the term intellectual property with patents. But as we have seen above, the term covers a very wide range of rights of relevance to industry. For present purposes these rights can be very briefly summarized to give an idea of their scope (see Table 1). They will be looked at in more detail in later chapters.

1.2.1 Patents

A patent is a legal monopoly which is granted for a limited time to the owner of a new invention which is capable of industrial application: in essence, a patent is concerned with new technology, in the form of novel machines, processes and substances. In most of Europe, the duration of a patient is now twenty years, but it differs in other countries, so that – for instance – it lasts only seventeen years in the USA and in Canada.

The price that an owner of an invention has to pay for the grant of a patent monopoly is to register the invention, to prove it to be novel and inventive and of a type which is patentable in law and to disclose it in an official journal in sufficient detail to enable any suitably skilled person in the same technical field to reproduce it. If it is properly drawn up the patent granted to the owner can in some cases protect even an underlying inventive concept which forms the basis of the patented invention itself and not just the particular example of that concept which the inventor came upon.

The owner of a validly granted patent may in law prevent others from exploiting or using the patented invention without his or her consent for the duration of the monopoly and seek damages if they are proved to have done so. But merely to have a

Table 1 *Summary of intellectual property in the UK*

NAME OF RIGHT	REQUIREMENTS	DURATION	PROTECTION
Patent	Inventions which are novel, inventive and capable of industrial application and are not otherwise excluded from protection by law	20 years	Monopoly
Registered design	Features of shape, pattern, ornament, which are industrially applied in an original design and are not otherwise excluded from protection by law	15 years	Monopoly
Copyright	(1) Original drawings of functional designs	Author's life plus 50 years	Protection from copies
	(2) Original drawings of non-functional designs	15 years from first marketing	Protection from copies
	(3) Original works of artistic craftsmanship: prototypes	15 years from first marketing	Protection from copies
Registered trade mark	Device, name, signature, word, letter or combination that is: (1) Name of company, individual, firm, represented in a special or particular manner (2) Signature of applicant or predecessor in business (3) Invented word (4) Word or words having no direct reference to the character and quality of the goods or services and not being geographical or a surname in its ordinary meaning (5) Any other distinctive mark, with evidence of distinctiveness	7 years then renewable for periods of 14 years at a time	Monopoly, subject to certain limitations in favour of honest concurrent users prior to registration
Plant varieties and seeds	New varieties distinctive from other breeds, uniform and homogenous in continued reproduction, whether wild or artificially induced	15–25 years	Monopoly

patent does not give the owner the right to use or exploit a patented invention: that right may still be affected by other laws such as the health and safety regulations, or the food and drugs legislation, or even by other patents where what has been patented is a mere refinement of someone else's earlier patented invention whose patent is still in force. It is in a sense a purely negative right.

The patent is in law a property right and it can be given away, inherited, sold, licensed and can even be abandoned like other property. But it is conferred by the state and it can also be revoked by the state in certain cases even after grant, and whether or not it has in the meantime been sold or licensed. It may, for instance, be found to be invalidly granted for want of novelty or inadequate disclosure of the invention. It may also be subjected to a compulsory licence in favour of another person if it has been inadequately exploited by the owner. These matters will be looked at later but a very important point to grasp from the start is that such patents are generally granted by the state only in respect of the territory governed by that state. There is no such thing as a world patent. The same is equally true of the other intellectual property rights dealt with in this book. All are national in character.

1.2.2 Registered designs and design copyright

Registered designs protect features of shape and ornamentation applied to articles produced industrially (as opposed to individual works of art, sculptures and the like). Some common examples include designs for the patterns of ornamentation on tea sets, designs for the shapes of furniture, coathooks, or household objects such as draining racks for crockery, for the surface of a golfball, toys, luggage and a whole host of similar objects. Registered designs create legal monopolies in much the same way as patents but they do not protect any of the functional aspects of the product nor the underlying technical concept.

Like a patent, a registered design confers a legal monopoly, but in this case one which can last for only fifteen years in the UK. Unlike patent law, the law on designs differs markedly in other countries. Like patents a design can also be revoked before expiry in certain cases, and compulsory licences in them may be granted in extreme cases. Like patents they have to be examined for both novelty and certain other legal requirements and they must be registered and published in an official journal for the monopoly to be granted.

Design copyright by contrast protects the shape of an article even where not industrially produced, not by conferring a legal monopoly in it but by protecting it only from copying – there is no protection from another person coming up with the same idea independently. Use of this design copyright is possible by invoking it to protect the articles indirectly: it in fact protects the production drawings on which the article was based or occasionally the prototype for the article. Again, the law of the UK on design copyright differs markedly from that of other countries.

UK design copyright is also wider than registered design protection in that it protects not only features of shape or ornamentation but also the appearance of many functional designs with no eye appeal which the law on registered designs cannot protect – for example, the shape of an engineering component can be protected by design copyright to prevent other manufacturers from making and selling rival versions of the same design in shape even though there is nothing new or original in the design of the component except for its shape, which was itself dictated by the construction of the rest of the machine, subject to limited exceptions relating to the supply of the spare parts market.

A design copyright may last much longer than the registered design, depending on whether the design to be protected by design copyright is also a registerable one or not. If it is then protection may be limited to only fifteen years whether or not the design was in fact registered; if it is not protection may last for up to fifty years after the death of the person who was the author of the drawing or plans or prototype upon which the article is based. Where both the registered designs protection and design copyright protection are available, each type of protection has advantages and a decision will often have to be made as to which form of protection to opt for. There may also be occasions on which there is a choice between a patent and design copyright. This will be discussed later in the book.

1.2.3 Copyright

Quite apart from its artistic connotations, use can also be made in the UK and also in most other countries of the law of copyright to protect many written materials of commercial value such as technical instruction (e.g. car maintenance) manuals. Any record of research and testing and of other written and similarly stored information may also be protected from copying or commercial reproduction. Almost anything requiring effort to compile, such as logarithmic tables, customer lists, catalogues, directories, or trade journals, is capable of protection by the law of copyright. It can also protect computer software, the reproduction of which can be a lucrative business and which it is now essential to protect. Copyright lasts for a long time, until fifty years after the death of the author of the copyright work, and it can be acquired automatically, without any formality, cost or registration in the UK and many other countries, and subject only to very low costs and simple formalities in other countries.

1.2.4 Trade marks

Trade marks identify your association with goods or services placed on the market by you and help to distinguish them from other similar goods or services. They do not protect the ideas or designs behind the goods or services themselves from imitation or even duplication, but merely prevent other traders from claiming or deceiving customers into believing that goods or services in fact produced or marketed by them were produced or marketed by you, and in certain cases from using your trade mark without your authority, for instance in an advertisement in which comparisons are made between goods bearing your trade mark and goods bearing someone else's trade marks. Thus, for example, a company whose own products have acquired a reputation of quality and reliability may rely on its trade marks, to prevent a rival or new company from marketing its goods under that trade mark or a confusingly similar trade mark in an attempt to benefit from a reputation which it has done nothing to acquire.

In the UK trade marks can be protected both at common law, in which case they are acquired merely through the course of time and by gaining a reputation in the mark, which may be limited in geographical area or range of product, or by registering a statutory trade mark, which will usually gain protection throughout the whole country in the range of goods or services for which it has been registered. The scope of protection also differs slightly depending on the type of trade mark acquired.

The trade mark differs from the other statutory intellectual property rights mentioned so far in that it is potentially eternal. To take one example, the UK's first statutory trade mark, the Bass Red Triangle, registered in 1876, is still registered and in use today. Most countries of the world do have well-established trade mark systems that differ in strength and character from place to place, but follow broadly the same principles. Trade marks do not travel well, in general, because of the importance of cultural factors in their popularity, and often it will be necessary to use different marks for different export trade markets, though this is changing.

1.2.5 Plant seeds and varieties

New plant seeds and varieties are protected in many countries not by patents but by a separate legal regime, which protects the reproductive material of the plant from commercial reproduction and exploitation without the consent of the owner of the registered material. These laws confer legal monopolies similar to conventional patents. Similar legislation exists in many other countries.

1.2.6 Trade secrets

Trade secrets are not strictly true intellectual property rights in law but in practice have the same commercial significance and are dealt with in much the same way. They are not in law property rights at all, but depend on the imposition of personal obligations on others. It is possible to protect a trade secret in many countries by imposing legal obligations on those who come into contact with the information not to disclose it or use it without authority, where that information is not generally known and is of a legally confidential nature. The means of doing so varies from country to country. In the UK it is by use of the law of confidence, sometimes supplemented by the law of contract.

Trade secrets may protect many things which lack the novelty required for patent or other protection, but they are equally often an alternative to such a patent or similar protection. A wide range of secret information from technical secrets such as formulae, know-how, and processes, to information about a firm's customers, employees, business information about its projected prices and sales strategy etc., can be legally protected in this way but only under certain conditions and in certain circumstances.

The law does not confer a monopoly in a trade secret of this kind, so it is not possible to protect trade secrets from some independent discovery by another person, but the unauthorised use or disclosure of information, if directly or indirectly obtained from the owner or licensee of a trade secret, can often be restrained, and the obligation can often be imposed on another person unilaterally and without their consent.

1.3 WHY BOTHER WITH INTELLECTUAL PROPERTY?

It will be apparent that intellectual property can in law protect a wide range of commercial and technical knowledge. And such intellectual property matters today more than it has ever done. One need only look at the UK's technology and licensing royalty figures in recent years to see this: their importance to the UK's balance of payments is very considerable (Table 2).

Table 2 *UK technology and licensing royalty figures*

YEAR	EARNINGS (£ MILLION)	EXPENDITURE (£ MILLION)	BALANCE +/−
UK overseas royalties on technology and mineral licenses			
1980	410.0	353.8	+ 56.2
1981	480.3	397.1	+ 83.2
1982	501.6	414.5	+ 87.1
1983	614.5	482.0	+132.5
UK overseas royalties on printed matter, recordings, etc.			
1980	77.5	43.9	+ 33.6
1981	95.6	51.0	+ 44.6
1982	66.7	49.9	+ 16.8
1983	86.1	63.8	+ 22.5

Sources: *British Business* published by the Department of Trade and Industry.

It is almost a cliché that the British are very good at inventing but very poor at exploiting what they have invented. In so far as this is true, and there is a distinction between purely scientific invention and design and technological innovation, a great deal of UK technology and design has been and still is being lost to competitors through a failure to take adequate steps to protect it from imitation, either through our own managerial carelessness or pure ignorance.

If any proof were needed of the importance of care and knowledge in this field, one need only look at the huge number of patents and other intellectual property rights sought by the Japanese, once thought of as mere imitators of other people's technology, but now world leaders in obtaining proper legal protection for their own newly developed technology and design: see Tables 3, 4 and 5.

Table 3 *Number of patent applications for 1984 by country*

Japan	285 314
USA	114 784
UK	63 354

Table 4 *Number of design applications for 1984 by country*

Japan	54 683
USA	8 739
UK	7 237

Table 5 *Number of trade mark applications for 1984 by country*

Japan	161 882
USA	62 600
UK	22 796

People do not spend time and money acquiring these rights for no good purpose. There are a number of good reasons why it is important to bother with proper legal protection of new technology and design; some positive, some negative. Some of the positive reasons are –

- Intellectual property rights help to protect you from free riders, counterfeiters and copyists, who take your research and development expenditure for free. But these rights also keep you ahead of legitimate rivals who may be working on a similar idea independently. This is because most intellectual property rights confer monopolies for a limited period which keep others out of the market or require them to spend time and money designing around a patent or other right, or to take a licence from the right owner and pay a royalty.

So with an intellectual property right such as a patent you may be able to get onto the market place with a new product free from competition while being able to maintain the high prices necessary to recoup R and D costs. Your competitors too have high R and D costs. Occasionally your idea may be such a market winner that other companies simply have to take out a licence at almost any price: Pilkington Float glass in the 1950s was a very good example of this phenomenon. A patent is not a licence to print money. But it does buy some degree of legal protection for your investment in plant and materials.

- Intellectual property rights are often valuable intangible assets in themselves, which appear on the balance sheet of the business and which may be sold and exploited in just the same way as a machine.

With an intellectual property right behind you, the fact that you cannot go into production with an invention does not prevent you from making money out of it, although it is true that most successful inventions have been produced by the right owner personally, with profitable licences arising only after the product first hit the market place.

Such rights can also enhance the capital value of the owner's business. It is a characteristic of the new technologies that a higher proportion of them are being developed in smaller companies than was the case in earlier times. Many of these smaller companies have bright ideas men who are also poor commercial managers. Even the most successful run into trouble from time to time, particularly as a result of over-rapid development and expansion or of cash flow difficulties. This makes them attractive take-over prospects for larger companies who want to expand into a new area and who want the whole technical team and production facilities concerned with the invention rather than the patent or other intellectual property right itself. But although they are willing to provide the finance for the operation they will want the legal security of the intellectual property rights – both technical such as a patent and commercial such as the trade mark which enhances the goodwill of the product. There is no point in buying the idea if the inventor who made it famous leaves to produce a better and competing version with the benefit of his or her own trade name.

Such trade mark protection can be crucial to a company's credibility or to its reputation for quality. The value of trade marks is shown by the numbers of infringements which are pursued every year, especially against importers of inferior foreign goods bearing prestigious luxury or quality good trade marks. To take one example, think of the importance of the trade mark 'Ferodo' to the British company making reliable brake pads for motor vehicles, in the light of the imports of defective brake pads from overseas, and the serious consequences for that company of large numbers of such defective foreign brake pads being sold under the trade mark 'Ferodo'.

The integrity of a company's trade mark portfolio is also central to many advertising campaigns and to the success of the many businesses using format franchising, in which a chain of independently managed stores is set up under the umbrella of licensed formats characterized by uniform trade marks, store designs and products. These operations now depend entirely on full trade mark protection for their financial viability.

- Intellectual property rights can make the process of technology transfer and especially the international transfer of technology to Third World countries, much easier in many cases. They can enhance the prestige and appeal of the products they protect, especially when first marketed, when the award of a patent or design may confer a certain aura of quality in the mind of the buyer, whether trade or consumer. More generally, the patent provides a good basis for the licence and is easier to quantify for royalties than know-how.
- Intellectual property rights are not as expensive to acquire as many people believe. Nor is advice on the possibility of protection as expensive as is thought by many business people. Intellectual property rights can also protect more than many people believe. Even a weak or a relatively narrow right may deter potential rivals from imitating the design or invention it protects or induce them to take out a licence for it, to avoid litigation.
- The publicity afforded to intellectual property rights can alert other companies working in the field to your technology and make them potential customers and licensees instead of pursuing their own research and development which would merely duplicate yours. The intellectual property rights you have are an incentive to them to take out a licence because, on the one hand, there is some security behind what the company buys, and on the other, there is less point in trying to invent the same thing when the second inventor may well not be able to exploit it because of your rights. The presence of a patent number on the product gives it the appearance of a quality guarantee to some people, and a patent portfolio gives a certain image and cachet to the company owning it – which may make it a much more attractive trading partner.

Quite apart from the loss of the advantages which intellectual property rights can give you, the main negative reasons for placing this much importance on intellectual property should be obvious –

- Other people's intellectual property can put you in second place in the competition by keeping your product off the market and in extreme cases you can be kept almost permanently in second place by the constant improvements to technology being patented so as to keep the competitive advantage in one camp. To be only two or three months behind in seeking protection for your rights can leave you with an entire research programme wasted and redundant, and with it many redundant staff.
- Other people's intellectual property can cost you money if you even accidentally infringe their rights. A lost infringement action is a very expensive albatross around your neck, quite apart from the stigma of being held to have infringed another's legal rights and being branded an imitator.

- Published details of intellectual property grants provide information of what others are doing and helps you in monitoring their activity. This is of growing importance with the recent improvements in on-line data about new technology and design. So ignorance of these sources of information deprives you of much free R and D results material of use to your own firm.

These factors must always be considered when you determine not only general company policy on patents and other rights, which may be aggressive or relatively passive – publication of the idea can always at least prevent others from patenting it and it is cheaper than seeking and defending a patent – but in every case when an invention or other design which it is possible to protect by intellectual property rights is devised.

But it is not intended to imply that intellectual property should be the focus of all your concern. For the essence of exploiting technology and design is to make and sell things or processes that work, either directly or under licence. All the intellectual property rights in the world will not help you to sell your deficient technology or poor designs and obviously there is no substitute for sheer competitiveness in the quality or price of what you actually make. It is a prerequisite of industrial success. The prime functions of research and development teams in any firm must be to research and develop new technology and designs.

Thus the point of using the intellectual property rights is to assist you in capitalizing on research and development work to the greatest possible extent. These intellectual property rights are not golden eggs. They do not guarantee you a return on the invention you have invested time, money and effort into creating. But they are useful commercial tools, which in certain cases can assist you in creating that return. They should be seen as such and valued as such.

2

Protecting trade secrets and know-how

2.1 WHY BOTHER WITH TRADE SECRETS?

Trade secrets were used long before the patent system came into being and remain an effective tool today. If the proper practical and legal precautions are taken by their owner, trade secrets can be kept and exploited for a long time, especially if reverse engineering of the product or process containing the secret is very difficult or if the physical means of concealment of such information is otherwise easy to maintain. The formula for Benedictine liqueur has, for instance, remained a secret for many centuries. It is still as valuable a piece of secret information as when the formula was first devised and indeed with the benefit of the accompanying trade mark is now probably even more valuable than when it was first made. So called 'black box' licences in which a piece of advanced technology is licensed to a licensee without revealing its mode of operation are still relatively common, where the value of the technology justifies it and it is possible to detect a licensee's unauthorized investigation of it.

But even where such physical protection is not so easy, it remains the case that where the means of exploiting the information is through relatively limited licensed distribution (as opposed to a mass market), use of the legal obligation imposed on the licensees to preserve confidentiality of the information can be of very considerable assistance and must not be disregarded as a potential instrument.

- The technology or design concerned may not be patentable, but it may still be sufficiently valuable to need some form of legal protection.
- Often a patented invention which has been disclosed to the public is surrounded by some associated patentable or even non-patentable material which is not required to be disclosed in the patent specification but without knowledge of which it is not possible to

exploit the published patent profitably. Under the new patent laws in Europe there is no need to disclose the best way of operating a patented invention, only one way of doing so. Many such patents are commercially simply not viable without this associated know-how.

- Even where an invention is patentable and will in the future be patented, inevitably a period of time elapses before the patent is applied for and it is essential to be able to protect that information by some legal means pending the application. Likewise, an invention may still be in an embryonic state and thus require further development before the time is ripe for patent applications to be filed. The use of the law of confidence (i.e. the law protecting secret information from unauthorized use or disclosure by others) is one such means of protection, a useful legal back-up to any physical security measures which may be taken.
- Infringement of published technology may in some cases be very difficult to detect, so that the use of trade secrets as a means of exploitation is the only practicable way of preventing unauthorized imitation. Equally, it may be easier to design around a published patent publication than around technology which is unprotected but difficult to reverse engineer.
- As against protection through copyright, trade secret protection has the advantage that it protects not merely the form in which the material is fixed but also the underlying ideas, so that in one sense at least the scope of protection can be wider than with copyright.
- As against patents, it is generally easier in the USA to obtain interim injunctions preventing use of the technology when trade secret protection is relied upon than when patent protection is relied upon. This is less true in the UK, where the law on interlocutory injunctions is different.
- The means of protection using trade secrets and know-how is virtually free, whereas patent or other protection may be out of reach by reason of its cost. This may be an important factor for the individual inventor lacking the resources for patent applications. Similarly, if the development is of little commercial significance it may not merit the expenditure which patent protection demands.
- The duration of protection is not limited to a set period of years as in the case of a patent; it is potentially indefinite. Benedictine liqueur is a good example. Conversely, the short period of time for which legal protection is needed in a fast-moving industry such as the microelectronics industry may mean that the time lapse and cost involved in patenting is such as to render patents quite otiose for the piece of rapidly obsolescent technology concerned.

There are disadvantages to trade secrets, as follows.

- The information may be very vulnerable to breach of the obligation by someone in whom it has been confided, and once so disclosed it may be difficult to salvage the position as a matter of practice whatever your apparent legal rights, especially in situations where your capacity for exercising physical security measures is limited.
- Trade secrets are not legal monopolies. The use of trade secrets does not prevent exploitation of the information by someone who subsequently discovers it independently, whereas the patent does protect the owner from later inventors. Moreover, a key invention which is likely to be central to a new line of inventions in the future and which will be a valuable master licence product will almost always justify patent protection. In general, the more you intend to exploit through licensing and the less through personal production, the greater the argument for some formal patent protection.
- The scope of protection afforded to trade secrets in respect of third parties is legally less certain than is the case for patents and designs and copyright, and for Continental Europe protection is undoubtedly weaker. In the USA there are some difficulties which arise out of the doctrine of the Constitution of the USA that State laws may not pre-empt Federal laws, that is, in any case of conflict Federal law prevails over State law. Trade secret laws are State laws and so in the case of any conflict with the copyright or patent laws, which are Federal laws, these State trade secret laws are to be ignored. In some cases, courts have held that

material which could not be protected by – for instance – the Federal copyright law should therefore also be denied the protection of State trade secret law. The position remains uncertain for the moment and therefore professional advice should always be taken if the use of trade secret protection in the US is contemplated.

- Should litigation arise, matters of proof of ownership of trade secrets and of breach may be more difficult to establish than for a patent or for a registered design or copyright, unless very great care is taken by the owner. This highlights the importance of following the advice on practical precautions to be taken when making use of trade secrets.
- Trade secrets often lack the prestige which published patents may have for potential licensees and in some Third World countries the necessary government approval of deals involving pure know-how is often more difficult to acquire than for deals at least fronted by a patent.

These are some of the factors which should be kept in mind when considering the use of trade secrets. It will always be a matter of individual decision whether trade secret protection is to be used instead of any protection which might be available through patents or similar means. There are no golden rules.

2.2 HOW DO YOU PROTECT TRADE SECRETS BY LAW?

The English courts hold that there is an implied obligation of good faith imposed on recipients of confidential information in certain circumstances not to disclose it or use it without the authority of the person who confided it. If there is a threatened 'breach of confidence' an injunction may be obtained to prevent it, and in some cases the owner of a trade secret may claim money compensation for any breach which took place before he or she had a chance to do anything to restrain it.

When trade secrets are being licensed these obligations, implied by the law in certain cases, can be very much strengthened by the inclusion of appropriate terms in the licence agreement and the obligations of employees and others coming into contact with trade secrets can also be considerably increased in their contract of employment or by similar agreements or notices. This means of protecting such trade secrets is employed in the USA and in most of the Commonwealth countries, sometimes also backed up by statutes dealing specifically with criminal penalties for the 'theft' of trade secrets, and although this form of protection is not available in precisely the same form in other European countries, in fact very similar protection through the use of contract and unfair competition law may still be invoked in many countries in both Europe and the Third World, although these tend to be weaker than the law of the UK and USA and of the Commonwealth nations.

Under English law, the courts have held that a person infringes the rights of another if he or she is proved to have used confidential information, directly or indirectly obtained from that other, without his or her express or implied consent. At this stage four points should be stressed:

(1) Trade secrets can be infringed not only by use but also even by mere disclosure to third parties.
(2) Such use or disclosure without authority can take place purely subconsciously, without any such intent to infringe the rights of another person.
(3) The motive for the unauthorized disclosure or use is irrelevant.
(4) It is vital to take practical precautions to back up the legal rights on which you

rely if trade secret protection is chosen: failure to do so may lead the court to conclude that you have waived your rights. This applies equally to the security measures you adopt with respect to outsiders and employees, and to the enforcement of any confidentiality agreements you may enter into and action against those in breach of them. The type of precaution to be taken is considered later in this chapter.

2.3 WHAT TYPES OF INFORMATION CAN YOU PROTECT?

The law on trade secrets is almost entirely judge-made: no statute governs it. Accordingly the scope of protection is to be found entirely in the decisions of the courts. The basic requirement for information to qualify for protection are that:

- The information is information which is not publicly or generally known in the industry or business concerned and which is communicated to the recipient for only limited purposes or received by the recipient in circumstances which objectively he or she should know impose a restriction on the uses to which that information may be put.
- The information is of a kind capable of being subjected to such limitations on use by the person seeking to restrain its use or dissemination by the recipient. Not all information can be protected simply by calling it confidential information. The types of information which can be protected in this way must be considered. In brief they include the following types:

(1) Technical information and know-how
(2) Business strategy information
(3) Private personal information
(4) Literary ideas, plots for films, etc.

- Confidential information may be made up of relatively mundane components put together in a way which makes only the totality valuable. Something that has been devised solely from materials in the public domain may still possess the necessary quality of confidentiality. It is perfectly possible to have a confidential document, be it a formula, a plan, a sketch, or something of that kind, which is the result of work done by the maker upon materials which may be available for the use of anybody; what makes it confidential is the fact that the maker of the document has used his or her brain and thus produced a result which can only be produced by somebody who goes through the same process. So information built up from publicly available components may itself be confidential if it requires effort to put it into a useful form, such as a list of names which individually are available to the public in the telephone directory but which have been assembled together in one useful list of customers.

2.3.1 Technical information: know-how

The major point about this category of information is that there is no requirement that it be potentially patentable or registrable as a design or even protectable as copyright. In principle virtually anything not generally known can be protected by law of confidence, although the courts will not protect mere nonsense or mumbo jumbo, however confidential.

2.3.2 Business strategy information

The courts have consistently held to be at least potentially confidential a wide variety of business information, including information about:

(1) A firm's customer list and their requirements
(2) Negotiated prices paid by customers
(3) A firm's manufacturers and suppliers
(4) Negotiated prices paid by the company
(5) Overseas buying agents dealt with by the firm
(6) A firm's new ranges, actual or proposed
(7) Details of a firm's current negotiations
(8) A firm's samples and finance plans
(9) A firm's current 'fast-moving' lines
(10) Take-over bids and stock exchange quotations
(11) Projections of future business prospects

2.4 PROTECTING ABSOLUTE AND RELATIVE SECRETS

There is no need for absolute secrecy to establish a right of confidence. This is because the information can still be relatively secret after limited and often necessary dissemination (as, for instance, to employees and licensees, especially if on a 'need to know' basis, during the development of an invention or the execution of a commercial process).

What is in the public domain, too, seems to be a variable concept. In the case of highly specialized information, for instance a formula for catalysing a process, it would not matter that the general public was wholly unaware of it, if all or practically all the specialist users of that process were familiar with it: it could be denied the protection of confidence. At the other extreme the case law can be generous as to what can remain confidential despite fairly widespread disclosure.

The limits on this generosity by the courts is well illustrated by a decision of the Court of Appeal.

EXAMPLE: *Sun Printers v. Westminster Press Ltd (1982)*
The chairmain of a company in financial difficulties sent a letter to one hundred or more people within the company including junior management and junior union officials, dealing with the company's future. The letter was passed to the defendant newspaper which intended to publish it. The Court of Appeal held in this case that the information it was sought to protect had been widely distributed and discussed and could no longer be regarded as being confidential in character, although things might well have been different if the plaintiffs had taken the simple precaution of stamping the word confidential on the front of the documents circulated.

2.5 WHO CAN YOU BIND TO KEEP TRADE SECRETS SECRET?

To bind someone to keep the information secret the courts require the information in dispute to have been communicated to (or possibly intercepted or otherwise acquired by) the defendant, directly or indirectly, in circumstances which impose an obligation of confidence. The main examples of persons who may be bound of relevance in our context are:

(1) Employees
(2) Consultants
(3) Parties to pre-contract negotiations

(4) Recipients of ideas and suggestions
(5) Licensees of trade secrets
(6) Sub-contractors
(7) Third parties, especially trade rivals

2.5.1 Employees and ex-employees

Leaving aside for the moment express contractual obligations of confidence, the courts have been faced with conflicting public interests in employee loyalty and the freedom of movement of workers in the economy.

(a) Obligations during employment

During the course of employment obligations of fidelity and confidentiality are interpreted very strictly. The courts have implied into contracts of employment an obligation on the employee not to disclose to others any confidential information communicated to him or her in the course of his or her employment by his or her employer, whether they be his or her trade competitors, trade unions or other persons, and not to make use of it for personal gain to the prejudice of the employer; nor to use that information for his or her own purposes so as to enter competition with his or her employer, even in his or her spare time; and to reveal to his or her employer any such information he or she picks up in the course of employment as an employee and not to pass it on to others. These specific duties form part of the general obligation of an employee to further the interests of his or her employer's business, in whose prosperity he or she is perceived to have a direct interest. But there are limits to this.

(b) Obligations after termination of employment

By contrast, once the employee has left the employment of the master, although certain obligations of confidence follow the ex-employee, the ex-employer has less of a hold over the ex-employee's subsequent use of much of the information acquired while in the former employment.

The rationale of this rule is that after the employee has left the employment of the master, the interests of the employee can no longer be seen as inextricably linked with those of the ex-employer. The employer clearly still has some limited legitimate interest in restraining the disclosure of certain information about the secret operations of his or her business. But this conflicts with the requirement of freedom of movement of workers recognized by the common law from medieval times.

The courts have sought to balance the legitimate interests of the employer and employee after they part, by distinguishing between information which the employee may and may not use. On one side of the line is placed information properly labelled trade secrets; on the other side the general skill and knowledge of the employee, built on experience and which the employee is free to use for his or her own benefit or for any employer for whom he or she may subsequently work. Clearly the obligation of confidentiality will be lower after the employee's departure than while the employee continues to work for the employer who gave him or her the information before his or her departure; but it is far from clear in theory or practice quite what this obligation is.

The courts have in practice drawn some distinction between specific technical formulae and more general working methods and practices which may become part of the general knowledge and experience of an employee (this is discussed in Chapter 7) but any distinction between information which forms part of one's general recollections and information which is the subject of deliberate memorization and recording by an employee is inherently unsatisfactory, since the court should no more restrain the employee with a poor memory than it should release from his or her obligations the employee with an encyclopaedic or photographic memory who can without any effort recall a complicated formula or process in sufficient detail to reproduce it. The reason advanced for the distinction appears to be the technical legal difficulty of framing an injunction sufficiently precisely to cover information not reduced to some definite form.

EXAMPLE: *G.D. Searle and Co. v. Celltech Ltd (1982)*
The defendants had hired one Mr Carey following his departure from the plaintiffs' employ, both firms being engaged in research into genetic engineering. Before leaving Carey gave the defendants the names of other employees of the plaintiffs and the defendants recruited them, all the employees recruited leaving the plaintiffs having given notice in accordance with their contracts of employment so that no question of inducing breach of contract arose. The Court of Appeal refused to grant an injunction restraining the defendants from using the information given to them by Carey. The majority, Lord Justices Cumming-Bruce and Templeman, felt that the names, aptitudes and specializations of the plaintiffs' staff were not *per se* confidential and that the plaintiffs had never sought to impose an obligation of confidence on their employees in relation to that information. In the absence of a restrictive covenant it could not be restrained. Lord Justice Brightman, though refusing an injunction, thought that names were capable of being confidential, but that the law had to be built up sensibly and in this context protection could not extend beyond names. It must have been significant in this case that the plaintiffs themselves undertook some publicity of the identity and expertise of their employees, with photographs of them and their names appearing in promotional material advertising the plaintiffs' skills and services. Lord Justice Brightman's willingness to accept in principle that the identity of the workforce may be confidential is surely correct, on the assumption that the employer does not act inconsistently with that intent:

> 'I am prepared to accept that the personnel manager of Emporium A, on transferring his services . . . to Emporium B, which is about to establish itself in business, could not properly provide Emporium B with the names of all the buyers working for Emporium A, in order to enable Emporium B to attract to itself from Emporium A a ready-made team of skilled buyers. The personnel manager's knowledge of the unpublished identity of Emporium A's buying team, learned during his service with Emporium A, may well be confidential information of Emporium A. . . '

(c) Enhancing the obligation of confidence

Faced with attempts to strengthen the position of the employer by express contractual undertakings, the doctrine of restraint of trade prevents employers from including in contracts of employment obligations which would unreasonably prevent their ex-employees from making use of their working skills for other employers. This is not the place to consider the doctrine of restraint of trade and the contract of employment in depth, which is considered in more depth elsewhere (see Chapter 7) but it will be clear that the drafting of such obligations is extremely difficult because if the employer attempts to introduce specific obligations there is a danger that the courts will interpret them very narrowly and the employee will escape all substantial restraint; if on the other hand the employer attempts a broadly drafted clause he or she runs the risk of having it struck down by the courts as being in unreasonable restraint of trade.

To an extent this difficulty can be mitigated by careful drafting of a number of restraints as severable (i.e. separable) individual restraints so that if some fall others may survive examination in court, but the attitudes of the court are not altogether predictable when it comes to interpretation of such clauses. The leading recent decision will provide a good example of the difficulty an employer may face in practice.

EXAMPLE: *Faccenda Chickens Ltd v. Fowler (1985)*
A chicken salesman left his employer to work on developing a business selling chickens in which he employed data on the pricing of chickens, quantity and quality demands, names and addresses of rounds, etc. The employers then sought to restrain him from doing so.

The Court of Appeal held that the scope of the employee's obligations was to be determined both before and after his period of employment by the terms of the contract of employment, express if there were any, or implied in the absence of any express terms.

During his employment the employee was under a duty of good faith or fidelity which varied with the nature of the contract and which would be broken if the employee made a list of his employer's customers or deliberately memorized them for use after his employment ended.

The implied term as to conduct after terminating his employment was more restricted than that during his employment. There was an obligation not to use or disclose information for secret processes of manufacture or designs or other special methods of construction, and other information of a sufficiently high degree of confidentiality as to amount to a trade secret. This obligation did not extend to information which was confidential only in the sense that any unauthorized disclosure would be a clear breach of the duty of fidelity. To determine whether any particular item fell within the implied obligation after employment was a question of fact in every case. Among the factors to be taken into account were:

(1) The nature of the employment. A high obligation of confidentiality might be imposed where the employment was in a capacity where 'confidential' material was habitually handled.
(2) The nature of the information. Information would only be protected if it could properly be classed as a trade secret or as material which was in all the circumstances of such a highly confidential nature as to require the same protection as a trade secret. A restrictive covenant would not be enforced unless it was reasonably necessary to protect a trade secret or to prevent some personal influence over customers being abused in order to entice them away.
(3) Whether the employer impressed upon the employee the confidentiality of the information.

(4) Whether the relevant information could be easily isolated from other information which the employee was free to use or disclose. The separability of the information was not conclusive but the fact that the alleged confidential information was part of a package and that the remainder of the package was not confidential was likely to throw light on whether the information in question was really a trade secret.

The Court of Appeal left open the question of whether any greater restraint could be imposed on an ex-employee from selling the information to a third party rather than merely using the information acquired in the course of employment to set up in business on his or her own account to make a living.

Nevertheless, it is very common for the employment contract to include terms restraining employees from disclosing certain categories of information both during and after the termination of employment, and, despite the problems of drafting, from the employer's point of view this is clearly the most sensible course, because it appears that the contractual or fiduciary obligations imposed on an employee are not exhaustive of the employee's duties of confidence. The implied contractual and equitable burdens of confidentiality survive the agreement of express terms, unless expressly excluded by agreement. Thus even very wide clauses, should they be struck down by the courts as being unreasonably in restraint of trade, have no effect on the implied contractual or equitable obligations of confidence which the courts now recognize. Accordingly the inclusion of such a clause can only help the employer – it cannot penalize the employer. It is also possible to include a number of separate and gradually wider clauses so that if some are struck down others will survive.

In practical terms it can be very difficult for an employer to restrain the employee, both on account of the difficulties of proof he or she may encounter and because of the risks that litigation will indirectly reveal even more than it was sought to keep protected. Hence the practice, in some fields of commerce, of minimizing loss by extra-legal means. Apart from basic security provisions such as disclosure only on a need to know basis and even personal searches of employees leaving company premises, if it is felt that a salesman or saleswoman is imminently leaving for a competitor, steps may be taken to remove him or her, perhaps even sacking him or her with pay in lieu of notice, and in the meantime sending in replacements to canvas 'his' or 'her' customers for the employer before he or she can begin to canvass them for the competitor.

In this context a related technical problem may arise. What is the scope of the obligation of confidence of employees who have been wrongly dismissed by their employer? Suppose the employees were under an obligation of confidence in relation to certain information governed by a term of their contract of employment. When that contract is repudiated by the employer and the employees accept that wrongful repudiation as discharging the contract of employment are they under any obligation of confidentiality at all, and if so is it that which would normally be implied out of the relationship of employer and employee or is it the more extensive obligation which the employees expressly undertook? It would seem that the repudiation of the contract by the employer still does not affect the equitable obligations owed by employees, so that even unjust dismissal does not free them to do what they please with confidential information which they would in normal circumstances be restrained from using after leaving the employment of their master voluntarily.

2.5.2 Independent contractors

The position with regard to independent or sub-contractors appears to be relatively straightforward. The use of such persons to manufacture specific components of a product or to undertake specific processes is convenient and in any case it is often economically preferable to concentrate resources and make use of those who can produce the item most efficiently. Where in pursuance of such an arrangement confidential material such as specifications and drawings is handed over to the contractor the courts are ready and willing even in the absence of an express agreement (which would be rare) to impose an obligation of confidence. The drawings etc. are supplied for a limited purpose only and must not be used for extraneous purposes. Distributors of the plaintiff's products are in a similar position.

2.5.3 Professional advisers, consultants

There are few difficulties in establishing an obligation on such persons not to make unauthorized use of confidential information, though precise analysis of why they are to be held liable in any given case may be confused since the principles of contract, the general doctrine of confidence and the principles of fiduciary obligations coalesce in the facts of most of the disputes. English law has labelled a vast class of persons fiduciaries: directors and other senior employees including bankers, legal advisers and some consultants being the classical examples in this sphere, but the label has been applied a long way down the ladder of seniority, indeed too far in some senses. The fiduciary will be liable to an action restraining him or her from using, misusing or profiting from confidential information on a basis quite independent from contract or the general law of confidence.

The starting point must be to identify what a fiduciary is. A fiduciary is one whose obligation is to advance or protect the interests of another, and the standard and scope of that obligation does not remain constant as between different types of fiduciary. This basic obligation has developed into two closely related rules – the fiduciary shall not make an unauthorized profit from his or her position; and the fiduciary shall not allow a conflict of interest to arise between himself or herself and the person(s) to whom he or she owes the obligation.

One manifestation of the rule is that fiduciaries must not purchase or exploit for themselves an asset which they should have purchased or exploited for the person(s) to whom they owe the obligation, nor must they seek to divert to themselves a business opportunity which would have been of interest to those persons, even if they know that for some reason such as a lack of finance it is impossible for those persons to make use of the opportunity themselves and it will slip away if not acted on. Only if a full disclosure has been made, the opportunity rejected, and permission obtained to proceed on their own behalf, may the fiduciaries seek to exploit the information personally. The usefulness of such a doctrine in this area of activity is manifest.

EXAMPLE: *Industrial Development Corporation v. Cooley (1972)*
The defendant worked for the plaintiff company in a senior position. It was intimated to him by a representative of a Gas Board that his company was not liked but that if he could free himself from the company he stood a good chance of winning a design contract. The defendant left the company on the pretext of ill health and was awarded the contract. He was held liable to the company in respect of the profit of that contract even though the company itself would never have been awarded the contract. He should have informed them of what he had been told and he was in breach of his fiduciary duties to the company.

Most consultants too or senior staff with some professional expertise will be placed in a fiduciary category by the courts. Such consultants, particularly business strategy consultants but also trouble-shooting technical advisers, are increasingly common means of introducing expertise into a company. The company may well need to provide confidential information to consultants to enable them to do their job. This raises little difficulty. But that information not known to the company but discovered by the consultants for the company in the course of their performance of the task allotted to them may cause more difficulty. In these cirumstances it seems that a test similar to that adopted in respect of employees must be applied in splitting up the information which becomes part of the general skill and expertise to be employed in the exercise of the profession and that which must be retained and not used which is peculiar to the client.

2.5.4 Licensees of trade secrets and know-how

As we have already said, a vast amount of novel technology is exploited by being licensed through the mechanism of confidential disclosure agreements, either solely or attached to a patent or copyright licence. And indeed almost every licence of a patent will of necessity include some confidential know-how licence alongside the main patent licence as part of the package which makes the licence commercially viable for the licensee.

Usually this is a relationship entirely regulated by contract, in the licence agreement itself, which will make full provision for the conditions under which the secret information is to be disclosed and used by the licensee. But in the absence of such a provision the courts have consistently held that information of a confidential nature revealed pursuant to a licence may only be used by the licensee for the purposes of the licensee.

The major difficulty will be to determine when this is so and many licences simply provide for a one-off payment giving perpetual rights in the know-how etc. If this is not desired it is generally advisable to make express provision in the licence agreement for the return of all documentation at the end of the licence with prohibitions on reproductions of it etc.

A further variation on the know-how licence is the so-called black box agreement, whereby the technology but not the knowledge of how it is to operate is licensed by leasing machinery the mechanics of which are concealed to the licensee, who undertakes not to interfere with the machinery or to seek to discover or permit others to inspect the machinery. Such licences clearly impose the obligation not to disseminate any information thereby put into the licensee's hands. In these cases the obligation is always backed up by the retention of ownership of the machinery itself. Where, as in the computer hardware industry, leasing gradually becomes replaced by outright purchase of hardware such licences become extremely difficult to police; this is especially true where second-hand markets develop.

2.5.5 Parties to pre-contract negotiations

In many cases negotiations for the assignment or licensing of know-how or other confidential information will be preceded by an express undertaking of confidentiality. Less commonly, negotiations will be conducted after the unilateral imposition of such an obligation of confidence by one of the parties which is accepted by the conduct of the other in continuing with the negotiation. Sometimes imposition of the obligation and revelation of the idea will be contemporaneous, as where an unsolicited idea is submitted accompanied by a letter imposing the obligation. Finally, the parties may enter into negotiation without a clear understanding on either side about the position should negotiations break down. Often of course negotiation does break down or the submitted idea is rejected. Later the recipient of the idea may be seen to be using what looks like the idea submitted and the confider will seek redress. Will he or she succeed?

The courts are clear that whenever information is communicated which has some commercial or industrial significance and it is passed on in a business context, the recipient is under a very heavy burden to prove that an obligation of confidence is not thereby imposed upon him. This places the recipient of unsolicited information in a very difficult position.

EXAMPLE: *Seager v. Copydex (1966)*
An inventor displayed an invention on a television programme called 'Get Ahead'. This interested a company who contacted him with a view to discussing the possible licensing of the invention. During these abortive negotiations the inventor voluntarily revealed another idea on which he had been working but the company was not interested in it. Some time later the company brought out a product virtually identical to the second idea. The inventor sued for breach of confidence. It was held that the company had forgotten about the idea at the time they went into production but had subconsciously used it. The idea had been communicated in circumstances which impliedly imposed an obligation of confidence on the company, not to use or disclose the idea without authority. The company had therefore broken the confidence and was liable to the inventor.

At least in Seager's case the company concerned had voluntarily entered into some form of negotiations, concerned with a product similar to that which the voluble inventor subsequently revealed. The extreme case of the unsolicited idea is that where a letter arrives which purports to reveal an idea and which expressly or impliedly imposes the obligation of confidence. There will of course be times when it is possible for the recipients to prove that the idea was already known to them, or that it was already in the public domain, perhaps in a patent specification, but the evidential problems may be considerable, involving examination of the R and D department's staff and day books, perhaps with much attendant bad publicity. It may also be the case that such communications are so careless that the courts will deny the necessary attributes of confidentiality to them. But the law is not certain on this point. Their concern about this problem has led a number of larger companies in the USA, who apparently suffer much time-consuming and irritating administration as a result of submissions, to take evasive action, with administrative procedures to eliminate these risks. These are considered below in Section 2.9.4.

2.5.6 Eavesdroppers, thieves, spies

Confidential information may be obtained without the voluntary or even accidental participation of the possessor, through industrial espionage or (less dramatic) simple leaks. The sources of leaks are many. They include, for instance:

(1) Talkative sales and research staff, especially at meetings, in publicity material, etc.
(2) Talkative contractors, component manufacturers, supplied with information essential to the proper performance of their task, who discuss it when persuading other potential clients of their technical capacity to take on similar work.

The recipients of leaks are equally diverse. As for the intentional recipients, professional spies come in many forms too:

(1) Temporary staff, employed as plants
(2) Readers of waste, carbon papers, etc. picked up from those who make use of unlawful means
(3) Permanent staff, especially those acquired through agencies
(4) Visitors on factory inspection visits, fake contract negotiators, etc., including some allegedly technology brokers, sales staff from other companies
(5) Market research organizations, advertising agencies, etc.

In the UK there is greater difficulty in acting against the mere eavesdropper or user of bugs which involve no trespass or crime. But in the USA even these people may be acted against. In a leading American case a photographer took photographs of the plaintiffs' new methanol processing plant as it was being constructed, which photogaphs could reveal the secrets of their new process. In taking these photographs no civil or criminal wrongs were committed but the plaintiffs were still able to prevent the use of the photographs.

2.5.7 Third parties in receipt of information

Not all recipients of secret information who subsequently make use of it are guilty of an intent to obtain an unfair advantage. A new employer, for instance, may be wholly unaware that an employee's apparently bright idea is in fact the bright idea of a former colleague in another firm. What if the new employer starts to make use of that idea in good faith, only later discovering its source? As in many cases the courts are faced with the difficult task of deciding which of two innocent parties is to suffer.

It seems that the courts will usually restrain the new employer from using the information after he or she discovers the truth, though without awarding compensation in respect of any use of the information before that time. Even this is rough justice on an employer who has invested a great deal of money in new plant based on the information and the courts will exercise a discretion not to restrain him or her in extreme cases.

2.6 PROOF OF BREACH OF CONFIDENCE

Where there is a substantial similarity between the technical information employed by a defendant and that disclosed to the defendant in confidence, or the products produced by a process closely resemble those which would be produced by the plaintiff's process, an inference of use may in appropriate circumstances be drawn by the court. A breach will be found even if only a part of the information was used if it was of any significance although the degree to which the confidential information affected the defendant's process or product may be material to the remedies obtained by the plaintiff.

2.7 LOSING THE RIGHT TO KEEP INFORMATION SECRET

In certain circumstances the plaintiff may lose the right to keep the information secret, even though he or she once enjoyed that right. This may occur because of publication of the information, because of consent, because of the public interest, because of independent discovery of the information by a defendant, or even because of the plaintiff's own inequitable conduct.

2.7.1 Publication

Disclosure to the public of the information may have taken place through a breach of confidence by an employee or other person, through an innocent third party, or even through the conduct of the original possessor of the information. To what extent does subsequent disclosure of the information release the person who has been subject to an obligation of confidence? Disclosure by the original possessor may have been deliberate or accidental: it would seem that this will release the other party.

Disclosure by a third party may deprive the secret information of the confidential nature that it had once possessed and leave the victim with at most an action for damages. But if the breakout of the information was only very limited it may still be relatively confidential and injunctions may be issued to prevent further dissemination.

Any publication of the information as part of a patent specification of course deprives the owner of the information of any rights to rely on obligations of confidentiality and only if the information is within the scope of the patent will the owner retain any control at all over its use except by contract with others.

2.7.2 Consent

Express or implied consent may release the other party: this consent may be given for limited purposes only, as where a firm authorizes a bank to release information for the purpose of providing a credit reference.

2.7.3 Public interest

It can be argued that certain information is of such a character that there is a public interest in its disclosure. The usual example cited is the inventor who claims to have discovered a cure for cancer but who will not disclose it. Should another who receives the information in confidence be able to disclose it? It has been said in an English case that the public interest could never justify the disclosure in breach of confidence of a valuable chemical formula. But such questions do raise grave questions. A recent instance is that of the Lion Laboratories breathalyser machine, where the public interest in the administration of justice and acquittal of innocent persons was argued to demand the disclosure of test reports on the accuracy of these machines which had been used as evidence of drunken driving by defendants in criminal cases. The courts may order the disclosure of such confidential information in a trial and such disclosure cannot be a breach of confidence.

2.7.4 Independent discovery

It has already been mentioned that being first in the field does not prevent someone else from exploiting the trade secret if they come upon it legitimately, for example, through independent discovery of it or reverse engineering in the absence of a contract not to attempt reverse engineering. If there is any serious prospect of this and patent or other protection is available to you, then of course it must be seriously considered as soon as possible.

2.7.5 Inequitable conduct

If the plaintiff seeks an injunction to restrain publication or disclosure or use of the confidential information he or she must remember that injunctions are only granted in the discretion of the court and that any inequitable conduct on his or her part may induce the court to refuse the injunction.

2.8 OTHER WAYS OF PROTECTING TRADE SECRETS

The use of the law of confidence as a means of protecting trade secrets does not exclude all other legal means. Although as has been made clear the option of taking out a patent or registered design will automatically exclude the possibility of maintaining trade secrets protection for the published material, trade secrets protection can supplement these forms of property right to a degree and confidence can itself be backed up by other causes of action in law. Two in particular may be used to back up the obligation.

2.8.1 Criminal law

In certain cases it will be possible to invoke the criminal law, though of course this may not prevent the break-out of information, but merely penalize the perpetrators. This will not arise out of the breach of confidence in itself but from use of illegal means of obtaining the information such as burglary, bribery, telephone tapping, etc.

Where use of the criminal law and – not something necessarily guaranteed – action on the part of the relevant authorities can be invoked, this may have the advantages of greater resources being directed towards the pursuit and detection of the guilty parties and the enhanced powers of investigation search and inquiry which these authorities enjoy. On the other hand, the present powers of courts to order compensation to victims of crime are of little use in the UK in this context and the victims may find their lack of control over the proceedings extremely inconvenient.

In any case, because of the law and the manner in which they operate, the opportunities for criminal sanctions over such activities by a professional are few, because a professional will usually take no physical property, but merely copy the information, if only to reduce the risk of detection of the leak.

EXAMPLE: *Oxford v. Moss (1978)*
The defendant was a University engineering student who had 'borrowed' an examination paper, copied it and put it back where he had found it. He was charged with its theft but acquitted. Theft requires an intention permanently to deprive another of the property alleged to be stolen, and the student had returned the paper. Secondly, the student had not 'stolen' the information on the paper either, not least because the definition of property in the *Theft Act 1968* does not extend to information.

Somewhat artificial prosecutions may sometimes be brought where computer hacking or other such electronic techniques have been used, employing crimes such as the abstraction of electricity under the *Theft Act 1968* or even the *Forgery and Counterfeiting Act 1981*. In other countries the possibilities of using the criminal law in such cases are rather greater and should not be ignored. In many of the States of the USA specific criminal statutes exist to deal with the question of abstraction of trade secrets, though it should be said that these have generally proved to be unsatisfactory in practice. Special provision also exists for the problem of computerized data banks which facilitate huge information gathering exercises.

In the UK specific legislation does deal with some telephone tapping, and such activities as industrial espionage may amount to a conspiracy to defraud where two or more persons are involved. Where some element of deception has been employed such as impersonation, and a pecuniary gain results from it, the chances of gaining a conviction for obtaining a pecuniary advantage by deception are also greater. Obviously specialist legal advice should be taken in such circumstances if private prosecutions are to be contemplated, and police assistance may be sought.

2.8.2 Copyright

Very frequently plaintiffs seeking remedies for breach of confidence can also rely on copyright where the information was put down on paper and copied or reproduced by the defendants. This is a useful back-up. Always have the trade secret well documented – and of course keep the documentation safe. On the one hand copyright is more durable in that it survives such a publication untarnished. On the other hand it provides less protection in that it is only the form in which the trade secrets are expressed that is protected and not the content of the information itself that is protected, as has already been discussed. The confidential drawings of specific components and the like are easier to protect through the use of copyright than manuals of procedures or processes. Certain difficulties may be encountered in the USA when attempting to rely on both trade secrets and copyright, as indicated earlier. Copyright is considered in more depth in Chapter 6.

2.9 PRACTICAL STEPS IN PROTECTING TRADE SECRETS

The law exists to help you but the best course of action is self-help to protect confidential information from any unauthorized use or disclosure. A number of practical steps may be taken to make best use of the available legal protection for trade secrets and similar information. If some of the following are not possible for one or more reasons, others may still be of use. Most are relatively cheap to introduce.

2.9.1 Employment contracts and after

Questions of the employment contracts of research and other staff are considered in greater depth later. Here it suffices to summarize one or two points. As has already been mentioned it is possible to enhance the already implied obligations of a contract of employment so far as the law of confidence is concerned, but these implied obligations should themselves be made express in any case both as a measure of precaution and as a way of bringing them to the attention of the employee when he or she joins the firm.

An additional useful precaution relates to the employee about to leave the firm. This involves the signing of a letter, a model of which is reproduced in the appendices, whereby the employee is reminded of his or her obligations both during and after employment and requesting that the employee should sign the letter and return it to the sender and disclose any inventions made during his or her employment of which the employer may not be aware and any information of a confidential nature to which he or she has had access, and to return any documents of a confidential nature in his or her possession. Even if the employee does not comply with the terms of the letter, his or her signature and return of it establishes the cause of action should any subsequent unauthorized disclosure on the employee's part be detected as well as assisting in questions of ownership of inventions and designs and merely to send it to the employee puts him or her on notice of obligations.

2.9.2 Visitors' books

It is good practice to maintain a visitors' book signed by anyone entering the firm's premises – and not just selected security zones. This book should contain a prominent notice which the visitors see before signing, whereby they acknowledge the obligation of confidence upon them in respect of information they may acquire in the course of the visit. A model notice appears in the appendices.

For legal purposes this is useful in so far as it expressly imposes the obligation of confidence on visitors. It does not stop the use of a false identity or other 'James Bond' style activities but it does impose some limited degree of supervision and enables a firm to a limited degree to establish who had access to information which has subsequently been leaked and to that extent operates as both a deterrant and as a means of preventing future break-outs by detection of the offending parties.

2.9.3 Restricted zones in research and production

These of course make sense, but how many firms maintain the regime properly? It is difficult for a junior employee to deny entry to a senior manager because he 'hasn't bothered' to get a pass for the day. Where any sensitive research is going on there should be proper provision for authorization and passes, combined with special provision for maintaining proper security by locking up research books etc. Senior staff should of course lead by example.

2.9.4 Pre-contract negotiations and submissions

These require consideration from two points of view.

Firstly, what of the submitter of an idea? The submitter may well wish to prepare the way for negotiation by obtaining an undertaking from the proposed recipient to maintain confidentiality. A draft letter for beginning this process appears in the appendices. The main difficulty the individuals face is the practical one that they will probably not get an undertaking and the imbalance of bargaining power leaves them in a weak position. For them the only tactic may be to turn up and preface their presentation with an express imposition of an obligation of confidence. That way the listeners must either stop the presenter from continuing and dissolve the meeting or, if they sit back, impliedly accept the obligation imposed upon them. At least the tactical advantage lies with the presenter in such a situation, particularly if a sufficiently vague but interesting outline of the nature of the invention he or she is trying to market preceded the detailed exposition prefaced by the warning.

Secondly, from the point of view of the company prone to receive submissions, a number of defensive measures may be taken.

(1) Firstly, it is possible to introduce a regime whereby all mail directed to the research division must pass through a non-technical department so that if any such submissions are made they can be returned immediately without being seen by a technical person, accompa-

nied by a letter explaining what has been done and inviting resubmission on a non-confidential basis, while assuring the submitter of the firm's good faith. A model letter appears in the appendices.

(2) Secondly, as a more general defence mechanism, a general notice stating the terms on which such submissions are received may be appended to technical publicity material, putting the public on notice of the regime applied in the company.

2.9.5 Confidentiality notices and staff training

If a document is confidential make sure that the reader is put on notice that it is confidential. A notice to that effect on the cover is of considerable assistance. Although it is difficult to persuade a court of the confidentiality of a document which is widely distributed merely because it bears the confidentiality notice, any document which receives relatively narrow distribution and which is one of only a few copies will probably be protected by this means, so long as it is treated carefully, and not left lying around for anyone to pick up etc.

Also make the staff understand why confidentiality matters and explain fully what their obligations are and why they matter. Employees who understand why they are meant to do something are more likely to do it. There should be an attempt to make the firm's personnel more security conscious.

2.9.6 Physical back-ups and personal searches

These understandably create a certain amount of employee resistance and always require a great deal of sensitivity and discretion, but they may be necessary in certain cases, to prevent unauthorized removal of documents to an employee's home. Much of the difficulty may be removed by adequate explanations in the training programmes advocated above. Such precautions may even mean preventing the employee from working on a project at home in the evenings. In many ways it is desirable to prevent this in any case, as discussed in Chapter 7.

Mere locked doors are a good initial barrier to the casual inquiry. Shredding machines for documents which have sensitive content are essential. If at all possible, do not sub-contract shredding work. Buy an internal machine and use it. Similarly, items such as once-only typewriter ribbons and carbon papers should be destroyed. Similar considerations apply to the waste from the laboratories. Ask yourself, where do your failed prototypes and samples go? Make sure that they are properly disposed of.

Another precaution worth taking is the practice of maintaining records of distribution of documents. If the worst comes to the worst it at least helps to track down the source of a leak if only to prevent further leaks and to take disciplinary action against that person; and knowledge of the risk of detection of itself can help to deter leaks. One important way of doing this is to determine who had access to the information, by keeping a record of the distribution of documentation through an initialling or similar system. Include a record of when a document was destroyed.

2.9.7 Disclosure on a need to know basis

One of the best ways of keeping information secret is not to tell anyone about it. This is of course impractical but disclosure on a need to know basis is not, and it is a valuable tool so long as it does not hinder the research programme itself. There is no need to brief fully sales people not directly concerned with the marketing of the product, and even those who are concerned may be given a lower degree of briefing sufficient to attract interest in the product before drawing in technical staff to explain more fully the specification of the product under properly controlled negotiation conditions, which include confidentiality provisions. Even the best sales people can become carried away in the effort to win orders and commission. What they do not know they cannot divulge.

Related to this is the need for vetting of two important outlets of information about a company's work: the company journal or staff magazine, and the publications of the research staff in papers and at conferences. These can raise extremely difficult personnel problems which merit close attention from the personnel officer.

Every company likes to boost its employees' egos with information about new developments and orders won, usually in the company magazine. These journals are an important source of technical and of commercial intelligence surveyed with great interest by other firms. Technical staff often write an initial draft which the publicity officers of the firm edit into readable form. Neither class of employee is particularly security conscious by tradition and some form of vetting procedure of such material, perhaps by a member of the patents department if there is one, or some senior officer of the company if there is not, may and should therefore be introduced and maintained.

As to scientific and technical publications, although this issue is raised in all establishments, it is a particular problem in universities. Academics depend for their continued employment, promotion (and career advancement through outside consultancy work) on their published research results. The pressure to publish is considerable. Some also pursue the ideal of knowledge for all and for free – although the acquisition of that knowledge has cost their employer a considerable amount in the provision of facilities. There may well be express obligations imposed on the university by the research contract funding the work. These matters are considered in greater depth elsewhere. If publication is to be held up there must be a procedure for gaining permission to publish.

Vetting has to be quick and there has to be a good reason for preventing the publication. Individuals have to be persuaded that they are being deprived of glory for a good cause, especially in those cases where progress with one's career is dependent on publication. This problem links up with the response to a new discovery on the part of management and the procedures for assessment of it. This is considered in greater depth in Chapter 7 together with a suggested procedure for dealing with the problem.

Linked to this is the question of how much someone has to know to perform their job efficiently. Division of knowledge among individuals can be an effective way of making the task of industrial espionage harder by diversifying the sources to which one must go to get the whole picture or simply making the risk of harmful leaks smaller.

3

Protection for technology by patents

3.1 WHAT IS A PATENT?

A patent protects a new technological development, a practical invention. In simple terms, it gives you an exclusive right in law for a period of time to prevent others from exploiting a new invention in the country where the patent is held. The monopoly it gives you can be sold or licensed so that you can share it with other people or even give them all your exclusive right in it for the whole life of the patent or for a more limited period of time. The price of this legal monopoly, apart from the fees you have to pay the State to acquire and keep it, is that details of the new invention must be published in an official journal which anyone may read and on the expiry of the patent anyone may exploit the invention without the consent of the patent owner.

3.2 HISTORICAL BACKGROUND TO PATENTS IN THE UK

It is often said that the earliest patent regime in the modern sense can be traced back to a decision of the Venetian Senate in 1474. In England the practice of granting monopolies in certain products and processes can be seen as far back as 1449 when one John of Utyman was granted a twenty year monopoly in a process for making coloured glass, and a form of patent system grew somewhat haphazardly out of the exercise of the royal prerogative to grant such monopolies. In fact, the very term patent betrays its origins in English history. It is derived from the phrase 'litterae patentes' or open letters. Letters patent were legal documents issued by the monarch and sealed in such a way as to be open for public inspection. They covered many administrative and public matters, including such privileges as the grant of a coat of arms.

Elizabeth I in particular made very great use of the royal prerogative to reward trusted servants, even giving monopolies in everyday staple products. but the increasing power of Parliament did lead to the gradual curtailment of royal prerogative in this as in many areas of public life, and in 1623 the Statute of Monopolies imposed a general prohibition on monopolies. The Statute did however recognize the very special characteristics of invention monopolies, which made them desirable so long as they were kept within reasonable limits.

From the time when the *Statute of Monopolies 1623* accorded Parliamentary recognition to the validity of grants of monopolies for inventions for a period of fourteen years so long as those monopolies were not 'contrary to the law or mischievous to the State by raising prices of commodities at home, or hurt of trade, or generally inconvenient', the patent system developed along a more or less consistent line until finally, after the *Patents Act 1977*, an inventor ceased technically to be granted letters patent and instead the inventor today receives a certificate of grant issued by the UK Comptroller General of the Patent office.

3.3 INTERNATIONAL PROTECTION THROUGH PATENTS

There is no such thing as a world patent. Patents are national rights and they protect inventions only in the country where they were granted. But it is possible to gain patent protection throughout most of the world by making a series of co-ordinated patent applications in individual states. You may of course only want such protection in the UK itself or in some of the leading industrial countries. A variety of routes to patent protection at home and abroad exist. Let us look at the possibilities and the decisions you will have to make in choosing between them.

3.3.1 Applying for UK patents

You may wish to seek patent protection in the UK alone, at least initially. It is possible to obtain a patent in the UK by a direct application for a national UK patent under the UK Patents Act of 1977, or via the routes provided under the European Patent Convention or the Patent Co-operation Treaty, which will be looked at below. But if you only want patent protection in the UK then it is quicker and cheaper to make a UK national application direct to the UK Patent Office. An initial reference to the UK Patent Office will in any case be necessary even if you do not want protection in the UK, because on grounds of national security any UK resident who wishes to apply for patent protection abroad must seek permission from the UK Patent Office at least six weeks before doing so unless the resident has already applied for a patent in the same invention in the UK.

If the Patent Office for some reason considers that the invention in question contains information whose publication might be prejudicial to the defence of the realm or to the safety of the public it may prohibit or restrict that publication. There are provisions made for compensation of applicants in such cases, and criminal penalties for failure to comply with this requirement.

3.3.2 Overseas patents using the Paris Convention

You may decide that you wish to seek more patent protection overseas. All patent systems insist on some degree of novelty for your invention to be patentable. If you had to file your patent applications everywhere where you wanted protection at the same time to avoid the possibility of the invention being held not to be novel in Country *X* because of its prior disclosure in a patent application in Country *Y*, you would have to commit large sums of money to a venture which you did not yet know to be commercially viable. This would be a big gamble and a waste of resources.

To help to relieve the burden of international patenting, an international convention, the Paris Convention of 1883, ratified by most of the countries which have a patent system, now enables you to wait for up to a year after making your application in the UK before you have to file patent applications overseas, while still being able to back-date those overseas applications to the date of the original UK patent application. The 1883 Convention also guarantees equal treatment for domestic and overseas applicants in the signatory states. This means that your first application in the UK cannot harm your chances of gaining equivalent protection in overseas markets for the same invention. The subsequent patent applications need not be identical with the first one to gain the benefit of the back-dating procedure, but only that part of a later application 'supported' by the content of the first will gain the priority date. The timetable in Table 6 will give an example.

Table 6 *International patenting priority: Bloggs and Co.'s application*

---

1 April 1986:
Bloggs and Co. file a UK patent application for their new road surfacing material involving a new rubber-like compound which has extremely good properties for tyre roadholding.

1 October 1986:
Bloggs and Co. write about their new invention in a trade journal in an effort to raise further development cash for the idea.

1 December 1986:
Bloggs and Co. discover that there is another rubber-like compound which works just as well as the one they described in the UK patent application but which they omitted to describe in that first application. They decide to keep it a secret for the time being.

30 March 1987:
One day before the expiry of the twelve month period permitted by the Paris Convention, Bloggs and Co. file a French patent application for their new road-surfacing material.

The application in France on 30 March 1987 is back-dated to 1 April 1986 and so is valid despite the earlier disclosure of the invention in a trade journal and in the UK patent journal, where the UK application will by now have been published. But they cannot obtain priority for the second compound which they omitted from the earlier application.

Of course the more countries you seek patents in, the more money you spend acquiring them, so unless you have unlimited funds some discretion clearly has to be exercised as to the countries in which protection is to be sought. This is a commercial rather than a legal decision. Which markets are most valuable to you?

Which countries are a threat to you? Can you cut costs?

3.3.3 Linking a series of overseas applications

If you do want overseas protection, and this will often be commercially essential, it is quite possible to make a series of national applications in those countries abroad where you have decided that you want protection. But this is of course going to be rather expensive: it almost inevitably will involve a fresh set of patent papers, a fresh set of legal advisers, of fees, and translations into the native language of the country concerned, for every country you choose. Even though it is possible to delay the expense of multiple overseas applications for up to a year after your UK application by making use of the 1883 Paris Convention priority period, the expense will eventually be met. If you want to make applications in a series of countries then the routes provided by either the European Patent Convention or the Patent Co-operation Treaty, may have advantages over making individual national applications in each country for which you seek patent protection. Let us look at these routes, which effectively link overseas applications.

3.3.4 The European Patent Convention (EPC) route

At first sight the European Patent Convention may seem to create a partial exception to the rule that there are only national patents. It is possible to file a single 'European Patent' application in order to be awarded patent protection simultaneously in a number of member European countries including the UK. Under this regime you file a single application designating those Member States of the European Patent Convention in which you want protection and the application is dealt with in a single office. The fees payable depend on the number of countries you have selected. Using this system it is possible to obtain protection through a single patent application in Austria, Belgium, France, Germany, Greece, Italy, Luxembourg, The Netherlands, Spain, Sweden, Switzerland (including Liechtenstein) and of course the UK. Although the procedure is unified you are still granted a series of national patents, each interpreted by the national courts of the country concerned, and each requiring separate renewal fees and enforcement. So it is quite possible, for instance, to gain patent protection through the European Patent Convention procedure in all the Member States but to find that whereas the courts in West Germany have held it to be invalid, the courts in France have upheld it.

This is of course a mixed blessing. Although it means that one adverse decision in a foreign court does not overturn the patent protection you have acquired in other countries, it does mean that if the patented invention is of any commercial value the adverse decision in, say, Germany, which will certainly have been noticed by your competitors (whose agents will look out for this sort of thing for them), will induce those competitors to contemplate challenging or even ignoring your patents for that invention elsewhere, with the result that you run the risk of a series of court actions in the various countries in which you took out protection.

Thus it is now possible to gain patent protection in the UK for an invention by two routes: a UK national application made through the UK Patent Office in London or a European Patent (UK) application made through the European Patent Office in Munich. But at this point is should be made clear that while it is possible to apply both for a national and for a European UK patent in respect of the very same invention the Comptroller General of the Patent Office has the power to revoke a UK patent obtained through the national patent route if both types are granted.

3.3.5 The Patent Co-operation Treaty (PCT) route

The Patent Co-operation Treaty became effective in 1978. It has been ratified by nearly forty states. Again, it does not create a world patent, but it does simplify the means of filing many patent applications simultaneously in a number of countries. It is in the end very expensive but time saving and it postpones costs.

As a national of or a resident in the UK, it is possible for you to file an international application under the Patent Co-operation Treaty in which you specify which of the Member States you want protection in and to send it to a single 'receiving office'. For UK nationals or residents this receiving office is generally the UK Patent Office so you submit an English language application.

This has the merit of simplifying the procedures and of delaying some of the expense as compared with filing individual applications in those countries right from the start. Initially only relatively small fees for each country designated in the application are charged by the office which receives the application. The receiving office sends the application to an 'International Search Authority', in the case of UK applicants usually the European Patent Office, which undertakes the initial steps in examining the patent application and then at about eighteen months after the application it and the search report of the International Search Authority are published by the International Bureau of the World Intellectual Property Organisation in Geneva. Copies are sent to the applicant and to the Patent Offices of every country designated in the application. Within two months of this you file the application at the national patent offices of those countries for which you still want protection, providing the fees, translations and other formal documents required by the national laws of those countries, so that from that moment onwards the patent application is treated in each country just like a normal, i.e. national application, but you will have avoided committing yourself to that expenditure for nearly a year and you will have had a chance to reconsider whether you want protection in all of the places you originally designated. Some countries also accept a so-called International Preliminary Examination by a central authority in place of their own national examination when deciding on whether to grant a patent.

The list of countries for which you can use the Patent Co-operation Treaty route is to be found in the appendices to this book. You will see that it is possible to specify the European Patent as one of the patents you seek, and combine the advantages of the European Patent Convention and the Patent Co-operation Treaty. The point of all this is that it enables you to postpone the vast expense of multiple applications until after a preliminary opinion about patentability has been given by the International Search Authority, when you can give up altogether or press on with a better idea of which countries you ultimately want to pursue protection in. You can therefore abandon some countries at minimal expense.

The Patent Co-operation Treaty (PCT) gives you a long time in which to postpone expenses because it is only on receipt of the patent application by the national offices that translations are required for further progress. Furthermore, the information gleaned by the International Search Authority is sent to the national offices which much reduces the cost of their search. But the PCT cannot eliminate the various national searches altogether, because the national patent offices have not yet all reached agreement on harmonized standards etc. and so the International Search Authority merely consults an agreed list of common matters, leaving the final decision to the national offices themselves. Moreover the various International Search Authorities used by the PCT system do not all use the same materials in undertaking their search, which differ according to the working libraries of the authority concerned.

A Patent Co-operation Treaty application can even claim the benefit of the priority date of an earlier national patent application under the Paris Convention. Thus, if you have first chosen to patent in the UK and seek a national patent at the UK Patent Office alone it is possible for you to seek later to obtain protection in a number of foreign countries still using the Patent Co-operation Treaty route if you act within the twelve months of filing your UK national patent application, for under those circumstances the PCT application can be back-dated to your UK priority date in just the same way as any other overseas national application.

3.3.6 The Community Patent Convention route

This Convention is NOT yet in force but it will be a significant development in international patenting which is intended to provide a genuinely European-wide patent covering all of the EEC and it will eventually replace the European Patent route altogether for the EEC Member States. The 1977 UK Patents Act already has provisions dealing with its eventual implementation in British law. When it comes into force the Community Patent Convention will provide a 'unitary' patent, i.e. one covering the whole of the EEC, and one which can be assigned and licensed for the whole of the EEC alone and not for individual member countries. It will not be possible to have an exclusive licensee or owner in the UK and a different one in, say, France. It is designed to avoid the possibility of division of the EEC into separate national markets by the use of patent rights, which has the effect of partitioning trade between the member states, contrary to the intentions of the Common Market.

Which route are you going to take?

Deciding on the proper course of action is not just a matter of mathematics, of calculating the respective fee totals of the routes available. Which route you choose will depend on a number of variables, including market forecasts, strength of protection, on available finance, etc. It will involve discussion with your patent agent, commercial and technical staff. There is no golden rule: all depends on the individual invention in question. Some of the factors you will consider in taking this decision will include the list set out here:

- PCT and EPC applications involve a more extensive literature search than those in the UK. This means that your patent application is more likely to get through in the UK but more open to subsequent challenge after grant. What you win on the swings you lose on the roundabouts, but, of course, it may be a consideration that to have an invalid patent is better than to have no patent at all.
- Patent Co-operation Treaty and European Patent Treaty applications may be cheaper than a series of national applications in a number of foreign countries for large numbers of patent applications in respect of the same invention or may at least delay some expenses until later. But where you intend to seek protection in only a few overseas countries, it may still be cheaper to use the individual national routes.
- Filing a UK national application first not only gives you a priority date but also provides a cheap way of getting a first official impression of the chances of success for your application, by giving it to you before you embark upon the expensive business of multiple applications and giving you time to react.
- Confining subsequent applications for the same idea to the scope of the initial application helps you to delay expenditure by making use of the full one year priority period. But a number of countries permit a degree of amendment and refinement upon initial filings without the sanction of losing the earlier priority date for the new material. The precise amount of leeway you have varies from place to place.

3.4 WHAT INVENTIONS ARE PATENTABLE?

Although, as the last section pointed out, patents are generally only national in their scope, so that for world-wide protection it is essential to make a number of applications, we cannot cover the patent laws of the whole world in a book of this size, as national patents inevitably differ slightly in scope and the range of inventions which are patentable.

Fortunately, through international treaties such as the Paris Convention and the more recent Strasbourg Convention and the European Patent Convention, there is now an increasing degree of harmonization of patent regimes in the world, so that differences which still exist are diminishing with the passing of time. We will first concentrate on European patent laws and then look at the USA, where there are some significant differences.

3.4.1 Patentability in Europe and the UK

In European Patent Convention countries, to be patentable, an invention must:

(1) Be novel
(2) Display an inventive step
(3) Be capable of industrial application

These basic requirements each apply equally to an application for a UK national patent, a European (UK) patent and to an application for a national patent in other countries which have signed the European Patent Convention. As we have said, very similar requirements apply to countries outside Europe but only the USA will be looked at in any depth.

(a) Novelty

The first requirement of patentability is novelty. An invention is considered to be new in law if it does not form part of the 'state of the art'. The state of the art is judged by looking at all the information, or 'prior art', which has at any time before the 'priority date' of your application been made available to the public whether in the UK or anywhere else in the world, by written or oral description, by use or in any other way. The description 'prior art' means technical developments prior to the application and from which the invention advanced.

This makes a huge amount of information relevant in deciding whether your invention is novel in law and the patent offices of the world keep vast stocks of the technical journals, references in books, magazines and newspapers, advertisements and the like, which their examiners search for anticipations of the invention for which a patent is sought.

For a UK patent, the 'priority date' of a patent application is usually the date of filing in the UK Patent Office, though in some cases, by virtue of the Paris Convention, where the applicant is a foreign resident who applied for protection elsewhere earlier, it is possible to rely on an earlier filing abroad within twelve months of the UK application.

As we have said, UK residents in this country have to file in the UK first to comply with national security procedures, at least six weeks before foreign filing. The 'prior art' therefore includes any matter contained in an application for another patent (either UK or European designating the UK) published after the priority date of the patent application in question if its priority date is earlier than that of the patent application in question. In the case of applications to the European Patent Office alone this applies to prior applications to the extent that they designate the same states. This has the unfortunate effect that the patents granted for various European states may differ in scope, even though they were initiated in the same patent office, the European Patent Office in Munich.

All this sounds very complicated. But it has a number of very simple practical consequences for those with an invention of any commercial potential.

- The significance of the priority dates is that your application can be defeated by an earlier patent application which has been filed but not published at the date that you applied so that you could not have had any knowledge of it at the time.
- Secret use by another person, for instance in secret experiments, which used to be a disqualification in some cases under the UK's 1949 Patents Act regime, is now wholly immaterial to the success of your patent application. It is not the first to invent but the first to apply for a patent in an unpublished invention who wins the race. There is therefore always a risk in deciding to keep a patentable invention secret and delaying application. If patents do not bother you but you want to exclude others from patenting the idea you have had you should publish it as soon as possible, preferably in a trade or research journal.
- Publication anywhere in the world is relevant to novelty – even if it was in a language which is very difficult to understand, and the inventor in the UK had never heard of it. So Japanese, Russian, even Mongolian publication or public use count. Of course, if you have not come across the earlier work in searching the literature when deciding whether to apply for a patent, the patent office may not do so either. But that very obscure information would if discovered later still be a ground for revoking your patent.

● Publication in documentary form – such as in a research paper – or mere public use of the invention elsewhere can be cited against your application. And, moreover, use of something which although it is not made available to the public as such is still easily discoverable from what is made available may defeat you – for instance a product made from a process which is easily deducible by reverse engineering will sometimes be held to reveal the process to the world, not just the product of that process.

● Information communicated to even one person without an obligation of confidence being imposed on that person has in the past been held to constitute publication. Communication of information to a wide range of people under an obligation of confidence will not but the courts will be sceptical of suggestions that there was an industry-wide secret. The importance of keeping potentially patentable material confidential until a decision whether or not to patent has been taken should be obvious.

● Mere potential availability will not amount to publication if no one has actually taken advantage of it. Thus where an abstract of a thesis was available which stated that the full thesis was available in London, but there was no evidence that anyone had ever taken up the offer to inspect it, this was held not to amount to publication.

● Accidental disclosure and disclosure the true significance of which the discloser is unaware is just as fatal as deliberate disclosure. This highlights the necessity of taking adequate measures to safeguard any research information which may lead to some patentable invention, even if its immediate application is not at the time of any apparent use.

● There are qualifications to the fatality of prior disclosure.

Firstly, disclosure by the applicant within six months before the date of filing is disregarded if it was made in the course of certain international exhibitions, if the applicant states this on his application and furnishes evidence of it. The Department of Trade and Industry certifies which of the many trade exhibitions count, but they are few and far between.

Secondly, it is also excused if disclosure takes place within six months before the application and it took place as a result of being unlawfully obtained or in breach of confidence from the inventor or from any person to whom it was made available in confidence from the inventor or because the inventor believed he was entitled to obtain it, or from such a recipient. For this purpose the 'inventor' includes the owner of the invention for the time being. Breach of confidence and the protection of trade secrets is discussed at length in Chapter 2.

● One form of prior disclosure which was at one time excused in the UK, but which is now not excused, is prior disclosure to a learned society. Nevertheless, today most academic disclosure is not in this form, so the problem is not so important as it once was, in the sense that most academic disclosure is now in a form which was never protected in Europe: the publication of research results in journals which may be read by anyone.

The European novelty rule, preventing patenting of material which has been disclosed already, particularly by academic disclosure in research journals, is a major difference from US patent law, which it is essential to understand and which is discussed at length later in this chapter.

(b) Inventive step

To be patentable the invention must not only be novel. It must also be shown to involve an inventive step. An invention involves an inventive step if it is not obvious to a person 'skilled in the art', having regard to any matter which forms part of the state of the art but disregarding unpublished pending patent applications. This means that in contrast to the legal position on novelty, the inventive step is not to be judged in the light of patent specifications which the person skilled in the art could not have known about, filed before but published after the priority date of the application under consideration.

The threshold of the inventiveness is difficult to lay down, being a question of fact in every case. The European Patent Convention seems to require, and the European Patent Office to apply, a much higher standard for applications than was the practice of the UK Patent Office under the 1949 Patents Act. Nevertheless, the difference will be one of emphasis and degree only, not one of approach and it will still be the case that, for instance, the invention, to be patentable, need not be an advance as such in the sense of being better than existing methods, it need only be a different method. Factors which will be taken into account will include the following.

- How close the idea is to the state of the art – for instance, whether other inventors came up with the same idea at about the same time, which might indicate that the invention was not particularly ahead of the state of the art.
- Whether the invention meets a long felt want and the manner in which the commercial world receives it, together with its commercial success and its degree of commercial significance.
- The technical advance of the invention itself (as opposed to its extraneous characteristics) is what counts.
- The degree of effort put into an invention is irrelevant. Thus, the inventive step may merely be an accidental and misunderstood discovery of a solution to a particular problem, a putting of existing technology to an unexpected use. It does not matter that once the idea has been thought of the means of putting it into practice is obvious, so long as the idea itself was not. But the mere analogous use of something known or the fresh application of a known technique will not be sufficient unless it required some inventive step to overcome the technical problems involved. Although for many years disapproved by the courts, the practice of adducing evidence of how the invention was come upon has always been common and now the courts appear to accept it for the purpose of assessing inventive step. Thus complete records of the research project may be crucial to the success of a claim. This is considered further in Chapter 8.

As we said above, what is obvious depends on what is obvious to the person skilled in the art. Quite what his or her (or their) level of knowledge and capacity is will vary from case to case. Points to bear in mind are the following.

- In the highly developed industries the person or persons skilled in the art may well be a multi-disciplinary team.
- The amount of knowledge the persons skilled in the art are endowed with has been in the UK at least subject to much disagreement in the courts. Some cases assumed that they are aware of all the literature which counts for assessing novelty; some that they are aware of all the literature but choose to ignore the more obscure of it; some that they are only aware of what diligent workers in their position would be aware of if they went on a search for literature in that problem area; some that they are aware only of what is commonly known in the trade concerned.
- In assessing inventive step the courts picture the persons skilled in the art as thoroughly uninventive and unimaginative despite their assumed great specialist knowledge of the trade in which they work. They are well-informed and highly trained dullards.

These requirements of inventive step may appear to be very difficult if not impossible to satisfy for all but the best inventions. In practice, however, they are not. Do not overestimate the standard you have to meet. A quick look through some of the more recent patents in your field will reveal that even quite simple things which with the benefit of hindsight are obvious have gained protection and scientists never cease to be amazed how low the standards really are. The golden rule is this: whenever you devise something which is of potential commercial value, consider calling in the patent agent to assess its patentability, or other possible protection.

(c) *Industrial application*

To be patentable the invention must be 'capable of industrial application'. An invention will be capable of industrial application in accordance with the requirements of the law if it can be made or used in any kind of industry, including agriculture. Note that there is no requirement that the invention should actually have been applied industrially.

(d) *Inventions which may not be patented*

The general requirements for patentability which we have just looked at – novelty, inventive step and industrial applicability – are supplemented by a number of specific exclusions from patent protection, which require consideration in their own right, and which reflect specific policy decisions which have been taken by governments on the desirable scope of patent protection. There are a few minor differences between the laws of some of the European Patent Convention Member States. The following is the UK position.

(i) Pharmaceuticals and methods of treatment

New pharmaceuticals are now patentable in most countries in Europe, including the UK, though they were not patentable for many years in Italy and are still not in some countries on moral grounds. But in the UK an invention of a method of treatment of the human or animal body is not patentable, even though a product consisting of a substance or composition is not to be prevented from being patented merely because it is intended to be used in any such method of treatment. Other countries do now permit the patenting of methods of treatment as such and the presumably policy basis for the exclusion is not wholly clear.

However, the 1977 UK Patents Act does provide that 'in the case of an invention consisting of a substance or composition for use in a method of treatment of the human or animal body by surgery or therapy or of diagnosis practised on the human or animal body, the fact that the substance or composition forms part of the state of the art shall not prevent the invention from being taken to be new if the use of the substance or composition in any such method does not form part of the state of the art.' This is a crucial provision for the pharmaceutical companies. It is now generally interpreted as meaning that the first but only the first medical application of a known substance is patentable. Various attempts have been made by patent applicants to circumvent this prohibition on the patenting of second and subsequent medical uses by various drafting devices. These have met with varying degrees of success before the Patent Office.

(ii) Theoretical discoveries

The law provides that a discovery, scientific theory or mathematical method as such cannot be patented. In other words, what we need is the application of the discovery to some practical end.

(iii) Literary, dramatic and other artistic works

The law provides that a literary, dramatic or musical work, or artistic work or any other aesthetic creation whatsoever cannot be patented 'as such'. It may be that where aesthetic creations are capable of industrial application, perhaps as industrial articles, they may then be patented. Generally, these are all perfectly adequately protected by the law of copyright and to a lesser extent by the law of designs, to which they should be confined.

(iv) Business and calculation methods etc.

The law provides that a scheme, rule or method for performing a mental act, for playing a game or doing business or a program for a computer, cannot be patented. Contrast the pieces of equipment for a new game which may be. The most discussed exclusion is possibly the computer program. This will be considered later.

(v) Presentation of information

The law provides that anything which consists in the presentation of information is excluded, e.g. in methods of notation, or of learning a language. But if the method is actually applied to a product or result, then on the basis of the old law they will qualify, being more than the presentation of information *per se*.

(vi) Immoral inventions

The law provides that a patent will not be granted for an invention the publication or exploitation of which would be generally expected to encourage offensive, immoral or antisocial behaviour; it is not enough that the behaviour would be prohibited by law.

(vii) Animal and plant life

The law provides that a patent will not be granted for any variety of animal or plant or any essentially biological process for the production of animals or plants, not being a microbiological process or the product of such a process. It is important to recognize the scope of this exclusion, which does not appear to exclude from patent protection many examples of innovation in the processes and products of genetic engineering. These matters are discussed in more detail later. For the time being it is sufficient to note that there exist requirements for deposit of specimens in the case of patent applications for microbiological species.

3.5 HOW DO YOU APPLY FOR AND OBTAIN A PATENT?

This section is concerned with:

(1) The procedure for obtaining a patent (Section 3.5.1)
(2) Requirements of a valid application (Section 3.5.2)

3.5.1 Procedure for applying for a patent

Of course, patent application procedures differ from country to country. In some countries there is basically no examination of the substance of a patent application and it is left to the courts to examine validity if someone else challenges this. In other countries there is a very rigorous examination system. The European systems are basically the same as that in the UK, but some other systems, and in particular the US system, do differ markedly from the UK system.

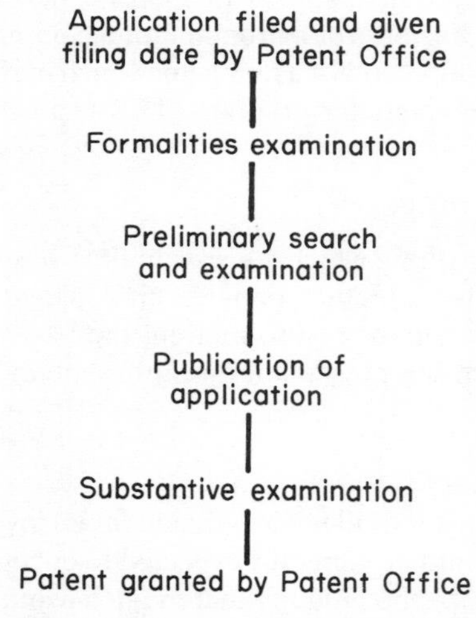

Figure 1 Outline of procedure before UK Patent Office

(a) Procedure for applying for a UK patent (see Fig. 1)

When applying for a patent in the UK you request the grant of a patent on an official form, paying an initial filing fee. With this request form you file a specification, which describes the technical content of the invention, plus any drawings which help to describe it, and with the full name and address of the applicant. This is then given a priority date and a number which it retains throughout its life, whatever its fate. At this stage the law permits you to apply the terms 'patent applied for' or 'patent pending' to the invention.

There is no need for the specification to be particularly sophisticated at this stage, so that even relatively embyronic inventions may be protected early on in their lives in cases of urgency. Nor need there be any 'claims', i.e. statements of precisely what it is that you are seeking monopoly protection for at this stage, merely a description of the invention.

The wording is still important, because it must form the basis for a later and precise specification which you will eventually have to supply and which is interpreted as a legal document determining the scope of protection you are to be given. Further, you cannot expand the description you initially submit to include material which was not there at first, even if you knew about it at that time or forgot to put it in, for any subsequent refinements must be within the scope of the original draft. Nor can your later claims cover any new material not in the original description.

You now have about twelve months in which to review the position, to test out the markets, to decide which foreign countries you will seek protection in, etc., before it is essential to make the next formal move. And there is now nothing to stop you from marketing the invention and giving it the greatest possible publicity once you have the priority date and number, though you may if you are very cautious still keep it secret for there is still some time left during which it is possible to withdraw the application before it is published and therefore retain the possibility of exploiting it as a trade secret in appropriate cases.

Within twelve months after the date of filing, you have to file a claims form stating precisely what it is about the invention for which you are seeking the monopoly; you must also file an abstract of the description and make a request for a preliminary prior art search and examination of your application, and pay the search fee.

If you are relying on an earlier priority date, you must pay the search fee and request the preliminary examination and search within twelve months of the earlier priority date but you may file the claims and abstract within twelve months of that date or one month of filing, whichever is the later date.

Although you cannot make the original application any wider you can of course abandon it and file a fresh application which can then claim the benefit of the earlier application's priority date, although only in so far as the new application contains material which was in the old application. If you do this and allow the old patent application to lapse you can recover some of the fee for the old application, in respect of work which the Patent Office has not had to do on it. If you do make a fresh application based partly on the old one but with some new element – for instance a better model prototype – which could not be added to the old patent application, you must include the number and filing date of the old application.

When the Patent Office receives these materials they will be checked and you are given an opportunity to rectify errors in the formalities required. If you fail to make these corrections in the time specified by the Patent Office examiner, the application fails.

It may be that you wish to use up almost all of the twelve month period seeking backers or testing the market or developing the invention further to assess its commercial potential. This is perfectly acceptable. If with the expiry of the twelve month period you have done nothing, the patent application expires. Otherwise once you have made your application for a preliminary examination and search in time two months can then go by before anything more happens.

About two months pass and you receive the Patent Office preliminary search report, a document listing any relevant prior act which the Patent Office examiner thinks ought to be looked at when examining your own invention, e.g. which might limit the possible scope of protection in some way, though the examiner will at this stage come to no formal opinion about the patentability of your invention.

The examiner will look mostly at the existing patent specifications and a good search by your patent agent in advance should usually have uncovered most of what the Patent Office examiner looks at – and the patent agent will have taken account of it when drafting the application – though it will not uncover unpublished pending patent applications with an earlier priority date. You will also be given the date of the proposed publication of the application. On the basis of this report you will have a number of options:

(1) Abandon the application altogether
(2) File amendments to description or claims or both
(3) Leave the application as it is

You can abandon and withdraw a patent application at any time before grant by informing the Patent Office in writing of your wish to do so. This withdrawal is irrevocable and immediate. If you act quickly enough in withdrawing you may be able to prevent publication even at this stage and perhaps be able thereby to retain the protection of trade secrets for the invention. But this is unlikely, especially if the request for search was left until quite late in the priority period. It depends on how far the Patent Office has processed it.

Similarly, if you have decided to amend the patent application by reducing its scope, these amendments will not necessarily get into the first published version of the application unless you act very quickly, for the same reasons. In each case it depends on whether the publication date of the application has yet been notified to you. If you make any amendments before receiving the letter notifying you of the proposed publication date, the original claims and description will be published with the amended claims (but not with the amended description).

You may instead delay any amendment altogether until after the publication, and if you are only made aware of difficulties after you receive the notification of the publication date, this will in effect be your first chance to do so. The advantage of getting your amended claims into the first publication of the patent application is simply that if the patent application subsequently succeeds, any conduct by another person between the first publication and grant which infringes the patent claims as first published, including the amended claims, may entitle you to an award of damages: but claims as amended which did not appear in the first publication will not entitle you to such damages. The relative merits of this as against amending claims at greater leisure must be weighed up.

If you decide to proceed with the application, as it is or amended, and you are not the inventor or the only inventor, you must then within sixteen months of the filing or priority date, whichever is earlier, file a statement of inventorship and of your right to grant, which the Patent Office will send copies of to all the persons named in the statement as inventors. This document establishes how you come to own the invention, if you were not the original inventor, by assignment or by being the employer of the inventor.

If you go ahead with the application either as it is or amended in some way, about eighteen months after the application, the specification of the application, as amended by you after the initial search must then be published by the Patent Office. Until that time no one gets to see or hear about the invention except the Patent Office officials, but it is published in exactly the form in which you wrote it and anyone can read the published text, together with the examiner's search report.

This is both good and bad from your point of view. No later patent application, by you or anyone else, in respect of that invention, can succeed because it is not novel any more. The search report may alert others to your application and induce them to oppose it on the basis of patents belonging to them cited in the report. If the application is thrown out for a formal defect or because you have not paid the fees on time, you have lost the chance of patenting it for good.

On the other hand, once this publication takes place any later action you may bring for an infringement of your patent when granted, although it cannot be brought until after the patent itself has been published, may be back-dated to the date of this first publication of the specification.

From this point on it is open to anyone to send comments and objections to your application to the Patent Office, which will be considered by the Patent Office examiner when the final, 'substantive', examination of the application is made. You will be given a copy of any such submissions and also be permitted to comment on them. But the third party who has made these observations has no right to reply to your comments or to attend any hearings. The examiner may make what use of them he or she wishes.

Up to six months after first publication, you have to file a request for a substantive examination by the Patent Office, much preliminary work for which was done by the examiner in the initial search report, but which will be updated and improved upon, and pay a further fee. If you fail to do so the application lapses, and because publication has taken place you are not even back to square one, but worse off than before you started in so far as trade secret protection is not possible. There is a provision for buying an extra one month's time in which to make the request, but no more. Some people advise you to file this request at the same time as the request for the preliminary examination for safety's sake. If you withdraw after the preliminary examination you can obtain a refund of the fee charged for the unused substantive examination. The substantive examination checks again on the patentability of the invention and your compliance with the requirements of the patent law on disclosure etc. This may involve some material which the examiner did not cite in the earlier preliminary examination.

About twelve months later an official examination report is issued by the Patent

Office which will often raise objections to your patent based on the prior art found in the search or in the subsequent substantive examination. In the light of objections raised by the examiner you may have to make amendments even at this stage of the proceedings.

This may involve a certain amount of negotiation between the patent agent and the examiner. Amendments may be made to the application to meet such objections or voluntarily in the light of further reflections or research you have done. But no new material may be introduced at this stage. You are entitled to make amendments in response to the letter of objection sent by the examiner to overcome those objections, but in all cases the Patent Office has a discretion to refuse to accept amendments for good reason. Each amendment you make requires the payment of an additional fee and must be filed on a separate form.

These are set time limits within which any such amendments in response to the letter of objection may be made, which if not complied with will cause the application to lapse. These will be determined by the Patent Office examiner, who may, for instance, give you a period of – say – six months in which to reply. If you cannot agree, the dispute may be referred to the Patent Office for a hearing with appeals to the Patent Court and even in exceptional cases to the Court of Appeal.

If you have overcome this stage successfully, the examiner will accept your application and your patent is then granted and published in its final form. All these stages must be completed within an overall time limit of 54 months. There is no possibility of any more extension beyond this time, so that if you do not get through all these stages by that time the application will lapse. Protection endures for a period of twenty years from the date of filing and so long as the annual renewal fees which become due are paid.

(b) Procedure for applying for a European patent (see Fig. 2)

Because of the national security provisions in force in the UK, which require you to obtain special permission to seek European Patent Office (EPO) or overseas patents if you have not filed a corresponding UK application at least six weeks before initiating any overseas applications, you will begin a European patent application at the UK Patent Office unless you are able to claim priority from an earlier UK application, in which case you may choose whether to go via the UK Patent Office or direct to the European Patent Office. Either way the procedure is the same once the application has been forwarded to the EPO. (If an application is held up by the UK Patent Office on the ground of national security the procedure is as for a UK application.) To be valid the EPO patent application must be forwarded from the UK Patent Office to the EPO within fourteen months. In the highly exceptional case of this not occurring the *UK Patents Act 1977* provides for the EPO application designating the UK to be converted into a normal UK national patent application.

You file the application which must be in English, French or German (unless the language of the patent applicant's nationality is different in which case you may file in your national language but you must later on provide a translation). We might expect a UK applicant to use English in order to save on translation costs. The same language is used in all the subsequent proceedings relating to the patent application.

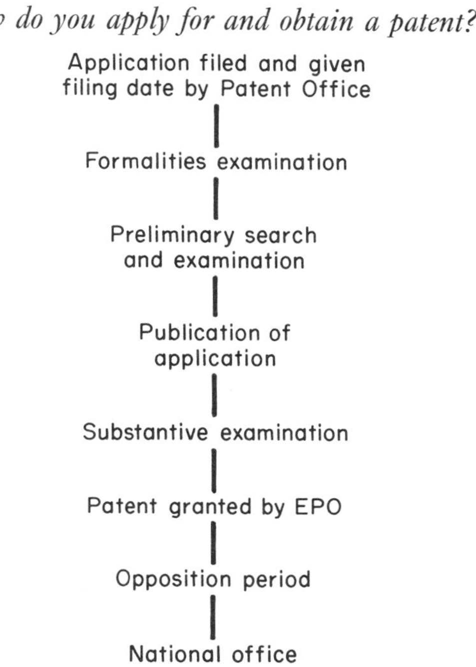

Figure 2 Outline of procedure before European Patent Office

A formal examination of the application is first conducted by the EPO Receiving Division at The Hague to determine that you have complied with all of the formal requirements for the application including translation, where necessary, and paid all of the requisite fees. It is at this point that applicants must designate which countries they seek protection in and pay the fees accordingly. At this stage too there must be a description of the invention, at least one formal claim, and an abstract plus any drawings, and the inventors must be designated. In the case of microbiological inventions there may also be a requirement to deposit a micro-organism if relevant.

While it is possible later to cut down the number of countries which you designate you cannot later increase them.

Any EPO objections concerned with formal requirements may be dealt with by the applicant who is given up to one month in which to pay fees and to make any necessary corrections. Only when all the formalities objections are met and fees paid will you be given a filing date, which may be important if there is no earlier priority date you can rely on.

A preliminary examination and search is conducted by the EPO Search Division in The Hague and a report is made on the application in much the same way as in the UK Patent Office. In the light of this report the applicant may amend the application once without consent and thereafter with consent but may not add material which extends the patent application's scope beyond that of the original application. At this point too any objections about lack of unity of invention may be met by filing a divisional application.

Eighteen months after the filing or the earliest priority date upon which the applicant relies, the application, search report and abstract of the application are published by the EPO. Following the publication of the application anyone may make observations on the application as in the case of a UK patent but that third person will not become a party to the proceedings.

The applicant now has a period of six months after the publication of the application in which to request a substantive examination and to pay all the fees this implies. On receiving this request, the EPO Examining Division in Munich next appoints a three person team to examine the application. Usually one alone of the team deals with the application but all three take the final decision. The examining team sends out a letter listing objections and giving a time limit in which any replies must be made. If there is disagreement, there is an appeal to a Board of Appeals, but no further appeal to the courts is possible.

The applicant who has overcome these obstacles must pay the fees for the grant and the printing within three months and must file translations into the other two official languages of the EPO, and the patent is then published in the original language plus translations. The EPO then grants the patent.

There now follows an opposition period lasting nine months in which third parties may oppose the grant of the patent. These oppositions are heard by a team of three in the Opposition Division, with appeal to the Board of Appeal, but again with no further appeals to the courts. The result may be to uphold the patent, to amend it or to revoke it, usually for all designated states, apart from the exceptional case where the objection is lack of novelty but the basis of this was an earlier unpublished application which did not designate all of the countries designated by the later application.

At the end of the opposition period the patent is sent to the national offices designated in the patent application, to be accorded the status of a national patent. Its treatment will vary slightly in the national office concerned. Most will, like the UK, convert the EPO grant into their national patent only when they have received a translation of the patent specification and claims into the official national language of the state concerned, if that is different from the official language used before the EPO.

Thereafter revocation of the patents granted is possible only country by country and there is no central revocation tribunal such as exists in the first nine months after the EPO makes the grant, so that any parties seeking the revocation for some reason of your European patents which had been granted in – say – nine of the Member States of the European Patent Convention might in theory require you to fight nine separate revocation actions, one in each country designated in the original European application.

3.5.2 Objections to a patent after grant

As we have seen, the fact that what you have invented is prima facie patentable within the criteria stated above is not sufficient for success. This is because the inventors or other applicants must confine themselves to claiming a monopoly in those characteristics of their invention which qualify for protection in this way and must disclose all that they seek protection for. This is reflected in several of the grounds for revocation of a patent granted by the UK Patent Office or through the EPO in Munich. Even if granted, a patent is liable to be revoked if it is invalid.

There are several grounds on which invalidity might be claimed in a revocation action brought by a third party. Grounds for revocation of a patent after grant are as follows.

- The invention is not a patentable invention. The types of invention which are patentable have been considered above.
- The patent was granted to persons who were not the only ones entitled to be granted the patent – this ground may only be invoked by persons whom the Patent Office finds to be entitled to the grant themselves, and may not be invoked more than two years after grant unless it is proved that any persons registered as proprietor knew at the time of the grant or of the transfer of the patent to them that the proprietor was not entitled to it. The question of who is entitled to apply for a patent and some problems of ownership of inventions are considered in Chapter 7.
- The patent specification does not disclose the invention clearly enough for it to be performed by a person skilled in the art, in other words it does not meet the requirements of adequate disclosure for a valid patent grant.
- The matter disclosed in the specification (i.e. that granted protection) extends beyond that in the application for the patent as filed, or in the case of divisional applications of the earlier application as filed.
- The scope of protection conferred by the patent has been extended by an amendment which should not have been allowed. Before grant a patent may be amended in certain cases so long as this does not have the effect of extending the disclosure of material originally disclosed in the application, even though that has the effect of extending the protection conferred by the patent when granted. But after grant, no extension of protection can be allowed by amendment, even if that does not involve any additional disclosures.

These possible objections to patent grants require us to look more closely at the requirements of a valid patent application.

3.5.2 European requirements for valid application

This section is concerned with the requirements of a valid application for a UK or European patent, and in particular

(1) What has to be disclosed to obtain a patent
(2) Limitations on what can be claimed in a patent

The problems can be summed up in this way. Even if an invention is patentable, defects of substance in the application may invalidate the grant. The patentee may meet objections from the Patent Office for disregard of any of the following three principles.

- The patentee must not claim a monopoly in more than what is described in the patent specification, i.e. the invention the patentee describes in the patent application must justify his or her claims for monopoly protection.
- The patentee must say enough about the invention to meet the publicity requirements of the patents legislation. The patent specification must thus disclose the invention in a manner which is clear enough and complete enough to be performed by a person skilled in the art.

- The patentee must not seek to protect more than one invention per patent application. Thus the specification must display unity of invention, i.e. must relate to a single inventive concept.

These principles govern the way in which the patent specification is filled in. The specification is the most important part of the collection of documents which form a patent application. The specification performs in essence two functions:

(1) It fulfils the duty of disclosure
(2) It delineates the scope of the monopoly claimed

To achieve this, every patent specification is divided into two important parts, called respectively:

(1) The description
(2) The claims

The patent application will also require an abstract. These must be looked at in turn. The same requirements apply to all UK and all European patents.

(a) The description

This part of the specification performs the function of telling the public how to perform the invention; it satisfies the requirement of disclosure which is the price of patent protection. It is often said that this is the most interesting part of the patent documentation to the scientist, because it shows him or her what the invention is all about.

In modern practice, the description begins with a simple title which indicates the nature of the invention, then goes on to explain the technical field, the background and the problem to which the invention provides a solution. It provides the essence of what the invention does and how it does it sufficient for someone who is trained in the relevant technical field to reproduce it, with examples of its application and preferably illustrative drawings to amplify these where applicable.

It is not essential to describe down to the minutest detail everything about the invention including things which any trained person in the field would take for granted. Under the 1949 UK Patents Act the applicant for a patent had to do more than merely give a clear and complete disclosure. He or she had (and in the USA, among a number of important Common Law inspired jurisdictions, still has) to disclose the best method known to him or her of performing the invention. This is no longer necessary in the UK or in Europe, following the 1977 Act and European Patent Convention.

(b) The claims

This is the most important part of the patent documentation from the point of view of the lawyer or patent agent because it defines the scope of protection sought or obtained. The specification must end with at least one or more claims which must:

(1) Define what the applicant seeks protection for
(2) Be clear and concise

(3) Be supported by the description
(4) Relate to a single inventive concept

This means that there must be nothing in the technical content of the claims which has not already appeared in the description or which is inconsistent with the description. It is by these claims that the patentability of the invention is ultimately to be judged.

Drafting the claims as widely as possible without overstepping the mark is a skilled business – too narrow and the patent will be worth little because it will protect little – too wide and the patent may be struck down or cut down for having trespassed into the known or the obvious, or simply for having claimed more than the description can justify.

It is common modern practice to put the central claim first and to make a number of dependent claims to applications of the central concept thereafter: some of these may be cut down during examination. In the EPC (but not in the UK) it is no longer possible as it was in the old legislation to include a so-called 'omnibus claim' in which the applicant sought to claim anything patentable revealed in the description which he or she has omitted to mention in the preceding claims. Omnibus claims are still possible in the UK.

Apart from these strictures, there is a final requirement of a specification, that of unity of invention. This requires that the invention for which the patent is sought must display a single inventive concept. The specification must not disclose more than one invention, unless there are two or more inventions disclosed which are so closely linked as to form a single inventive concept. The reasons for this rule are basically administrative. The rule:

(1) Ensures that you cannot gain protection for two inventions when only paying for protection for one
(2) Makes searching the prior art in the official patents journals easier when new applications for patents are made

You can divide your application if this objection is raised, by filing a divisional application, which is thence treated as two (or more) separate UK patent applications, each requiring its own set of fees etc., so it is not fatal. Furthermore, although lack of unity of invention is a reason for you being refused a patent grant by the Patent Office, unlike the other objections which may be raised against an application, for example unpatentability or inadequate disclosure, if the Patent Office does not raise this objection and so a patent is granted, no one can later seek its revocation on the ground of lack of unity of invention.

(c) The abstract

Every application must include an abstract of the invention in respect of which protection is to be sought. The purpose of the abstract is simply to inform others of what current applications are in existence. It is filed on a separate piece of paper and gives a short (less than 150 words) summary of the description, with one of the drawings if any taken from the original description to illustrate the invention.

(d) Amendment of patent applications

What can you do if, although you have a patentable invention, because of some defect in the way you have drafted the patent application, your application has failed to meet the standards set by the Patent Office examiner? It is possible to salvage a defective patent application or even a granted patent which comes under challenge, by amendment of the specification. But this has strict limitations, because excessive amendment is a ground for revocation of a patent too. Amendment may arise at three stages in the UK system:

(1) Application by the applicant before the patent is granted
(2) Application by the patentee after the patent is granted
(3) An order of the court or comptroller of patents in proceedings after grant in which the validity of the patent is put in issue

Note that in any of these cases the decision to permit any amendment, which is discretionary, may be challenged by interested third parties, apart from if filing a divisional application to overcome the problem of lack of unity of invention. Furthermore the 1977 Act prohibits any amendment which discloses material which extends beyond that disclosed in the application as filed or the patent as granted.

It may be possible to extend the protection you claim by drafting wider amended claims in the application before grant, but they can only be extended to cover material which is already to be found in the description of the invention as it was when first submitted for examination. It is not possible after a grant to re-draft the claims to make them wider, even if the same matter is disclosed in the invention's description. But correction of any purely clerical mistakes obvious on the face of the document is possible.

This points to the importance of getting as much right as possible first time around. For this the person who drafts the patent application, who will usually be a patent agent, will need to be as fully informed as possible. See Section 3.6 below.

3.5.4 Obtaining patent protection in the USA

The patent laws of the USA diverge significantly from those in Europe, particularly with respect to the law on priority. But although their terminology is very different the USA requirements for a patent of novelty, non-obviousness and usefulness are in essence the same as those for the UK.

Patent protection is afforded in the words of the Act to 'any new and useful process, machine, manufacture, or composition of matter, or any new and useful improvement therefor', subject to the conditions of the USA's Patent Act. There are certain judicial exceptions to patentability which broadly speaking correspond to those codified in the UK Patents Act and certain laws passed by Congress relating to inventions of security significance related to the atomic and space programmes of the USA. A similar power exists to that in the UK to prevent the publication of foreign filing of inventions prejudicial to the national security of the USA, again subject to criminal penalties for non-compliance and compensation for loss caused by any ban.

How do you apply for and obtain a patent?

The Patent Act protection in the USA endures for a maximum period of seventeen years only, subject to some limited possibilities of extensions in the case of pharmaceutical drugs.

The major and crucial difference between USA and European patent law is that the USA law protects not the first to file a patent application in respect of an invention, but the first to invent it, so long as he or she has not previously abandoned the invention. Only the true and original inventor(s) may apply, or their assignee.

Thus, for the inventor to be able to apply the invention must be novel. But novelty is construed very differently in the USA from in Europe because of the first to invent rule. It requires you to look at who invented first and at whether he or she had ever abandoned the invention. Under the USA law an invention is not novel if it:

(1) Was known or used by others in the USA before the date of the invention which is the subject matter of the application by the applicant, or
(2) Was patented or described in a printed publication in the USA or in a foreign country before the date of the invention which is the subject matter of the application by the applicant, or
(3) Was described in a patent granted on an earlier application for a patent by another applicant filed in the USA before the invention by the present applicant for a patent

In addition, even if you are the true and original inventor of the invention for which you seek a patent, you will be statute-barred from obtaining one if you have already abandoned it. You are held to have abandoned it if more than twelve months before the date of your application for the patent you have:

(1) Patented or described the invention in a printed publication in the USA or in a foreign country
(2) Publicly used or sold the invention in the USA
(3) Applied for foreign patent protection granted before the filing date of the USA application

The effect of this is that inventors in the USA have a period of one year after their first publication or sale of their invention during which they may still file a valid USA patent application. Equally, if you are the inventor, the fact that someone else publishes details of the invention between the date you invented it and the date, less than twelve months after that person's publication, that you applied for patent protection in the USA, will not deprive you of patent protection. The position is therefore very different from that in Europe where these events would have instantly deprived you of any chance of patent protection unless you could invoke the six months period of grace for so-called evident abuse. Nevertheless, caution in disclosing the invention is in practice still a good idea.

By contrast, foreign inventors, applying for a USA patent, cannot claim any invention date earlier than their date of filing for a USA patent or a priority date conferred on them by the Paris Convention of 1883. They cannot like the US inventor make reference to laboratory books to establish an overseas invention date earlier than this first filing. This is severely prejudicial to overseas applicants and much favours competing American applicants, especially in a race to patent. But any publication within twelve months before making a USA application will not prejudice that application, so that the time bar rules apply equally.

However, not everything about the USA system is more generous to the inventor than in Europe. A major difference is that whereas in the European system unpublished patent applications of which you could not have known may be cited by the patent examiner against you in respect of novelty, but not obviousness and inventive step, in the USA these prior publications are used to assess obviousness as well as novelty. There is also a rapidly developing doctrine of fraud on the US Patent Office, considered below, whereby lack of full and candid disclosure to the Patent Office of prior art (i.e. technical development prior to the application) of which the applicant was aware, may lead even after grant of the patent to substantial sanctions.

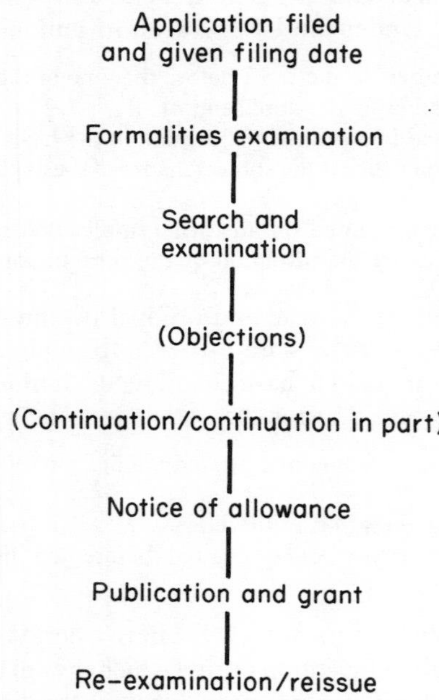

Figure 3 Outline of procedure before US Patent Office

(a) American filing procedures (see Fig. 3)

Unlike the European system, only the inventors may apply for a patent even if they have subsequently assigned the invention or their employer is entitled to it. The application is filed along with specification and a declaration of inventorship, and a filing fee is paid, whereupon the patent application is accorded a number, though there is provision for such assignment of a number before the fees and declaration have been received. Once the application has been received by the Patent Office, the applicants may put the wording patent pending or patent applied for on the product and market it, the idea being to discourage competition.

When filing the application, you should remember that the Patent Office asks the applicant who files an application to include in it or to accompany it by an Information Disclosure Statement in which any material prior art of which the applicant is aware is listed. A failure to disclose with candour material which you are aware may be relevant may well lead to rejection of the application, cancellation of any patent grant obtained thereby, denial of remedies for infringement by others, liability for defence fees and even criminal or antitrust liability.

The US Patent Office then examines the application for satisfaction of all formalities and assigns it to an examining group and then to an individual examiner.

The examiner then takes an initial look at the application and may object that the patent application claims more than one invention, in which case he or she will impose a so-called restriction requirement, in effect the equivalent of a European lack of unity of invention objection. The examiner may instead impose a species election requirement, where he or she feels that the applicant has claimed alternative forms of the same invention, requiring the applicant to elect just one of those forms for consideration. Where such a restriction has been imposed, the applicant is given a month in which he or she may ask for reconsideration of the case, giving reasons for the request, or elect which claims to pursue; but in any case the applicant may at any time during the course of the first application make a new divisional application, so as to claim the abandoned claim in a new and separate patent application.

At this stage the examiner will make the first so-called Office Action in which he or she either:

(1) Issues a Notice of Allowance under which the patent will be granted and the applicant moves straight to grant. All consideration of the merits of the patent application is terminated and no further amendments can be made of right to the application.
(2) Lists his or her objections to the patentability of the invention, of which there usually will be some. There is a three month period in which to respond to any such objections, though extension of this by a month may be bought as of right and subsequent extensions of up to six months maximum from the examiner's objections if good cause can be shown. This is the latest stage at which the applicants can fulfil their obligation to bring prior art of which they are aware to the attention of the examiner and escape potential liability under the doctrine of fraud on the US Patent Office.

Following the applicant's response to objections, the examiner will issue a second Office Action in which he or she chooses one of three options:

(1) Issues a Notice of Allowance by which the patent is to be granted, in which case all consideration of the merits of the application is terminated and no further amendments may be made to it or claims added as of right.
(2) Goes through a repeat of the procedure if any new art or new objections have been raised in the course of the applicant's response.
(3) Reaches a so-called final decision rejecting some or all of the claims, if no agreement can be reached with the applicant as to amendments or restrictions. In any response to this Second Office Action the applicant may seek to make certain amendments to meet the patent examiner's objections, perhaps by restricting the scope of the claims or dropping some of them, whereupon the examiner may withdraw objections. But following a final rejection of all or some of the claims, the examiner will at this stage now issue what is called an advisory action setting out his or her position, required amendments, etc.

Apart from these options, the examiner may find that someone else's published or pending patent covers substantially the same field as the one under his or her consideration, in which case he or she may declare there to be a patent interference.

This is designed for the case where two people invent more or less the same thing at about the same time, and the examiner is not certain who is entitled to a patent grant if any is to be granted and the patent interference proceedings which may follow are intended to determine this entitlement. The issue will be resolved in favour of the first inventor to reduce the invention to practice or the first inventor who exercises reasonable diligence to reduce it to practice even though the second inventor in fact beats the first inventor to it.

These interference proceedings may be very complex and place great importance on adequate records of trials, laboratory notes, etc. Even after a patent is granted it is possible to find yourself embroiled in one of these disputes.

Assuming any patent interference issues to be resolved, after receiving the advisory opinion, the patent applicant has one month in which to pursue one of the following options:

(1) Amend to comply with the advisory opinion
(2) Appeal against it
(3) File what is called a 'continuation' or a 'continuation in part' of the first application
(4) Abandon the application altogether

These may require some explanation as they are somewhat different from the European patent procedures.

(i) Appeal

A notice of appeal is followed within two months by a document specifying the grounds of the appeal. These are met by the examiner's response which may concede points or press objections and any new objections may be countered by the appellant. A team of three from the Board of Appeals hears the appeal. The decision of the Board of Appeals may take anything up to two years to be delivered.

Further appeals may be made to the Court of Customs and Patents Appeals or if there is new evidence which the Court cannot consider, then to the District Court in Columbia. In exceptional cases there could be an appeal from Columbia to the US Court of Appeals and thence to the Supreme Court or direct from the Customs and Patent Appeals Court to the Supreme Court: that would involve phenomenal expense.

(ii) Continuation

A continuation is a refiling of the original application, with the benefit of the original filing date, which enables changes and amendments to be made to the application which could not be made in the original proceedings because they had gone too far. This also gains you time and apart from a certain amount of additional expense and delay you have lost nothing.

A continuation in part is a new filing in which you include new material not present in the original application, including broader claims and descriptions, not just refinements or narrower claims. Again you retain that original filing date for the matter which was present in the first application, and a later filing date for the new material, that of filing the continuation in part. This enables you to abandon the original application.

The only difference so far as examination of the continuation in part is concerned from the normal procedure is that if the new claims are essentially those presented in the earlier application, and that earlier application has been finally rejected by the examiner, the procedure may in the discretion of the examiner be truncated so that the first examination and issue of objections is taken as the final rejection stage.

Where a series of patents are to be taken out as the result of a continually evolving research and development programme it is possible to avoid the problem of earlier applications from counting as prior art against later applications from the same programme, by filing the initial patent and following it up with a series of applications for divisional, continuation or continuation in part applications. In this way an issued patent cannot count against these later applications.

Following an issue of a Notice of Allowance by the examiner, the applicant has a period of three months in which to comply with any outstanding formal matters and pay fees after which the patent is printed and then published, which is the date of its grant. Protection then endures for seventeen years from the date of grant unlike the European system which takes the date of the filing as the relevant date.

Examination of the patentability of an application may also occur after grant of the patent. We should now look at this possibility, which may arise, in the guise of re-examination or of reissue.

(iii) Re-examination

At any time after the grant of a patent in the USA any person may request re-examination of a patent and cite new prior art not previously cited by the Patent Office. This may challenge the validity of the patent on the grounds of novelty or obviousness in the light of a printed prior publication. The Patent Office has, however, a complete discretion whether or not to accept this request, and it will only hear it if it is thought to disclose a real issue. There is no appeal against the decision of the Patent Office.

If the Patent Office takes up the issue the patentee is given an opportunity to make observations and thence the case proceeds as if it were a normal examination of a patent application with the original complainant taking no part in the proceedings other than by receiving copies of the papers passing between the patentee and Patent Office.

This may result in the confirmation of the patent as it is, or as amended to take account of the prior art, or its complete cancellation. It is not possible to widen the scope of the patent beyond that of the original grant in these proceedings.

(iv) Reissue

If after the grant of a patent it is discovered that there has been some omission in citing prior art, not through fraud on the Patent Office but through some mistake, nor through deliberate omission of broader claims in order to gain the original patent; or if there has been some other defect or inadequacy in the patent as granted, the patentee may offer to surrender the patent and apply for a reissue patent.

If the purpose of applying is to seek broader claims, this may be done within two years of the grant of the original patent. If the object is to narrow the claims in the light of new prior art, this may be done at any time during the life of the patent.

It is possible to file for reissue exactly the same patent claims. If the Patent Office thinks there is nothing wrong with it the original patent stands. If there is something wrong this will be recorded on the original patent but if the patentee can amend the old claims adequately to cover the prior art, the reissue patent will be granted. This might be done for instance to gain a preliminary and cheap view from the Patent Office of the original patent in the light of newly discovered prior art before embarking on expensive patent infringement litigation.

The view of the Patent Office will not bind the court hearing a subsequent action but it will be useful ammunition and a potential adversary may be invited to participate in these proceedings rather than waiting to fight a full later infringement action. If the potential adversary refuses this may be cause for a prejuducial award of damages. But all that an adversary can do in the reissue proceedings is to lodge a protest against reissue, cite prior art and explain briefly in writing its relevance. No right to be heard is allowed in the procedure, which from his point of view is unsatisfactory. You cannot use the reissue proceedings in an attempt to get round possible fraud on the Patent Office problems: the Patent Office examiners will look closely at the documentation to detect any such attempt.

Finally it should be noted that if a patentee has discovered that one or more of his or her claims is invalid he or she should disclaim it before entering into any litigation enforcing the patent. This simply results in that claim being disclaimed on the Patent Office record; no other examination of the patent results. The importance of this is that if in subsequent litigation a claim is found to be invalid there will be a sanction of costs.

3.6 WHAT TO TELL THE PROFESSIONAL ADVISERS

To do their job in preparing a patent application and prosecuting it before the Patent Office, and for that matter to do the rest of their job properly, the patent agent and other advisers need full and accurate information from the client. To get your patent agent and other advisers to work for you efficiently you have to give them the right information. What sort of information should you be taking to your patent agent when considering the patenting of some new invention?

In drafting a patent application, the nature of and requirements of any patent specification are such that as much information as possible about an invention should be given to the patent agent. Patent agents can always leave out unnecessary information: but they cannot exercise any discretion at all about information of which they have no knowledge. You can be sure that, as professionals, they are not going to disclose it to others or run off with it themselves.

The patent agent in drawing up patent applications has to tread a very careful line between not saying enough to get a valid patent and saying too much which may help competitors unnecessarily particularly about the know-how necessary to make the invention work properly. It is the height of folly to withhold information just because you do not want publicity for the details – that can lead to loss of protection altogether. You cannot claim the benefits of patent protection without paying the cost.

The following is a list of the information you need to take when consulting a patent agent about an invention.

- An explanation of the history of the invention, where you got the idea from, how you developed it, any early failures and possibly prototypes, with all your laboratory note books, etc. if possible. This will help the patent agent to explain the inventive step which it is necessary to establish to obtain the patent, and it also increases his or her understanding of the invention so as to maximize the skill with which he or she can draft claims and specifications for it.
- What you think is the central part of it, the most inventive element or most useful aspect, together with what other similar prior inventions you know of or have developed the idea from or improved upon. If you have developed an improved version of your competitor's products, admit it, be totally honest. It is vital to be such so that the patent agent can define your invention properly in making the application and avoid excessive claims which might be struck down.
- A detailed description of the best way of putting the invention into practical use, results of your tests and trials, etc., including all the failures and defects.
- Alternative ways of using the invention, and the substitutes for parts of it – i.e. will one chemical compound do as well as any other in the process, is there an optimum size, etc. It may be worth drafting the patent widely enough to cover less satisfactory alternatives – if this is possible – to prevent rivals from marketing a less satisfactory competing product which because of its defects might bring the whole genre of product into disrepute.
- Both after an initial search and during the course of the patent application it is important to respond quickly and accurately to queries which the patent agent may have, to help the patent application on the way and to save you money. Thus the client should in particular keep the patent agent informed of any new developments or improvements or other changes made to the invention and any rivals which appear etc. When the search reveals prior art or if the Patent Office cites prior art, details of which the patent agent should pass to you, the technical staff involved in the invention should read it for themselves and make any comments they may have on distinctions between the art cited and their invention which may be of help to the agent in dealing with objections.

Some of the questions the patent agent asks you may require you to rethink the invention, to go back and do further experiments to determine whether substitutes of a certain kind would work as well, even to cut down the scope of what your invention is claimed to be, from possible future applications to known feasible applications. Patent agents are not opponents, but friends. Their objections are not personal affronts to the inventor but attempts to ensure from the start that they know the maximum legitimate boundaries of the invention and the optimum and desired scope of protection so that they can draft the strongest possible application and defend it against attack by the Patent Office or third parties.

Although in emergencies patent specifications can be drawn up extremely quickly, the maxim 'act in haste, repent at leisure!' applies with force. Patent agents have many clients with demands on their time. The average medium technology patent may be expected to take anything up to two weeks to pass from the in-tray to the out-tray on the patent agent's desk, unless there are exceptional factors which require greater speed. To this must be added time taken in responding to the patent agent's enquiries, making the checks mentioned above and so on.

3.7 WHO TAKES DECISIONS ABOUT PATENTS?

This is an important question. Your patent agent cannot tell you whether to patent or not: he or she can only advise you of your chances and give you information as to cost etc., so that your firm's decision is an informed one: your firm must make that decision.

Decisions about patents are commercial not just technical and have to be dealt with as such. In the smaller companies it may well be practicable for the managing director – who is often a technically qualified person in charge of commercial matters – to reach a decision. In larger companies the chain of command may not always be conducive to rapid decision-making, especially where the high cost of overseas protection is in question and there is a knowledge gap between the technical and commercial staff. Laymen's reports can often help in this. The matter is dealt with in depth later.

For now one point alone may be stressed. Do not assume that what you have is or is not patentable. Always consult the patent agent. There are many myths about what can and cannot be patented and many former truths which, with reforms of the patent laws, are no longer truths. The golden rule is this: if you have an idea which you think is worth protecting in some way, consult your patent agent: he or she will tell you if it is patentable or if some other form of protection may be available. That is an important part of the job. If you are in regular contact with the patent agent and the idea in question does not require more than a quick computer search you may not even be charged a very large sum and the advice could save you a lot of money in the longer term. Search costs could be as little as £30.00, though extensive searching can cost more.

3.8 PLANNING A PATENT APPLICATION CAMPAIGN

The first essential before taking the decision whether or not to seek patent protection for something is to make a search of the existing patent literature to determine patentability and feasibility. It is most usually a relatively cheap exercise and it will answer many doubts and save unnecessary costs.

Equally there is clearly no point in seeking any patent protection if the invention appears to be quite impracticable or obviously unprofitable for the foreseeable future. But it is common advice that if you are in doubt you should patent – or at least initiate an application – in the UK. Such a move is relatively cheap and it buys you insurance and time while you think out your next move.

This advice is in general sound. There may be reasons why it does not apply. The invention may be simply impossible to reverse engineer so that any explanations in a patent specification can only harm you. You may reach the conclusion that all you want to do is to be able to use the idea yourself without excluding others, in which case the mere fact of publication as soon as possible is both cheap and effective in attaining that objective. But once you have taken the decision to apply for a UK patent, and assuming that you have decided that, on the basis of an initial opinion taken from a search of the existing patent specifications, the invention is likely to be patentable, a number of other decisions must also be taken:

(1) In which other countries do you want protection?
(2) Which patent agents are you going to use?
(3) What time scale are you working on?

Many conflicting pressures will influence the decisions you make:

- Costs multiply with overseas protection. Can you afford and do you need extensive overseas protection? If the product takes off and a potential export market or licence market develops, you may rue the day you did not bother with that overseas protection. On the other hand, you may feel that it is only worth protecting the idea in your existing well-developed overseas markets. And, clearly, the market for certain technology will be greater in some parts of the world than others. Any genuinely commercial piece of technology will probably merit patent applications in Europe, the USA and Japan, assuming that the firm's finances permit this.
- Another factor may be the purely commercial one of balancing the cost of patent protection against the strength of patent protection in each of the countries under your consideration. You may, for instance, consider a patent strategy in which priority is given to patenting in those countries where the level of protection is high but the costs of acquiring patents is relatively low, that is, most of Europe and the Commonwealth, leaving more expensive and weaker regimes out as expensive luxuries. The choice of overseas protection may become fairly routine if you seek protection regularly because the same considerations are likely to apply each time.
- The applications, if any, must be filed before you disclose the invention publicly by selling it or exhibiting it or by testing in an unrestricted part of the factory or on the open road. There may be pressures for rapid commercial development of the idea from the sales and related departments of the company, against pressures for delay from R and D and other technical teams in the company.
- If there are rival producers they may be in a race to prevent you from gaining the monopoly even if they are not in a position to gain it themselves. In one well-known case, Du Pont de Nemours and Co.'s (Holland's) Application (1971), a number of companies were working on a new carpet fibre involving fibre with a trilobal cross section. A patent application in respect of an artificial fibre made by Du Pont was successfully defeated by publication of information about the fibre in the British Nylon Spinners magazine *Nylon Outlook* just six days before the patent application's priority date. Time can be of the essence in a highly competitive field.

By the same token, there may be a need to acquire more information by further experiments or tests so as to be able to formulate a stronger patent application. In every case a balance has to be struck between getting in an early patent application and getting in a strong application. You may be heavily influenced by the apparent activity or lack of activity by trade rivals in the same technical field.

- On the other hand, once you have initiated a patent application, a set time scale is imposed upon you in which to complete the application, so that if the idea is still capable of more complete development or its proper scope is not fully developed it may be unwise to apply too early, lest one obtain a weaker or narrower patent than that which would become possible with just a short time delay. This is a particular problem for those with on-going research programmes in a new field, though it can sometimes be mitigated by applying for follow-up improvement patents.
- Time can become a particular problem where overseas patents are to be sought and the process of application (which involves the preparation and checking of translations, posting of documents, preparation of applications in a form suited to the demands of the overseas system, taking account of developments in the meantime) may eat into the Convention priority year.

In reality you may have only a seven to eight month period before action becomes urgent and a final decision must be taken. There is a limited escape route if you do run out of time. It is sometimes possible to file a further overseas application outside the priority deadline which although it will only enjoy a later priority date and cannot benefit from the priority of the original application, will not be deprived of novelty because even though the twelve month period is expired, the original application has not yet been processed far enough to have been published. By making use of this route, which is a second best, you may be able to salvage foreign protection as much as eighteen months after the priority date of the original application.

3.9 WHAT RIGHTS DOES THE PATENT GIVE YOU?

Once you have obtained a valid patent, it gives you only the right to exclude others from using your patented invention or dealing in it in certain ways; the right to exploit it yourself is only subject to other laws and other people's patents. The scope and limitations on this monopoly in the UK law should be looked at more closely.

3.9.1 General points

The first half of this question has already been dealt with in part by looking at the claims and their interpretation. These determine the range of protection for the invention in terms of the forbidden technical territory. But the question of the nature of infringing activity and the duration of these monopoly rights must now be considered.

The 1977 Act provides a complete code of what now constitutes an infringement of the patentee's rights. Outside this code, acts which damage the patentee's interests in his or her patent will not constitute an infringement of it, whatever other cause of action he or she may have. One or two preliminary points may be made before looking at the scope of infringing acts.

- Whether or not there has been deliberate copying is in principle immaterial. The fact that the patent infringement was purely innocent, such as where the defendant reached the same result independently from the patent holder, is no defence, although it may be relevant to the limitation of liability for damages. Nevertheless, the court is likely in practice to find the presence or absence of deliberate copying by the defendant influential in determining the scope of the plaintiff's claims.

- So long as the activity complained of takes place within the scope of the plaintiff's claims, there is an infringement even if the activity of the defendant has added something to the original patented invention and contributed things which are in themselves inventive. Even one who has taken out a so-called 'improvement' patent requires the consent of the original patentee to exploit it; hence the proliferation of cross-licensing where the price of the licence to use the original invention is consent in return to use the improvement.
- The acts which infringe a patent holder's rights in the UK are set out in the 1977 Act: the provision is intended to cover both European patents (UK) as well as UK national patents. However, it should be noted that the section is also intended to have the same meaning as the equivalent provisions in the Community Patent Convention (CPC), which will have (when it comes into force) its own tribunal for infringement actions, and the text of the CPC is strikingly different in a number of respects, to the extent that a number of disparities could emerge between at least the UK interpretation of infringement under British and European patents and the CPC tribunal's view of things, if the Community Patent Convention is introduced. This should be borne in mind.
- There are two groups of infringing acts, called primary and secondary infringements respectively. The distinction between them is that primary infringements involve the direct use of the patentee's invention, whereas secondary infringements involve supplying others with the means of committing primary infringements.

3.9.2 Primary infringements

Under the *Patents Act 1977* s60(1), 'Subject to the provisions of this section, a person infringes a patent for an invention if, but only if, while the patent is in force, he or she does any of the following things in the United Kingdom in relation to the invention without the consent of the proprietor of the patent':

(1) Where the invention is a product, the person makes, disposes of, offers to dispose of, uses or imports the product or keeps it for disposal or otherwise.
(2) Where the invention is a process, the person uses the process or offers it for use in the UK when he or she knows, or it is obvious to a reasonable person in the circumstances, that its use there without the consent of the proprietor would be an infringement of the patent.
(3) Where the invention is a process, the person disposes of, offers to dispose of, uses or imports any product obtained directly by means of that process or keeps any such product whether for disposal or otherwise.

3.9.3 Secondary infringements

Under the *Patents Act 1977* s60(2), it is also an infringement of the patentee's rights if a person, while the patent is in force and without the consent of the proprietor, supplies or offers to supply in the UK a person other than a licensee or other person entitled to work the invention with any of the means, relating to an essential element of the invention, for putting the invention into effect when he or she knows, or it is obvious to a reasonable person in the circumstances, that those means are suitable for putting, and are intended to put, the invention into effect in the UK.

There is an exception to this secondary liability. Under s60(3) of the Act this shall not apply to the supply or offer of a staple commercial product unless the supply or offer is made for the purpose of inducing the person supplied or, as the case may be, the person to whom the offer is made to do an act which would constitute one of the primary patent infringements.

But equally, one should note finally that under s60(6), the persons entitled to work the invention, to whom a supplier of the means of working it may supply without infringing the patent under s60(2) do not include for the purposes of that subsection persons who are exempted from liability for working the invention without the consent of the patentee merely because they fall within the exemptions for private experimental or medical prescription use under s60(5), which we will consider immediately below.

This means that suppliers of patented material to experimenters infringe the patent right even if the experimenters with that material do not do so. No commercial activity within the scope of the patent, even if linked with research, is permitted without the consent of the patentee.

3.9.4 Exceptions to liability: permitted acts

There are certain acts in respect of a patent which are permitted even without a patentee's consent, as follows.

(a) Experimental use

Under s60(5) of the Act, any act which would otherwise be an infringement of the patent will not be so if it is done privately and for purposes which are not commercial or if it is done for experimental purposes relating to the subject matter of the invention. This exception for experimental use is of course essential for rival companies to seek ways of circumventing and improving upon the patent and there are no limitations on the purposes of these experiments so long as the product of the experiments itself is not disposed of or used for profit. Contrast the making of the infringing product to use it for teaching purposes or selling the infringing product to another who wishes to use it for his own experiments, both of which will infringe the patent.

(b) Medical prescriptions

The extemporaneous preparation in a pharmacy of a medicine for an individual in accordance with a prescription given by a registered medical or dental practitioner or dealing with a medicine so prepared is exempt from actions for infringement.

(c) 'Convention' ships and aircraft

Certain ships and aircraft are protected from interference by foreign countries on the pretext of containing some equipment or other feature constituting an infringement of patents when lawfully flying over their territory etc. by a series of international conventions.

(d) Exhaustion of patent rights

The doctrine of exhaustion of rights maintains that at a certain point the owner of a patent has had all he or she is entitled to have from the patent grant and that he or she cannot make use of the law to extract too much from it to the detriment of the rest of the community. This will be looked at much later in the book when considering the implications of competition law on the exploitation of patents and other intellectual property in the UK and EEC. It usually means that the patentee loses the right to control the use of the patented product after he or she has sold it.

(e) Secret prior use

Under the old 1949 UK legislation one who had begun to use the invention before its application priority date was not protected if the invention was subsequently patented by another; the prior user's only remedy was to seek revocation of the patent on the basis of secret prior use. Under the 1977 legislation secret use is no longer a ground of invalidity but the prior user is given a limited degree of protection from action by the patentee. In the 1977 legislation, s64 of the Act provides that where a patent is granted for an invention a person who in the UK before the priority date of the invention does in good faith an act which would constitute an infringement of the patent if it were in force, or makes in good faith effective and serious preparations to do such an act, shall have certain rights. These rights, however, are very limited. The section permits the prior user to do or continue to do that act personally; if the act was done or preparations were made for it in the course of a business, the prior user is permitted to assign the right or transmit the right on death or authorize his or her partners for the time being in the business to do it; and the prior user is permitted to dispose of the product so that the person to whom it is disposed may deal with it as if it was disposed of by a sole registered proprietor. The section does not permit the prior user to do a number of very important things.

Firstly, the scope of activity permitted to the prior user is to use the invention just as it was before the priority date: nothing by way of improvement or modification within the scope of the patent is open to the prior user, so he or she is time-locked, de minimis (very small matters) apart perhaps. If the prior user was making for hire he or she may well not be able to make for sale without falling outside the scope of protection afforded by the section.

Secondly, the prior user cannot license the product or process, use being confined to personal use and use by his or her partners and successors in title to the business where that business is a corporate body. Thus, although no limitations on the scale of the prior user's operations is imposed by the section, in practical terms his or her output will be limited by the scale of his or her operation and to increase it he or she must sell that part of his or her business, lock stock and barrel, to another to make that other the successor in title to the business.

Thirdly, if at the time of the priority date, the prior user had not yet gone into sales of the product or process, he or she may not do so unless he or she can claim that the activity he or she had undertaken constituted serious and effective preparation for the sales.

(f) Implied consent

All the infringement sections require that for the defendant to be liable he or she must have acted in one of the ways specified without the consent of the registered proprietor. This consent need not of course be express. The exhaustion of rights doctrine which was mentioned above is in a sense an artificial extension of implied consent. But there are other cases of implied consent.

There is for instance an implied licence to the purchaser of patented goods to effect repairs to them, even if that means refurbishing patented parts, though this does not extend to actually reproducing the patented product. But the borderline between repairing a patented product and reproducing it may be very blurred where new components have to be made for it. The patentee is taken to impliedly license repair, but not reproduction. Moreover, the licence to repair is transferable; the licence can be relied on by any subsequent owner of the article and by any contractor engaged to repair the item for its owner, and whether the product was originally obtained from the patentee or his licensee.

Of course, if the patentee sells the product without any restrictions as to its use, he or she loses all control over it. But, moreover, it would seem that purchasers of a patented product who were at the time of purchase unaware of any restrictions on the use of the product imposed by the patentee should be free to use it as they please. But it is not essential that the precise details of the restrictions should be known to the purchasers to bind them: to know that some exist, perhaps by means of a notice attached to the goods, and to have the means of discovering more will suffice. The same principles almost certainly apply to restrictions imposed by a licensee on products made under licence.

If the purchasers are aware of the restrictions and infringe them, they infringe the patent, unless they can rely on the provisions of s44 of the *Patents Act 1977*, which contain a list of forbidden licence terms. Normally, the breach of a restriction as to use of a patented product or process licensed by the patentee will constitute an infringement. However, by s44(3) of the 1977 Patents Act it is a defence to infringement proceedings that at the time of the infringement there was in force a contract relating to the patent made by or with the consent of the plaintiff or a licence made under the patent granted by the plaintiff or with the plaintiff's consent which contains a condition or term void under s44.

A number of contract terms in relation to patents are forbidden by the Act and the Act renders them void. These will be looked at in more detail later in the book, in considering restrictions on licensing of intellectual property rights.

3.9.5 The duration of protection

In Europe patent protection now lasts until the end of the twentieth year from the date of the application being filed. There is no possibility of any extensions of this term. In the USA, however, the usual seventeen years runs from the date of grant and it may in some cases be extended for pharmaceutical patents to compensate for the lengthy time the USA Food and Drugs administration takes to approve their use.

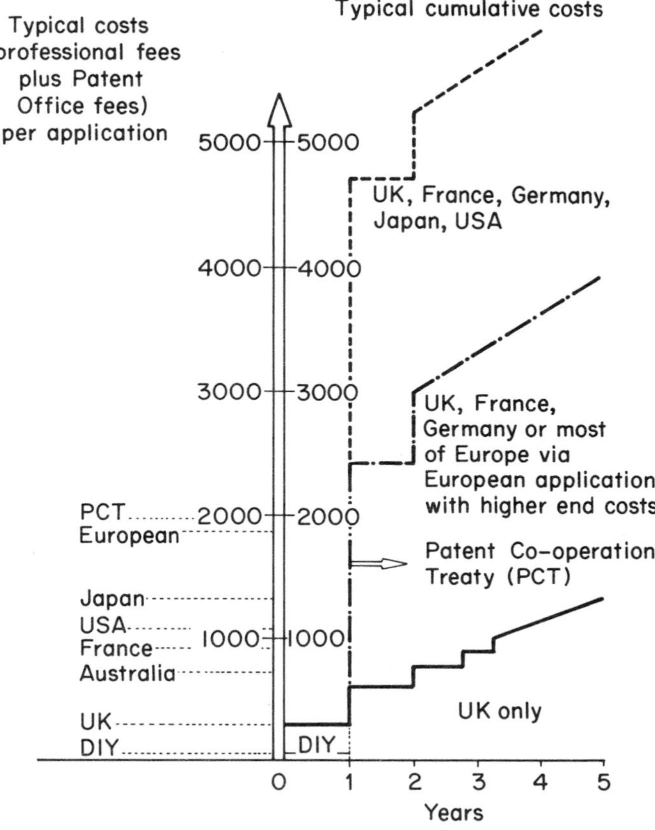

Figure 4 The costs of patenting

3.10 THE COSTS OF PATENTING (See Fig. 4)

It is impossible to give accurate figures, since so much depends on the facts of the individual case and inflationary tendencies are strong in this field as in any other. There are two factors:

(1) Professional fees
(2) Patent Office fees

Professional fees will be made up of patent agent fees and where appropriate translation fees. Most UK patent agents charge on the traditional hourly basis, but with variations depending on the complexity of the work. Patent applications are time-consuming matters. The cost of employing a patent agent in a highly complex new field will obviously be greater than for a simple mechanical device because of the greater technical expertise needed. Usually it will be possible for an estimate to be given.

UK Patent Office fees are easier to set out. Table 7 gives an indication of the fees payable.

Some Patent Office fees in the UK (as at May 1987)

		£
Filing fee		10
Preliminary search fee		85
Substantive examination fee		100
Amendment before grant fee		24
Amendment after grant fee		50
Patent renewal fees at:	4 years	78
	5	84
	6	92
	7	100
	8	112
	9	122
	10	134
	11	146
	12	164
	13	182
	14	200
	15	220
	16	240
	17	260
	18	290
	19	320

3.10.1 Financing a patent in force

In the UK and most other countries once you have the patent you have to pay renewal fees to keep it in force. Fees get progressively larger as the patent gets older. Again, for each country in which you have a patent this is going to require a regular decision from someone. Options may include those to:

(1) Fail to renew or abandon the patent to cut costs
(2) Keep on paying and keep the patent yourself
(3) Find a buyer to take the patent
(4) Reduce fees by allowing 'licences of right'

Licences of right are discussed in Chapter 14. In essence the UK Patents Act permits you to cut renewal fees in half by declaring that anyone may apply for a licence in the patent on paying a royalty which in the absence of agreement is to be decided upon by the Comptroller General of Patents.

Factors which will condition this decision will be:

(1) How much royalty money is coming in on the patent
(2) What legal obligations you have to licensees
(3) What market potential there is for the product
(4) What prestige the patent has in the industry
(5) Whether there are any obligations to work it

The question of working requirements, which exist in some countries, whereby the patent is lost if it is not used locally, for example by local manufacture of the patented product rather than by mere importation of fully assembled products covered by the patent, will be considered later.

3.11 REACTING TO OTHER PEOPLE'S PATENTS

From the point of view of an inventor thinking of protecting his or her invention by patents, if in the course of a search or application for a patent or in surveying current technology the inventor comes across another person's patent which seems to cover his or her own invention, there are two ways of looking at the situation: either as an obstruction to ambitions, or as a short cut to the technical development sought. And from the point of view of a person with an existing patent, or keeping abreast of recent developments, a new patent application in his or her technical field may be seen as a challenge to his or her own patent's commercial power or as a possible complementary piece of technology of which he or she could make good use. Either way, there are a number of options in reacting to a patent or patent application which may get in your way. Which course of action you take is always going to be a business decision in the light of any professional advice as to the strength of the other person's patent, and any commercial advice as to the importance of the patent, the willingness of the other party to fight or deal, etc. Let us look at them.

- Ignore the patent and hope it will go away. In the short term this is cheap and effortless. It is in effect a calculated risk in which you balance what your chances are of being caught out against the chances of escaping unscathed. Factors which will influence the decision will include the length of time the patent has to run before expiry, the strength of the patent, the cost of the alternatives and the power of the patentee.
- Seek a licence from the patentee to use it. You may want or have to go further, buying an assignment of the patent or even buying out the company which owns it lock stock and barrel, particularly if there is useful associated know-how and staff.
- Design around it. This involves finding an alternative means of achieving the same technical result without infringing the scope of the existing patent. It is a skilled job both from the point of view of the designer and the patent agent. Of course, most patent specifications are drafted as wide as possible so as to minimize the possibilities of this happening, and the practicable alternatives available outside the scope of these claims may be few and far between.
- Challenge the patent. It may be possible to find an arguable case for invalidity based on prior art or some other cause. If the patent has been amended during its application stage to overcome some Patent Office objections these may prove a promising starting point for a challenge. Alternatively the doctrine of fraud on the Patent Office in some countries may prove to be arguable. This can be a costly exercise in the short term but save a lot of money in the longer term if it induces a settlement or the grant of a licence which would not otherwise be granted by the patentee, anxious to preserve the credibility of his or her patent, especially if that patent is central to his or her profitability, as is the case with a number of smaller one-product firms in the newer technology industries.

4
Designs

4.1 WHAT IS A DESIGN?

An essential part of the marketability of many of our industrially produced items, even very mundane and very functional items, resides in their eye-appeal and thus visual design has become an integral part of most consumer products. Where the efficiency and performance of rival products is virtually indistinguishable and where costs of their production and marketing are very similar, design is the focus of competition, supported in some cases by heavy advertising to create an image associated with the manufacturer's own product line, whether it be one of efficiency, prestige, or whatever.

The commercial importance of good design both of the product itself and of the packaging it comes in is clear from the success of names such as Terence Conran, Laura Ashley, Kenneth Grange, Michael Peters and David Mellor. But good design has been a traditional weakness of British industry. If you have not got a good design there is not much point in protecting it from imitation because few are going to be silly enough to copy it. By contrast good designs which attract public attention do attract an equal amount of plagiarism.

The best designs have lasted a long time, and they therefore need a considerable period of protection. But even shorter term design trends require some degree of protection to enable the recoupment of the costs of producing them. So non-functional design is a valuable aspect of a product. In many industries a degree of imitation of a popular design is not only expected but also accepted, for instance, in the fashion industry, as a necessary part of the trade, so long as slavish exact imitation of the more expensive by the cheaper is not involved, and even thought a healthy practice. Indeed, the lawyer's views of the problem of industrial design may in some regards diverge markedly from those of the industrial designer. But the line has to be drawn somewhere. The law is not particularly successful in drawing it but the firm with a new design of some commercial value has to cope as best it can.

4.2 HISTORICAL BACKGROUND TO DESIGN LAW IN THE UK

Design protection laws date back as far as 1709 in the UK, but although some limited copyright protection was extended to textile products before then, the legal registration of designs in the UK dates back only to 1839 when registration was introduced in response to a demand for registration as a means of saving industry from unfair overseas competition.

The history of legal protection for designs in the UK is a monument to many policy changes, constant tinkering with statutes and private lobbying by interest groups, all of which have combined to produce a regime which is a complex and anomalous, even unpredictable mess. A series of new Acts developing registered design protection continued to be passed through the nineteenth and early twentieth centuries.

Finally, in 1949 the Registered Designs Act was passed as a parallel measure to the 1949 Patents Act and this remains in force today. If this were all, things might be relatively straightforward, but there is a parallel source of protection for designs in the shape of copyright law which much complicates matters.

4.3 MEANS OF PROTECTING DESIGNS IN THE UK

There are now two basic ways of protecting industrial designs in the UK:

(1) The *Registered Designs Act 1949*
(2) The *Copyright Act 1956*

The two Acts differ both in the range of designs which they protect (although there is a considerable overlap) and in the nature and duration of that protection. They also interact in some cases. Today, registered designs appear very much the poor relations of patents and design copyright in the minds of both manufacturers and professional advisers, and they rank among the least understood of all the UK's intellectual property rights. Certainly domestic firms seem to make relatively little use of the UK's *Registered Designs Act 1949* compared with many overseas companies, who are more accustomed perhaps to having only registered design laws in their own countries and even though UK firms do maintain a high profile in the registration of designs overseas and use fully all the protection it affords to designs.

Nevertheless there is still a fairly high level of demand for the provisions of the Act, for a variety of reasons: some UK dependent territories, for instance, still give protection in those territories automatically on prior registration in the UK, and the use of a registered design number on a product may confer on it a degree of respectability in much the same way as a patent registration number.

4.4 INTERNATIONAL PROTECTION OF DESIGNS

As in the case of patents, the Paris Convention of 1883 guarantees in the Member States equal treatment for domestic and foreign applicants for registered design protection and confers a priority period for applications, but this time of only six months from the date of the overseas application relied upon.

From the point of view of the UK manufacturer a UK registered design application gives him a six month breathing space for obtaining equivalent registered protection in several overseas countries. But a UK resident with a registrable design cannot in general make an overseas design application unless the resident has first applied at least six weeks before that for a UK design registration or has otherwise obtained permission to do so, under the national security provisions. There are provisions for compensation for suppressed designs and criminal penalties for failure to comply with the requirements of the Act apply.

Since the form of copyright protection afforded to designs in the UK by the Copyright Act is not generally available elsewhere and would in any case mostly arise automatically no such problems arise with unregistrable designs protected through copyright law in the UK. There are no means for co-ordinating several design applications as there are for patents. So such overseas campaigns are far more difficult to implement administratively. Again, as in the case of patents, to benefit from Convention priority a design application relying on an earlier priority date must be supported by the earlier application, i.e. no new design feature may be added to the UK application. This may be a real problem where the cultural characteristics of a design were devised with one country in mind and it is desired to make slight adjustments to attune the design more closely to the different cultural environment of the overseas market for which registration is to be sought.

4.5 REGISTRATION OF DESIGNS IN THE UK

Registration under the *Registered Designs Act 1949* confers a monopoly in the registered design for at most a period of fifteen years. The first question is to ask what types of design come within the ambit of the 1949 Act. To be registrable under the Act, a design must be:

(1) Within the definition of design under the Act
(2) Not otherwise excluded from the scope of the Act
(3) New or original

4.5.1 The definition of design

The definition of design under the Act permits the registration of both two-dimensional designs, i.e. the pattern or ornamentation applied to a plate or a dress, and three-dimensional designs, i.e. the shape of a vase or teacup and saucer. The admission of parts of an article, if they are manufactured and sold separately, permits spares and components to be registered as well as whole articles. Designs which are purely functional and which lack eye-appeal (not merely visibility) will not be registrable, though there is no question of imposing any requirement of good taste or artistic merit. This is the major difference from the scope of design which is protected by the law of copyright, where even merely functional designs may be protected lacking in any eye-appeal whatsoever. Some examples are as follows.

(1) A design for an electric terminal connection which was only used housed inside a washing machine and was never seen by the consumer buying the machine was denied protection.
(2) A design for the inside of an Easter egg was accepted for registration even though it was not visible until the egg was broken open to be consumed.
(3) A design for the display panel of a digital watch was accepted for registration although it

was visible only when being used for its intended purpose.
(4) A design for the form of computer print-out paper which was conceived purely to make the print-out easier to read and had no other attractions was denied registration on the ground that it was purely functional, as well as being excluded as primarily printed matter.

No 'method or principle of construction' is to be registrable as a design under the Act. In other words, application must be made for some specific application in concrete form of the underlying idea, rather than the general idea itself, which is too indefinite, and is often also open to the criticism that it is purely functional. Again an example may make this clearer.

EXAMPLE:
A design was registered comprising a picture of a basket, the claim being for the pattern of the basket, represented in a particular arrangement. This was held to be an invalid registration on the basis that a wide range of baskets all different in pattern could be made using the method of construction employed

4.5.2 Designs excluded from protection

The Act provides that there shall be excluded from registration under the 1949 Act designs to be applied to any of the following articles: works of sculpture other than casts or models used or intended to be used as models or patterns to be multiplied by any industrial process; wall plaques and medals; printed matter, primarily of a literary or artistic character, including book jackets, calendars, certificates, coupons, dressmaking patterns, greeting cards, leaflets, maps, plans, postcards, stamps, trade advertisements, trade forms and cards, transfers and the like.

Thus many commercial items are excluded from the scope of registration but not from protection under the law of copyright as such, which is considered later in the chapter. Other grounds for refusing registration also exist. A number of classes of item may also be refused registration, even though they are prima facie registrable under the Act. These are self explanatory:

(1) Where words, letters or numerals appear in the design but are not of the essence of the design, they shall be removed from the representations or specimens; where they are of the essence of the design the Registrar may require the insertion of a disclaimer of any right to their exclusive use.
(2) Where a portrait of any member of the Royal Family, or a reproduction of the armorial bearings, insignia, orders of chivalry, decorations or flags of any country, city, borough, town, place, society, body corporate, institution or person appears on a design, the Registrar may require a consent to the registration from such official or other person as appears to the Registrar to be entitled to give consent and without such consent the Registrar may refuse to register the design.
(3) Where the name or portrait of a living person appears on a design, the Registrar will be furnished, if he or she so requires, with consent from such person before proceeding to register the design. In the case of a person recently dead the Registrar may call for consent from his or her personal representative before proceeding with the registration of a design on which the name or portrait of the deceased person appears.
(4) In addition, of course, the Registrar retains a discretion to refuse registration and has a specific discretion to exercise in the case of applications contrary to law or morality – such as offensive, pornographic, immoral designs.

4.5.3 Requirements of novelty under the Act

If the design is of a kind which is prima facie registrable under the Act and not excluded it must still be new or original. In other words it must be novel. Novelty is to be judged by looking at the relevant prior art available in the UK – not as in the case of patents extending to material which is publicly available anywhere in the whole of the world – at the priority date of the application for registration. In cases where the applicant can rely on an application for protection of the design made in another Paris Convention country within the six months before the UK application, novelty will be considered as at that earlier date. Otherwise, the date of filing in the UK is the relevant date.

Thus prior art to be considered is drawn from a much narrower body of information than is the case for patent applications. No design can be registered which is the same as a design which before the date of the application for registration has been registered or published in the UK or differs from such design only in immaterial details.

But not merely registration but also any prior publication will frustrate a registration application. Publication to the general public, or disclosure to others without the imposition of any confidentiality will defeat a subsequent design application. But prior publication or use outside the UK is, unlike the case for patents, wholly immaterial to the registrability of a design.

In exceptional circumstances prior registration or publication in the UK will not bar an application:

- Where the disclosure by another was a breach of confidence, the owner's subsequent application will not be prejudiced so long as the applicant acts within good time to establish lack of consent to the publication in breach of good faith.
- Acceptance of a first and confidential order for goods bearing a new or original textile design intended for registration will not prejudice the application.
- Where the registered proprietor of an existing registered design wishes to extend its application to articles of a kind for which it was not previously registered, he or she may do so by using an associated design registration notwithstanding the earlier publication and use of the design. But such an associated design registration expires at the same time as the original registration so it is not possible to extend protection indefinitely by periodic minor modifications. This is considered separately below.
- Where the publication took place at a designated exhibition whose opening took place within six months before the application, that publication will not of itself invalidate the application. Such exhibitions are few and far between.
- Communications to a Government Department for consideration of the merits of the design will not prejudice a later application.
- Prior use of the design where it has artistic copyright under the 1956 Copyright Act as an artistic work does not invalidate an application: only if the use of it has taken the form of commercial dealings with articles bearing the design or by industrial application of the design, in both cases with the consent of the owner of the copyright work, will that use count against the applicant. The design will have been applied industrially when it is applied (a) to more than fifty articles all of which do not together constitute a single set of articles or (b) to goods manufactured in lengths or pieces, other than hand-made goods. This sounds difficult but an example may be of assistance.

EXAMPLE:
If a painting which enjoys artistic copyright is for the first time applied to decorate an industrially produced product, for instance a vase, ten years after it was first painted by the artist, the fact that this painting has already been sold in the UK does not prevent the mass produced vase bearing reproductions of that old painting from being registered as a new or original design.

Novelty is only a matter of degree and the mere application of an old design never before applied to the type of article of manufacture in question will suffice, although the scope of protection will be very narrow in such a case. Just how little novelty there need be can be illustrated by the following example.

EXAMPLE:

The subject matter of a registration was a design of a spoon whose handle bore a design derived from a photograph of Westminster Abbey. This was held novel. The Act does not require novelty in an idea but only in the application of the design to the article for which it is registered. The novelty is not destroyed by the design being taken from a source common to mankind.

The large amount of litigation surrounding novelty has given rise to a number of guidelines on determining whether a design is novel for this purpose:

- The design which has to be considered is the design as it appears in the finished article, not in isolation from it.
- The design must not only be new but must also have a degree of novelty which is substantial in relation to the prior art. The introduction of mere ordinary trade variants into an old design cannot make it new or original.
- The design has to be looked at as a whole. The features of shape, configuration, pattern or ornament are not mutually exclusive. All the features which make up or contribute to the final result may be old in themselves but if in combination they produce a novel visual result the design is novel and registrable.
- Only part of the design need be novel in order for the whole to merit registration (and of course if that part is made and sold separately it may qualify for protection in itself even if it is not sufficiently important in relation to the whole of the article to gain registration for the whole).
- In determining novelty Design Registry practice is to ignore colour combinations, a practice apparently endorsed by the judiciary, but there may well be some exceptional cases where colour does matter on the facts.

In practice the Registry is particularly generous in conceding to an applicant the benefit of the doubt in these cases. This is good and bad. It makes it easier to gain registration but leaves the presumed validity low and open to a challenge in the courts which may be more expensive for the right owner than an initial rejection.

4.5.4 Registration of associated designs

In exceptional cases it is possible to acquire a registration for a design which has been disclosed to the public by publication or use in the UK in respect of the same or similar articles. This is where the design for which registration is sought may be associated with an earlier registered design in respect of the same or similar articles owned by the applicant for the new registration.

For example, if you have registered the design of a new range of items which has acquired some popular appeal, you may at any time during the existence of the registration seek to register the same design for other products despite publication and use of the original design in the UK. Thus you may apply for registration of the design in respect of a trinket or other objects to which the design may now be thought to be applicable and worth protecting. This may be of value where the original design gives rise to a spin-off industry of souvenirs and associated materials such as one finds with consumer products sold on the back of a motor racing team and marketed on the basis of an image of glamour or excitement.

Equally, you may seek a new registration for the same article in respect of which the original registration was obtained but for a design which is basically the same as the original design and incorporates improvements or other changes insufficient to alter the character or identity of that original design, perhaps where subtle changes in mood or public taste make this desirable. Nevertheless, associated designs confer relatively limited additional protection on their owner:

(a) Duration

The associated design expires when the original 'master' registration expires or is not renewed or is found to be invalidly registered. This means that it is not possible to use the device of registration of associated designs to perpetuate an original design's protection beyond its normal life.

(b) Ownership

Associated designs may be acquired only by the owner of the original design. If you do not own the design which is the master you must acquire it before you seek protection for the associated design.

(c) Scope of protection

The associated design is given the same amount of protection as the original design. Because it has to be one which does not alter the character or identity of the original to acquire this status it will not expand the concept of the design to be protected by very much, being largely confined to minor trade variants, but it may still be of some use in covering enough territory to catch a design which might just have escaped infringement of the master design. The principal benefit lies in extending the range of goods to which legal protection is afforded.

4.5.5 How do you get a registered design?

The procedure for applying for a registered design is much simpler than is the case with a patent. But, as for patents, the procedural requirements of the Act are crucial to the success or failure of an application for registered design let alone its scope of protection. As the proprietor of the design, you request the grant of a registered design, on an official form, depending on the type of design you are seeking, e.g. textile or non-textile, etc., and pay an initial filing fee. In making an application the applicant must lodge an application form of the appropriate kind and comply with all of the Design Rules 1984 whose most important requirements are:

(1) Representations of or specimens of the design as appropriate (where the design is three dimensional, a two-dimensional representation of it for the register is sufficient) which present an accurate and complete picture of the design. These will have to be made up of a number of views if the object is three dimensional.
(2) A statement of the features of the design for which novelty is claimed (unless the design is to be applied to a textile article, wallpaper or lace, for which no statement of novelty is required by the law), that is, which one or more of the features of shape, configuration, pattern or ornament are claimed to be novel. The Designs Registry cites as an acceptable example: 'The features of the design for which novelty is claimed are the shape and configuration of the article as shown in the representations', or, where only part of the entire visible shape is claimed, 'The features of the design for which novelty is claimed are the shape and configuration of the part(s) of the article as shown coloured blue in the attached representation'.

The statement of novelty is important because as in the case of a patent infringement case, in later litigation it may well play an important part in the court's interpretation of the scope of the design. The statement of novelty plays exactly the same role as the claims in a patent specification.

The Design Rules 1984 require a statement of novelty in most cases, and it is in any case sensible because in the absence of a statement of novelty the courts have held the claim to be for the entirety, with the result that to prove infringement by another substantial taking of the whole will have to be established, not merely taking of the particular design feature which really constitutes the special attraction of the design. By the same token, in drafting the statement of novelty it must be remembered that what is not claimed as novel is free for others to take and reproduce with impunity and therefore the statements of novelty tend to be made as widely and generally as is prudently possible.

The possibility of more than one design claim being registered in respect of the same design has already been indicated, but the limits of this should be noted, namely the need to avoid inconsistent design applications and the need to make co-ordinated applications to avoid your earlier application from being cited as prior art against the later ones.

You might, for instance, have a vase whose shape and pattern was new. You could apply for a single registered design in which you claimed that the novelty of the design lay in the shape and pattern of the vase, in which case novelty and infringement would be judged by examining those two characteristics in combination and looking at their overall effect, or you could apply for one in which you claimed the novelty of the shape and another in which you claimed the novelty of the pattern applied to the vase, in which case each registration would be considered only in respect of that feature claimed. To take the latter course of action might in some cases expand the scope of protection enjoyed by the ensemble of shape and design when compared with the single claim which the former course of action would entail. It is a matter for professional advice in all cases.

In seeking a registered design you can only apply to register a design in respect of a specified article. There must therefore be a series of registrations for a range of articles to which you intend to apply the same basic design, but there is some mitigation of this where you have articles of the same type which form a set such as cups and saucers etc. in a teaset, for which you may take out a registration as a set. The Registration for a number of articles as a set gives the applicant exactly the same degree of protection as if he or she had obtained a registration for each object separately, but it is of course cheaper in most cases. The decision of the Registrar as to whether what you have is a set is final.

The fact that you have conceived a design in the abstract which you may wish to apply industrially in the future is insufficient for registration. You must thus register for a specific application. Unlike the case of registered trade marks, it is not possible for you to 'reserve' a design for future products and classes of product not yet devised, although to a degree the law of copyright can enable you to achieve a similar end; but when extra products are devised the Act does enable you later to take out extra design registrations as and when the designs materialize through the use of associated design registrations.

On receipt of the application, with at least some informal representations of the design and the full fee, a formalities examination takes place to see that the application is formally in order. If it is the application is then given a priority date and a number.

The date of the application again constitutes an important matter because apart from the question of the commencement of design protection, if the application procedures have not been completed within twelve months of the date of the initial application, the application is lost, unless a request for extension of time is made and granted, which will give the applicant an extra three months in which to complete everything. The Registrar may on giving notice and on such terms as he or she thinks fit, 'enlarge' the time available for other proceedings too.

If the application is found to be directed to more than one article the Registry may permit the submission of a divisional application on payment of additional fees, credited with the original filing date for priority purposes. But it is under no obligation to do so and could turn the application down.

Once a priority date has been accorded it is then possible to begin marketing the design without the danger of losing registration. But unless you are under great pressure of time, this may not be too sensible unless the design has already been incorporated into a final production version of the design. If you have only reached the stage of a prototype which you begin to show to the trade, and the trade reaction dictates that some degree of modification is desirable, once the application has gone in no further modifications can be made to that and if you are forced back onto a fresh application with a later priority date you may find that your own prototype has deprived you of the opportunity of registering anything at all. If instead you press on with the original registration and the final production model differs markedly the chances of protecting it effectively with what is in essence a registered prototype design may be severely diminished.

A search and substantive examination of the application takes place to determine its registrability in the light of the requirements of the Act and the prior art. Because technically unpublished registration applications with an earlier priority date do not constitute in themselves prior art, an examiner who wishes to cite such an application may instead delay examination of your application until after the grant and publication of the earlier design, which may then benefit from its earlier priority date to defeat yours. The search is in effect confined to the existing design registrations plus a number of the recent periodicals available in the UK, a very small amount of literature compared with the patent searches which have to contend with prior art from overseas as well as from the UK.

If the examiner raises objections, these together with citations of any relevant prior art are sent to the applicant who may apply for a hearing. Once that application is accepted it is registered and only then is it published unless there exist some grounds for preserving its secrecy in the interests of defence. It usually takes effect from the date of application.

It should be noted that unsuccessful applications for registration are never published. This means on the one hand that your application for registration of a secret design cannot prejudice your position with regard to confidentiality, so that if you yourself have not published the design in any way the application does not lead to publication; but on the other hand it also means that a prior application does not of itself constitute prior art against any later application.

4.5.6 Constructing the claim to protection

It is important to bear a number of factors in mind when formulating an application to maximize the scope of protection from a registered design.

- A new product may possess both two-dimensional and three-dimensional features which are registrable – for instance the teacup may be novel in shape and also bear a new ornamental pattern. Procedural requirements make it important to state clearly which if not both characteristics you claim protection for, since an excessive claim to protection for the whole may lead to loss of protection for both characteristics if one is not new and original; and likewise failure to claim more than one aspect of the design denies you protection for the others which may be registrable and turn out to be the real selling point.

If both elements are new but one is thought more vulnerable than the other the applicant would risk losing both if a single application were made and in such circumstances it is generally desirable to make two separate applications, one for each novel characteristic. Moreover if a claim is made to both shape and pattern, there will be no infringement unless both are taken. However, beware attempts to register different aspects of shape of the same design. You cannot, for instance, obtain two registrations for the same design of table, one claiming novelty for the legs, the other for the table top, the legs and the top not being sold separately. You must combine the two in the application. Only shape and pattern can be split.

- Although you must register a design for a whole article or for a part sold separately and not merely for an inseparable part or unseparated part of an article, the Registry tolerates claims and statements of novelty which emphasize the novel part of the entirety – though the courts still have to consider the entirety of the article in hearing infringement claims.
- It will often be possible to overcome the method of construction objection by confining the claim and statement of novelty so as to extend only to the design in question and not any wider. If wider protection is sought and cannot be protected by confidentiality, consideration must be given to protection either under the *Design Copyright Act 1968* or, if the principle of

construction is sufficiently inventive, to even patent protection. The key to whether the claim is too wide seems to be whether it is possible to get several different appearances which all embody the general features which are claimed; if so, then those features are too general and amount to a mode or principle of construction.

4.5.7 Duration of registered design protection

In normal circumstances, the initial period of protection for a registered design is a period of five years from the date of registration initially and it is possible to obtain two further periods of protection of five years each if application for extension and the appropriate payment of fees is made with the prescribed time limits. There is provision for procuring the renewal for up to six months after the expiry of the initial or second period of protection, on payment of an addition fee, but after then the registration will lapse. There is no further discretion to extend the times set out in the rules. In the same way as patent agents will run a patent annual fees list to remind clients of payment due, so they will do so for registered design renewals.

In practice many UK registrations are not renewed beyond the initial five year period simply because design is a thing of fashion and fashion renders many designs obsolescent so that the cost of renewal would not be justified by production and sales levels.

Exceptionally, however, the period of protection may be cut down to below the fifteen year maximum by prior use of the design. If at the time the design was registered there was a corresponding design in relation to an artistic work which enjoyed copyright under the 1956 Act and the design registered was registrable only by virtue of the earlier artistic work not having been put to commercial use, then protection under the 1949 Act ceases at the same time as the copyright in that work under the 1956 Act expires if that is earlier than the general fifteen years from registration.

4.5.8 Scope of registered design protection

If valid, a registered design confers a monopoly. The owner has the exclusive right in the UK to make or import for sale or use for the purposes of any trade or business, or to sell, hire or offer for sale or hire, any article in respect of which the design is registered, being an article to which the registered design or a design not substantially different from the registered design has been applied, and to make anything for enabling any such article to be made, whether in the UK or elsewhere.

As in the case of a patent, the monopoly right is only a negative one. It entitles you to exclude others from infringing the scope of the monopoly but does not entitle you to market the product if other laws happen to preclude this. If for instance the product is banned under some consumer safety legislation the fact that it is a registered design is immaterial.

Because it is a monopoly right, no copying need be proved for an infringement action. An infringement will occur wherever another person does one of the things mentioned in the UK without the authority of the registered proprietor. The main problem for the court to resolve is the question of whether the design alleged to infringe is substantially different from the registered design. The closer the registered design to the prior art in the field, the narrower its field of protection from subsequent designs. Attention must be paid to the scope of the claims made in the initial design application. Special attention must also be paid to the 'selling points' of the design.

But most importantly of all, the requirement that for infringement to occur the design must be registered for the relevant class of goods must be stressed: this is because outside the classes of goods for which the design is registered and subject to any rights you may have under the *Copyright Act 1956* rivals will have a free hand. Thus, if you have only registered your design for, say, table linen and someone else uses it on crockery, you have no action against him or her under the Registered Designs Act. There is, however, unlike some of the older designs statutes, no obligation to mark the items as registered to give rise to a cause of action.

4.6 COPYRIGHT PROTECTION OF DESIGNS IN THE UK

The use of copyright as a means of protecting the design of new products is an underestimated option.

(1) It is often possible to protect the design of an industrial product even where it is not patentable, by the device of protecting the copyright in production or other drawings on which they are based. This is so even if the product is entirely functional and it lacks any visual appeal, such as many engineering components like exhaust pipes, turbine blades, etc. So copyright is often a viable alternative to patent protection.
(2) Equally, it may be possible to protect a design which is unregistrable under the Registered Designs Act because of its functional or other characteristics or lack of the requisite degree of novelty.
(3) Thirdly, even if design registration is possible copyright protection is not excluded and may be available as an alternative. Where such copyright is available, the scope of protection is in some ways wider than that of the Registered Designs legislation and provides a cheap and easy method of quite extensive protection.

A design may attract protection through copyright in one of two ways:

(1) Drawings of it, such as blueprints, may attract protection as an artistic work, irrespective of their artistic quality. Alternatively photographs, engravings, paintings, or sculptures on which the design is based, may be found to confer copyright protection, again irrespective of artistic merit.
(2) A prototype of it may be produced which qualifies as a work of artistic craftsmanship.

The law changed in 1956 and in 1968. Different and much less favourable provisions govern designs made before those dates. But only the law relating to those designs made after 1968 will be looked at here, because of the diminishing relevance of the older regimes with the passing of time and the concern of this book with planning for the future.

4.6.1 Drawings, sculptures, paintings, engravings and photographs

In protecting designs through copyright usually we will be concerned with drawings since most new products begin their life as a blueprint on a draughtsman's board: but in the electronics industries increasing use is being made of photographic methods in the initial stage of production. Sculptures may be relevant in the toy industry and in protecting fancy goods, jewellery and the like.

Protection of designs of products will be possible through the drawings which precede them because it is an infringement of copyright in a drawing to produce a three-dimensional object which appears, to persons who are not experts in relation to objects of that description, to be a reproduction of the drawing on which that object was based. A reproduction of the photograph or engraving will also be a potential breach of copyright with the same caveat as to the non-expert.

The most important heading in practice is that of drawings. What designs may be protected by copyright in this way? Copyright in the UK arises automatically on the reduction into a fixed form of an original drawing. This may be hand drawn by a draughtsman or draughtswoman or produced in the form of a print-out from a computer-aided design project. Thus, the major initial obstacle to protection of industrial products is that there must be a drawing on which one can rely which is proved to be original.

In fact this is not much of a problem for most drawings (or for that matter photographs or engravings) for there is no requirement of novelty such as that required by design registration or patent legislation. Indeed, the basic requirement of originality really means that it must have taken effort to produce and has been more than merely traced from a pre-existing drawing. It appears that the requisite degree of effort for drawing to acquire copyright is low. If, using your knowledge of standard components and their appearance, you draw one, that drawing attracts copyright. If you trace a drawing of one from a standard text book, that drawing does not attract copyright. Secondly, there is no requirement of any artistic merit in the drawings. Even relatively simple rough drawings may attract copyright. And purely technical drawings such as blueprints qualify too, which convey nothing but purely technical information about the dimensions to which a particular engineering component must conform.

It would appear that when you rely on an original drawing in which copyright subsists, two difficulties of proof immediately raise their head.

(1) The problem of proving the chain of causation from drawing to alleged infringing product.
(2) The question of whether a non-expert will be able to understand the drawing sufficiently to recognize that the product is a three-dimensional representation of it.

Copyright in a drawing will be infringed by its reproduction in any material form, which includes reproduction in three dimensions, if it would be apparent to a lay person that the resemblance between the two exists. Copying does not have to be total but merely partial. And where the two products are very similar an inference of copying may be found by the court in the absence of strong evidence of independent creation and even subconscious copying still constitutes an infringement. But obviously any similarities which are substantially dictated by the substance or functions which the object has to perform are less important in assessing copying than any non-essential or unusual features.

But copyright does not protect the underlying idea, however novel, of the product but only the manner in which that idea has been expressed. Furthermore, in the case of nuts and bolts and the like, even if the drawing of them did acquire copyright, proof of copying of the drawing would still be very difficult. This is because in an infringement action the plaintiff must show copying and, although where an identical or very closely similar result is reached by a defendant the court may draw an inference of copying, where the object drawn is very commonplace, it will be virtually impossible for a plaintiff to prove that the defendant took inspiration from the copyright drawing rather than the commonplace object.

It will always be of considerable assistance if you can prove that the defendant had ready access to the product alleged to have been copied, for instance that the defendant had bought one. Secondly, although it is not possible to use expert evidence to the effect that one article is a copy of another, it is possible to bring expert evidence as to which features of the plaintiff's product are functional and which not, and which aspects are substantially dictated by external circumstance and which discretionary, to point the judge's attention to those aspects of the design for which the prima facie inference of similarity would or would not be copying of the original.

A defendant can escape liability if he or she can show that a non-expert would not recognize that his or her three-dimensional product is a representation of your two-dimensional drawing. The non-expert is permitted to look at the drawings alongside the allegedly infringing product and can look at all the relevant drawings together and together with any explanatory literature; and indeed, have a sectional object for comparison with a sectional drawing and interpret the drawing in the light of any instructions on them designed to enable a technician to put them into production. This can be a problem with highly technical production drawings which are difficult for lay people to understand.

It must be emphasized that only three-dimensional reproductions can ever benefit from the non-expert defence. Thus, a competitor's production drawings based on your own product would not benefit in this way, for example if one of your staff copied company production drawing without consent and took them to a rival from which the rival produced them it would be possible to prove infringement even if only an expert could interpret the two sets of drawings. So, if it is feared that action on the basis of the finished article is impossible, one based on the defendant's own production drawings may succeed, where no such defence can be available.

4.6.2 Works of artistic craftsmanship

The law has made it much more difficult to found one's claim to copyright on prototypes not falling within the heading of drawings, for to qualify for copyright they must be found to be works of artistic craftsmanship. The crucial difference here is that the prototype must display some characteristics of artistic merit, unlike the mere drawings on which it might have been based. Two examples may illustrate this:

EXAMPLES:

A knock-down prototype settee which could not be sat upon and had been created only to give a better impression of what the final product would look like could not be relied upon as a work of artistic craftsmanship and therefore could not be used as a basis for a copyright infringement action.

A dress made up by two textile workers who were following precisely the instructions and drawings of the dress designer could not qualify as a work of artistic craftsmanship because there was no artistic input into the dress by those who actually made it. Even if there had been, the author of the design on paper did not participate in the actual execution of the design into a model dress and therefore would not have enjoyed any copyright in it.

By contrast, a skilfully made first cast of – for instance – a vase, or a plaque, or a piece of furniture which could be used and which was to be imitated by machine through the use of robotics would be protected in this way. Equally a prototype piece of jewellery would qualify for protection under this heading.

The requirement of artistic input is thus an important restriction. It has been said that 'Many small and medium-sized companies produce new designs right up to the stage of production planning, that is a finished prototype which has been tested and perhaps even shown in the trade, before ever they make a single drawing (if indeed they ever make a drawing at all). This may be because in engineering matters where adjustments and alternatives are to be made after tests, it may be the most economic and practical way of going about a problem' (Drysdale and Hands).

Where this is so the restrictions placed on the protection of prototypes make it important to change design practice at least to the extent of following up new prototypes with detailed drawings so as to acquire the higher level of protection afforded by the artistic copyright in drawings. The costs of development may be increased, but if the increases can be borne they should be borne. It is money well spent.

4.6.3 Scope and duration of copyright protection

Obviously the major infringement is simply to make a copy of the design or as the Act puts it to reproduce the item in any material form. But the scope of design copyright is such that it can trap a large number of people involved in the manufacture or distribution of infringing items. In particular, it is an infringement intentionally to sell, let or hire, or by way of trade to offer for sale or hire or by way of trade exhibit in public any article the making of which was to one's knowledge an infringement or would if imported have constituted any infringement had it been made in the country to which it was imported.

It is also an infringement to distribute the articles by way of trade or so as to affect prejudicially the owner of the copyright if one knows of these facts. The significance of these other means of infringement, so-called secondary infringements, lies in the much greater remedies which may be available for them. For knowing infringements involving commercial exploitation of the infringing designs may lead to damages being awarded which represent the full retail value of the infringing items, making little or no allowances for deductions in respect of its costs. They are an important reason for relying on copyright in court. Imagine the case of a cheap and nasty design of, say, an owl, which is adopted for a diamond brooch. The design of the owl may in itself be of little value. But if the design is copied by another jeweller and the copies are sold, the measure of damages payable by the seller in respect of those sales will be the full retail value of the brooches sold, including the diamonds, the gold and other raw materials going into the products, and making no allowances for the cost to the seller or the raw materials, cost of labour, distribution costs, etc.

The duration of copyright in these circumstances varies with the facts. It depends on whether the design for which copyright protection is sought is or is not also a registrable design under the Registered Designs Act.

The protection of copyright from three-dimensional reproduction of a two-dimensional and non-registrable design is the life of the author plus fifty years, even after the design has been industrially applied. So a purely functional design such as an exhaust pipe may be protected for a very long period of time, equivalent to the drawings on which it is based. There is however a restriction on the protection available where reproductions are made by others in order to repair products of which the copyright part was a component. This is examined later.

Where a design is registrable once it has been applied industrially by or with the consent of the owner of the design, the duration of copyright protection is only for fifteen years from the date of its first industrial application so that no greater protection is enjoyed than would have been for the registered design; after the expiry of the fifteen year period it is no longer a copyright infringement to do anything which would before the end of the fifteen year period have constituted a breach of the monopoly in the registered design or any associated designs and articles under the Registered Designs Act, though the artistic copyright in the drawing as such endures for the normal term of the life of the author plus fifty years.

4.6.4 Acquisition of copyright in designs

Copyright protection arises automatically and without formality. But because of the lack of a formal registration system any design copyright claim will be dependent on bringing adequate evidence of ownership of copyright in the material it is sought to protect. This clearly requires careful administrative action in ensuring that all the required documentation is available and properly ordered. In bringing a design into production the following precautions should always be taken in relation to production drawings:

- Make sure that copyright in any drawing or model, in which copyright may subsist, is vested in you. This is easily forgotten, but if outside draughtsmen or draughtswomen are used as is often the case in a smaller firm, title in the drawings may belong to the outsider. It may be necessary to have a formal assignment of all such drawings either before or after they are executed and this will require certain legal formalities under the *Copyright Act 1956*, sections 36 and 37, so any such document should be drawn up by a legal adviser. This will be important too where you have employed an external source of parts for your product. Where outside production sources are used care should be taken that you enjoy copyright in their production drawings etc., otherwise you could find that you were stopped from changing your original component manufacturer for some cheaper or more reliable sources, by the original manufacturer's copyright in the parts you want manufactured.
- Make sure that all original production drawings are kept up to date with the development of the product so that each time a new development is decided upon it is reflected in a new set of original drawings. Some drawings based on a prototype, i.e. preceded by it, may not always be held to be original, and it has been argued that it depends on whether they are mere two-dimensional translations of the prototype or are works which required considerable effort, inspired by the prototype, though the better view is that they are all sufficiently original to merit copyright protection. In the same way, a full re-drawing of an earlier drawing will attract copyright: a mere tracing of an earlier drawing will not.

- Make sure that all drawings are correctly dated and signed by the draughtsman or draughtswoman drawing them, recorded and stored safely: they may be needed decades later. Make sure that they are the originals and not photocopies, etc.
- Make sure that the date of the first marketing or publication of the design as a commercial product is recorded and evidence kept of that date. This will establish the period of protection by copyright so that you are not unable to prove that copyright still subsists in the work.
- Make sure that the link between the drawings and the product is clear. If there is a diversity of product on the market it may be difficult to link a series of different drawings to those particular products. Always keep samples of the products derived from the drawings to make this task easier.
- In designing your drawings, bear in mind the law's requirement that a non-expert should be able to recognize the defendant's product to be a three-dimensional reproduction of your two-dimensional drawings. You should ensure that your drawings so far as possible will be intelligible to a non-expert, by including, for instance, pictorial views of the assembled product as well as technical blueprint drawings, with sectional illustrations and explanatory memoranda attached.
- Although it is strictly unnecessary in the UK, it is still good practice to make an express claim to copyright on the drawings and accompanying notes etc., as well as the final product both in prototype and as marketed, most conveniently by applying the © symbol plus the year in which copyright arose and the name of the copyright owner. This may also in fact be of some assistance in relying on copyright protection in some of the overseas countries. If the product concerned is such that for technical or aesthetic reasons it is impracticable to apply such markings, then at least the claim can be made in any accompanying literature or labels, on the packaging of the product, as well as in trade literature such as catalogues.
- If you have applied for and been refused a UK registered design for the material concerned, it may nevertheless attract copyright and indeed you should keep evidence of this refusal since it may constitute evidence that the design is not of a kind which is registrable as a design, and is therefore prima facie entitled to copyright protection for the much longer period of life of the author plus fifty years. This leads us to the next question, that of the relationship between registered design and copyright protection.

4.7 MULTIPLE PROTECTION OF INDUSTRIAL DESIGNS

It is very common for a single product to be protected by patents in respect of its patentable content, design copyright for its functional but non-patentable components, registered designs for visual aspects of the design such as a casing or display, and trade marks for the manufacturer's symbols and other marks. This is quite feasible. But what of multiple protection for the same element of the product? Can you, for instance, protect the casing both by copyright and by registered design?

4.7.1 Relationship between the 1949 and 1956 Acts

What is the position where a person seeks to rely on copyright protection under the 1956 Act for a design which is registrable under the 1949 Act? At one time the legislature had adopted an 'either or' approach to the question, but in 1968 in the Design Copyright Act the legislature abandoned that approach and allowed a person with a design registrable under the Registered Designs Act instead to claim th protection of artistic copyright under the 1956 Act; but this protection would last only fifteen years. So as we have seen once a design was found to be registrable and had since been applied commercially its protection endured only for a maximum fifteen years, whether registration or copyright were relied on from first marketing.

By contrast, the present position of a design which, although not capable of registration under the 1949 Act, has been applied to articles of manufacture and mass produced on them has been very controversial. The position now is that non-registrable designs enjoy the full term of automatic copyright protection of life of the author plus fifty years. Some have seen this as anomalous but it appears to be the correct view of a very complex statutory provision in the 1968 Act.

4.7.2 The relative merits of the two Acts

If you do have a choice between copyright and the registered design and do not want to seek both, then in deciding which mode of protection to seek, the owner of a new design may bear in mind that a brief comparison of the two statutes reveals just how little benefit the additional protection of the *Registered Designs Act 1949* provides for someone with the possibility of *Copyright Act 1956* protection and a well-documented claim, when bearing in mind the additional cost of acquiring the registered design, though this may well change if the reforms discussed in Chapter 16 are implemented.

(a) Duration

The length of protection under design copyright for a registrable design is no shorter than for a registered design and where the design is not registrable is markedly longer in duration.

(b) Damages

Herein lies one of the great advantages of actions under the *Copyright Act 1956* which almost inevitably provides a better measure of damages, a distinction uniformly criticized as anomalous, but one which gives much greater damages than a registered design claim.

(c) Compulsory licences

No provisions exist for the grant of compulsory licences in the *Copyright Act 1956*, such as are found in the *Registered Designs Act 1949* and *Patents Act 1977*, which give the design owner much greater scope for refusal to supply spares except on exorbitant terms, with the limited mitigation of the right to copy for repair purposes found by the House of Lords in the case of British Leyland v. Armstrong (1986).

EXAMPLE:
British Leyland PLC v. Armstrong Patents Co. (1986) Armstrong produced off-the-shelf

exhaust pipes to fit the Morris Marina car, which copied the originals. The House of Lords found that the owner of a car has the right to repair it by the most economical means at his or her disposal, and that he or she need not specially commission the part (as had before been thought) but may use parts supplied off the peg. Manufacturers may make spares in anticipation of orders for this purpose without fear of infringing copyright in the parts, though they could not set out to make parts for purposes other than for repair, nor can they claim them to be genuine original manufacturer parts or use the trade mark of the original manufacturer.

(d) Range of protection

Although the *Copyright Act 1956* requires copying for liability and the *Registered Designs Act 1949* does not, a decision of the House of Lords substantially diminishes the importance of this distinction: in LB (Plastics) v. Swish (1979), it was held that even where the alleged infringer has never seen a component, the copyright in which he or she is alleged to have infringed, if that part existed before a part into which it fits, and the alleged infringer designed his or her replacement part by reference to the dimensions of the part into which it fits, this still satisfies the causal link between the original drawing of the component and the making of the part, since all its dimensions are still dictated by that drawing. This of course places the alleged infringer in an invidious position: which came first, the component alleged to be infringed or the part into which it fits? (in the case itself, the drawer or the carcass?). Furthermore the present judicial practice is to infer from similarity the prerequisite of copying, a difficult inference to rebut. Again, the non-expert defence which might have been thought to make infringement actions in copyright more burdensome for the plaintiff has in practice been reduced to something of a paper tiger. Thus the real difference between freedom from copying and monopoly, though it remains, has shrunk.

(e) Range of items protectable

Drawings for any articles, even the most trivial or functional may be protected under the *Copyright Act 1956*: only those objectively novel or original designs with eye appeal earn protection under the *Registered Designs Act 1949*. This difference really makes itself felt in the field of engineering components where manufacturers who for years were unaware of their rights have made extensive profits as a result, although the British Leyland decision (above) threatens these profits.

(f) Overseas protection priority

A registration under the 1949 Act gives priority for certain designs and utility model registrations in many of the overseas jurisdictions and is a means of obtaining automatic protection in some countries, of which the most important commercially is Hong Kong.

(g) Prestige and proof of infringement

This is often cited as an advantage of design registration. It is true that the problems of establishing the date of your creations and chain of title are much fewer in the case of a design registration, and it may be that the prior search of the register and the official recognition of novelty by a registration number may have more effect. The conferring of a registration number may be very prestigious and assist sales of the product or operate to scare off competitors from closely similar designs.

(h) Licensing transactions

These are equally affected by these factors, with a record of the transaction and the associated benefits which come with it. The only partial mitigation which one can raise against this if it is decided to or is necessary to rely on copyright, is to make use of the facility of Stationers' Hall Registry as a means of reliable record but this has no legal effect as such and it is not necessary if you can preserve the evidence yourself.

(i) Cost of protection

Of course this is much greater in the case of registered designs, but it is still one of the 'economy grade' intellectual property rights.

The appeal of using copyright protection rather than the registered designs is obvious, particularly in the light of the fact that the measure of damages obtainable under the 1956 Act is markedly greater than that available under the other intellectual property statutes, and the acquisition of initial protection cheaper, less cumbersome and quicker. The attraction stretches beyond design, to make copyright an attractive alternative to patents in some cases.

4.8 FOREIGN DESIGN LAWS

There is possibly more diversity in industrial design laws of the world than is the case for any other intellectual property right, with some merely requiring deposit of copyrighted material, others having design registration systems close to that of the UK (for example in Australia, New Zealand and South Africa), and yet others with more unusual systems. The use of an overseas design agent, usually an associate of a UK agent, is essential in those cases where one is seeking overseas protection for a design. The important thing to remember is the necessity of acting within the six month priority period afforded by the Paris Convention. The requirements of applications vary to a much greater extent than is the case with patents and so it is not so easy to achieve rapid translations of applications.

The strength and duration of protection in these countries also varies markedly when compared with the relative consistency of patent protection. This and the cost of obtaining overseas protection which cannot be reduced in the same way as for a patent may combine to induce a greater discretion on the part of the design owner in choosing in which countries to seek protection, though the same factors apply as when considering overseas patent protection, possibly with the added consideration of the strength of the local law a more prominent factor than for patents.

4.8.1 Design patent protection in the USA

The USA design patent law provides a means of obtaining a so-called design patent by registration in much the same way as in the UK. The requirements of the law are however somewhat stricter than those in the UK and also the scope of protection somewhat narrower. The result of this is to make USA design registration even less attractive than in the UK. To be patentable in the USA a design must be:

(1) Novel
(2) Not obvious
(3) Ornamental
(4) Embodied in an article of manufacture

Novelty is determined in the same way as for a normal utility patent. Non-obviousness is clearly a more difficult criterion to apply in the context of design than for a patent and to further complicate matters the courts in the USA have disagreed on whether the test is what is obvious to the ordinary intelligent person or to the ordinary designer who is practising in the design field in question, and who will presumably be much better informed. Whichever it is it is a stricter test than that applied under the UK Act of 1949. The third requirement, that the design should be ornamental, means that functional aspects of design may not be registered under the design patent. The final requirement that the design must be embodied in an article of manufacture again corresponds to the requirement of the UK Act. And, as in the UK Act, the feature of the design to be protected may be one of configuration, shape or ornamentation or even a combination of these things, but it must be embodied in an article of manufacture.

An application for a design patent may relate to the application of the design to one type of article of manufacture only. Design patent protection endures for a period of fourteen years. It is generally felt that the requirements of the patent specification and claims make the scope of protection for a US design patent rather narrower than that enjoyed by a UK registered design and accordingly some advisers do not recommend the expenditure of money on acquiring the protection of the Act.

4.8.2 USA copyright protection for designs

Since 1954 it has been clear that in the USA the fact that an article registered with the Copyright Office as a work of art is later applied industrially (in the case in question, *Mazer v. Stein* (1954), a statuette was used as a lamp base design) does not deprive that work of copyright protection. Thus, in addition to design patents, there is some scope for use of the American law of copyright to protect your designs, though this is more limited than in the UK, in so far as the protection afforded does not extend to purely functional designs, but only to those which may be judged to be a pictorial, sculptural or graphic work.

Thus, in summary, under the USA Copyright Act of 1976, a design which comprises non-functional features of a pictorial, sculptural or graphic work and which can exist independently from the object to which they are applied, may be protected by copyright on complying with the requirements of formality and registration prescribed by the law of copyright in the USA.

This means that it is quite possible to protect through USA copyright law such commercially valuable designs as jewellery, ornaments, carpets and textile designs and sculptures and to acquire thereby a far longer period of protection than that conferred by the design patent, provided that what you seek to protect is not purely functional.

As in the UK, the protection of the USA law of copyright extends only to copying and no monopoly is conferred. Duration of copyright in the USA is a period of the life of the author plus fifty years for all works created on or after 1 January 1978.

4.8.3 Acquisition of copyright in the USA law

The USA law has certain formalities and certain registration procedures which must be complied with in order to acquire or enforce copyright. The requirements are considered later in Section 4.12.

4.8.4 Conflict of copyright and designs in the USA

It is not always possible for one to combine both copyright and design patent protection for the same apparently registrable design, since the US Copyright Office will refuse copyright registration to any design in respect of which there is an existing design patent, though the Patent Office will not refuse protection of a design patent to a design which has already been made the subject of copyright registration even though it remains registered with the Copyright Office.

4.9 WHAT TO TELL THE PROFESSIONAL ADVISERS

Much of the advice given in the chapter on patents applies equally to this area. Again it is important to consult the agent at an early stage, as soon as any physical embodiment of a design which it is felt may be of some commercial significance is attained, either in the form of a drawing or model. On some occasions the use of a design registration or copyright protection will be an alternative to or adopted instead of a hoped for patent claim. So where it is appropriate a patent agent may be consulted on both matters at the same time or even just asked for advice on what if any form of protection may be suitable or available in law. Again, the choice is a commercial decision informed by the professional advice you receive.

In approaching a patent agent to consider design registration or merely to check up on the position with regard to copyright protection, the owner of the design should bring with him or her as much of the following material and information as possible consistent with an early consultation:

(1) All drawings, photographs, models, work books, etc. used in creating the design and an explanation of the history of the design.
(2) What features of the design, either technical or visual the owner thinks to be novel or otherwise important.

(3) Any technical or functional aspects of the new design, materials in which it is to be reproduced, etc.
(4) Any legal or other, for example technical or performance, requirements to which the designer has had to conform in formulating the design specification.
(5) The articles to which the design has been or is intended to be applied and intended production methods.
(6) Any relevant prior art of which the owner is aware, including any earlier registrations taken out by the owner.
(7) Any existing or proposed variations to the design which it may be desirable to protect.
(8) The source of the design, for example an artistic work etc. from which it has been drawn.
(9) The authors of the design and their relationship if the authors and the persons seeking protection are different persons, together with any documents which purport to confer ownership of the design on the person seeking protection.

If you do not have all of the information at the first visit, much of it will appear in the course of time and should be passed on to the agent handling the matter, including details of any new improvements, any prototypes, variations, etc. which you may later develop.

The patent agent will be able to give you advice on the best way of presenting the drawings etc. to comply with the requirements of both the Acts we have been considering and on overseas protection. The patent agent will advise on the safest and best courses of action for acquiring the best level of protection. What the agent will not be able to do so well as in the case of patents is to tell you what your chances of success are to be. For it is a fact that searching for prior art is much more difficult in the case of a UK Registered Designs Act application than is the case for a patent. It is in effect essential to make use of the Designs Registry officials to do this. The reason is that there is no comparable classified designs journal such as exists for patents and a general search by an ordinary member of the public is impossible in practice. Instead the only real option is to apply for a search to be done by the Designs Registry, sending a copy of your design and requesting a search for prior art. This search is even more limited than that conducted when examining an application for novelty, since it covers only those registrations going back about twenty years. It is of more use in determining whether a design is likely to infringe someone's registered design than in deciding whether an application is worth making.

4.10 REACTING TO OTHER PEOPLE'S DESIGNS

Competition often requires a manufacturer or designer to follow a trend. How can he or she do so without infringing copyright the existence of which he or she has little opportunity of confirming? This raises the question of designing around an earlier design while retaining its basic character to appeal to the same emotions or tastes as the original. On other occasions the problem may be the purely technical one of wanting to produce spare parts and compatible parts for the design of another.

- From the point of view of manufacturers who are commissioning designs from a consultant or other third party and who want to follow some design trend without getting their fingers burnt, the only safe course of action apart from use of common sense and observation is to procure in any consultancy commission agreement a contractual warranty that the work produced does not knowingly infringe another's copyright and to reserve a right of indemnity if it does. If they are caught out, an immediate withdrawal of the infringing design will help to reduce damages not least because the measure of damages for innocent infringement is markedly lower than for knowingly trading in infringing goods.

- From the point of view of the designer who is commissioned to design following a trend and who bears in mind that intentional plagiarism is unethical and an instance of unprofessional conduct, it is legitimate to use the earlier design as inspiration for underlying concepts and ideas. What is not permitted is to use the earlier design as a short cut to executing one's own version or implementation of that underlying concept or idea. Designing around something is quite permissible but may be very dangerous for it is difficult to say that a design which began life as the original and is changed to appear different from it has gone far enough unless the effort of doing so has introduced sufficient independent content to amount to as much effort as starting from scratch oneself. It is a question of fact in every case. But better to start with your own fresh design working from the mandate of the design brief and compare the two designs at the end of the day, than to start with the original design and try to design-in any differences sufficient to escape liability for copying.
- Spare parts and compatible designs manufacturers may appear to be worst affected by these developments in the law, but even they have some scope for action in the light of the British Leyland decision already discussed above.

Firstly they may be able to benefit from the right of an owner to repair the goods which includes a right to reproduce components designed to fit into the main machine.

Secondly, it may be possible to rely on the way in which the parts were created originally to defend a claim. This is explained most clearly by the authors of a lawyers' text on copyright, Laddie, Prescott and Vitoria on the *Modern Law of Copyright*, 1980, Butterworths, London (paragraph 3.57), using the example of a nut and bolt design, as follows: '. . . To take a simple illustration, suppose there exists a nut and bolt of original design, and a competitor desires to make and supply nuts to fit the bolts. Necessarily there will be an exact correspondence between the female threads, and so the competing nuts are bound to resemble the original nuts as to at least a substantial part. It does not follow however that the law of copyright will interfere with the manufacture and supply. Going back to the time when the original nut and bolt were designed, it is more likely than not that the bolt was designed first, the shape of the female thread being dictated by the design of the bolt. Therefore, if the intending competitor makes his nuts to fit the bolts, and does not simply copy direct the nuts of his rival, the chain of causation is broken. He does not infringe the copyright pertaining to the nut, for his rival will fail to prove copying. Probably he does not infringe the copyright pertaining to the bolt, even though there does exist a chain of causation leading from the former, because there is no sufficient objective visual similarity between his nut and the other man's bolt – the two things look quite different and (the non-expert defence) will make his position even stronger.'

4.11 THE COSTS OF DESIGN PROTECTION

So far as copyright protection in the UK and many other countries in which copyright is available is concerned, the only costs will be those of preparing suitable drawings or prototypes and keeping them in accordance with the advice given above, though in some countries there will be additional costs of registration and appointing local representatives. So far as registered design in the UK and in other countries is concerned, much the same comments apply as in the case of patents, with the professional fees being a variable but on the whole much lower than where a patent is being sought. The UK fees for registration and renewals can again be set out more easily (Table 8).

Table 8 *Some registered designs fees in the UK (as at May 1986)*

	£
Application for single textile design	12
Application for set of textile designs	19
Application for single article design	42
Application for set of article designs	81
Renewal of registered design for years 5–10	99
Renewal of registered design for years 10–15	146

4.12 NON-INDUSTRIAL COPYRIGHT

Non-industrial copyright is simply acquired in the UK at no expense to the owner. It is an under-rated and an important source of protection for business from its unfair unwanted competition. Non-industrial does not mean non-commercial: the entertainments industries have enormous turnovers, highly dependent on the use of literary dramatic artistic and musical copyright and on the copyrights in films, sound recordings, broadcasts, etc. But these are not dealt with here. We are here concerned only with non-industrial copyright as a back-up to protecting new technology and design and not the wider aspects of intellectual property in business.

The types of valuable commercial material for which copyright is a valuable source of protection outside plans for industrial designs and computer software material, dealt with elsewhere, include:

(1) Process and maintenance instruction manuals
(2) Franchisee manuals, accounting and employee data
(3) Databases, including customer lists, catalogues
(4) Advertising copy and related trade materials
(5) Additional protection for trade mark designs
(6) Product packaging design, specification papers
(7) Research reports, employee assessment papers
(8) Project proposals, costings and estimates letters
(9) Certificates, coupons, quotation or order forms
(10) Designs for store shop fronts, decor, etc.
(11) Film recordings of tests, experimental runs, etc.

Copyright cannot stop industrial espionage. The use of trade secret protection and physical security measures alone can help you in that regard. But it can sometimes help you to mitigate the consequences of such industrial espionage. And it can help to prevent direct reproduction of materials which might dilute the effect of your own publicity or image of your everyday working materials such as manuals and the like.

Copyright can provide significant additional means of protection to trade marks even where trade mark registration falls short of assisting you. Of course, as we shall see, the duration of copyright is limited, and trade mark protection is potentially eternal, but this does not detract from the uses to which copyright can be put while it is in force.

EXAMPLE
Schweppes v. Worthington (1985)
The defendants produced a shampoo bottle using a very similar label to the plaintiffs' drinks

bottles labels 'Schweppes', but using the name 'Schlurppes'. This was held to infringe the copyright in the label design, a parody being an infringement as much as anything else even though there was not confusion between the two labels.

4.12.1 International protection through copyright

Almost every country in the world has a copyright regime and extends equal treatment to foreign works as to domestic authors' works, either under the Berne Convention, in which case there is no requirement of registration or any other formality, or under the Universal Copyright Convention, which does require some degree of formality and sometimes registration. Thus in most countries protection of a British work will extend either automatically, or after only a minor requirement of formality, the biggest market in which formality is required being the USA. The position is, however, very complicated inasmuch as the original Berne Convention in particular has been subjected to so many amendments, not all of which have been ratified by all the member states, that the precise regime applicable to an author abroad always depends on the combination of countries involved, and the date of the work under consideration. In every case it is necessary to check the position.

4.12.2 What material can copyright protect?

Most written material of any originality can be protected. The Copyright Act provides that a literary work includes any written table or compilation. Other than that little help is offered by the statute. But the work must be of more than minimal substance to acquire copyright as a literary work, and mere words or slogans, however inherently original or inventive they may be, will lack that necessary degree of substance for copyright protection as literary works under the Act.

EXAMPLES

Rigby v. Henley Ltd (1928)

The plaintiff had offered by telephone the first refusal on the slogan 'Drive a bargain home from Henleys'. The defendants refused payment but used the slogan. The court held that even if written down it was too insubstantial to gain any copyright protection.

Exxon Corp v. Exxon Ins Con Int Ltd (1981)

The plaintiffs sought to protect by copyright the word Exxon, a registered trade mark and their business name which another company had taken for itself. The court held that although much research had gone into its formulation and that it was literary, original and a work, it was nevertheless not an original literary work because it conveyed no information, it provided no instruction and gave no pleasure.

The Act protects paintings, sculptures, drawings, engravings and photographs, irrespective of their artistic quality. Thus a vast range of useful trade and advertising material, display material and the like may be protected in this way, which may be crucial to the success of get up in format franchising operations. But again, despite the term artistic work, these do not have to be of an artistic nature and so include maps, plans, and charts. Similarly, sculptures appear to extend to wall plaques, medals and the like. Drawings include any diagram, map, chart or plan or similar work, not being a photograph, even of purely technical material, as we have seen in industrial design protection.

EXAMPLE

Tavener Rutledge v. Specters (1959)

The design of a fruit drop box used by the plaintiff was imitated by the defendants. They had in fact used different colours and a different trade mark, but the pictorial design and physical arrangement of the sweets in the box was the same. The plaintiffs succeeded in an action for breach of copyright.

Again there are limits on this.

EXAMPLE

Merchandising Corp of America v. Harpbond (1983)

Protection was sought for a new design of the make-up used by the popstar Adam Ant on stage, which a poster-artist had reproduced simply retouching a photograph to update the poster to reflect the new appearance of the popstar. It was held that the make-up did not attract protection under the Act.

Works of architecture, being either buildings or models for buildings, attract copyright protection, quite apart from their plans. Wherever a new plant is designed by an architect and it is envisaged that any material alterations or some identical buildings might be wanted it is as well to check that the contract with the architect does enable you to do these things, whether or not he or she retains copyright in the plans or building itself. Similar considerations apply to advertising agent's work.

This might be of considerable importance where a franchise chain was to include outlets designed to have the same appearance whenever possible, such as those in some fast-food chains involving portable or mass-produced, ready-made component buildings. Of course, other considerations enter into the design of formats of this kind, such as local planning laws and regulations. In addition of course, uniforms of staff, cutlery, menus, etc. can all be protected by copyright in this way.

All of these works require originality, but not only does this vary in degree, but also there are a number of ways in which a work may be original. Copyright is not concerned with the originality of ideas but with the expression of the thought, and in the case of literary works with the expression of that thought in print or in writing etc. The originality which is required is therefore related to the expression of the thought. The Act requires not that the mode of expression be original or novel, but that it must not be copied from another work – that it should originate from the author. It is for this reason that mere compilations and selections, etc. may themselves attract a copyright even though they are collections of other work, if they also require a degree of original effort of some kind. It is a question of fact in every case. As a very rough guide what is worth copying is worth protecting. This is reflected in the protection afforded to commercial databases.

EXAMPLE

Whitaker v. Publishers' Circular Co. (1946)

A list of current publications won copyright not because of its originality in selection but because of the great comprehensiveness of the list which required much effort and research.

But the protection of catalogues and tradesman's lists etc. is limited to the form of expression and not to the selection itself.

EXAMPLE

Purefoy Engineering Co. Ltd v. Sykes, Boxall and Co. Ltd (1955)

A rival firm who copied the selection of stock listed an the plaintiff's catalogue was not held liable for copyright infringement of those parts of the catalogue in respect of which it could not be proved they had copied the presentation of that information.

4.12.3 Foreign copyright

Virtually all the British Commonwealth has a UK model Copyright Act, including Hong Kong, Australia, New Zealand and Canada, and the law is thus basically identical to that discussed already in the preceding sections.

In the USA copyright protection is basically the same as in the UK, arising automatically and covering basically the same materials and giving basically the same rights. But there are certain formalities to be observed in order for a work to qualify for the fullest protection in the courts, whether or not the work concerned is one which has been or is intended to be published by its author. These can be summed up as the following requirements for full protection:

(1) Deposit and registration of the work must be effected
(2) Copyright notices must appear on the work

Deposit and registration

An unpublished copyright work may be registered by depositing a copy with the Copyright Office and filing in the appropriate form and paying the appropriate fees for registration. It is necessary to deposit two copies of those works which it is intended to publish and they should bear the proper copyright notice in such manner and location as to give reasonable notice of the claim to copyright. If a work has been published bearing the proper copyright notice it can be registered at any time after publication in the same way, but only a registration taking effect within five years of the initial publication will enjoy a presumption of validity.

Failure to publish the work with the proper copyright notice means that you have a period of five years in which to file a registration and make reasonable efforts to apply proper copyright notices to all those copies of the work so far distributed, on the expiry of which registration is impossible.

Although registration is optional, deposit is compulsory apart from certain exempted categories of work and a fine may be levied if within three months of publication of a work bearing a copyright notice there has not been a deposit of two copies of the work with the US Copyright Office.

The benefits of registration are twofold. Firstly, that a certificate of presumed validity is issued, and secondly, that if registration is not effected within three months of the first publication of the work no award of damages or costs may be made in respect of a proved infringement of the work which was begun after publication and before registration. Finally, where the work is an unpublished work, if there has been no registration of the work no damages or costs will be awarded in respect of any infringement which was begun before the date of the registration.

The copyright notice

The proper form of the notice is as follows:

(1) The word COPYRIGHT or its abbreviation COPYR. or
(2) The Universal Copyright Convention (UCC) symbol ©;
(3) Followed by the date of publication
(4) Followed by the name of the copyright holder

(Note that in some countries one requires in addition to the UCC symbol the statement 'all rights reserved' (basically, the South American states). This formality should be observed too for safety's sake, where literature is to be published in those areas.)

If this notice is not present, the copyright in the work is not lost if it was omitted only from a few copies distributed to the public or registration is effected within five years of publication and reasonable attempts to rectify the omission on all copies after the omission is discovered or the omission occurred in violation of an express requirement in writing from the author that the copies published should bear the notice. But in the absence of a notice any person making unauthorized copies of the work may be held to be an innocent infringer and therefore be held immune from an award in damages. In practical terms it will be advisable always to place a readable copyright notice on all the printed materials. The Act does permit one to omit the year date from the notice where a pictorial, graphic or sculptural work, with accompanying text matter if any, is reproduced in or on greeting cards, postcards, stationery, jewellery, dolls, toys, or any useful articles.

The Western European copyright laws (and those of Japan which are based on Western European 'civilian' legal systems and in particular West Germany's) are in principle not dissimilar from those of the UK, and also give copyright automatically and without formality, but they do nevertheless differ in some notable respects from those referred to so far.

5
Trade marks and names

5.1. WHAT IS A TRADE MARK?

A trade mark is a symbol or name or device or any combination of these used by traders to distinguish their goods or services from those of their competitors. It goes further than that, however. It acts as a quality guarantee in the mind of the consumer who associates the trade mark with goods or services of consistent quality at a particular level, with goods or services of a particular price level or prestige, and acts as a shorthand link in identifying products in advertising media. From the point of view of the trader, the trade mark he or she uses is a focus of the goodwill surrounding his or her business and products. Legal protection of the trade mark both serves to protect the public from deception and protects the trader from unfair competition and use of his or her reputation by trade rivals. The UK law of trade marks is complex and whereas a full understanding of the law of patents and designs is of major importance to management in new technology and design, the law of trade marks may be dealt with in a rather less detailed way. This is not to underestimate the importance of trade marks, but merely to reflect the fact that they will occupy the specialist laywer rather more than the non-lawyer.

5.2 HISTORY OF TRADE MARK PROTECTION IN THE UK

Unregistered trade marks have long been and still are recognized by English law. Registered trade marks became available in the UK in 1875. Since 1875 the law has been changed markedly in a series of Acts of Parliament until the present law contained in the *Trade Marks Act 1938* as amended in the *Trade Marks (Amendment) Act*

1984 which for the first time extended the protection of registration to trade marks for services and not merely for goods. Before then, the law on service marks was governed entirely by common law, which had protected all types of trade mark from very early times. And even after over a hundred years of trade mark registration, the protection afforded by the common law to trade marks is so comprehensive that it is still relied on in many cases in addition to the statutory protection of the Trade Marks Act.

5.3 INTERNATIONAL PROTECTION OF TRADE MARKS

In this field as in others attempts have been made to protect marks on an international basis. But trade marks are different from the patent or design rights considered in earlier chapters. Often they have little value overseas in the form used in the UK and overseas trade mark protection may require an entirely different type of mark to meet the requirements of a different culture. Where a trade mark is valuable overseas four initiatives at international co-ordination merit some discussion.

5.3.1 The Paris Convention 1883

Under this Convention any person who has applied for a registered trade mark in the UK or another member country may rely on that application for a period of six months to gain the date of the first application as a priority date over other applicants in any other Convention states to which he or she applies. The earlier application must support the later one if it is to confer a priority, in the sense that it must not be for a narrower class of goods than that specified in the subsequent application.

In practice this is of little importance because prior use or publication of a trade mark does not as it would in the case of a patent application prevent you from obtaining a registration. Its only use is where in response to your application to register a mark in one country a third party seeks to prevent you from also registering in another country by making a pre-emptive application there personally.

5.3.2 The Madrid Agreement 1891

This is an agreement made under the 1883 Paris Convention, but the UK is not a party to it. Under the Agreement it is possible for a trader to register a trade mark in his or her own country and then to acquire the right to it in all the signatory states by depositing the registration at an International Office in Berne. Once this is done the mark is valid in all those states unless there is a conflicting mark on the register of that state already or it is for some reason not registrable in that country, but in such a case the mark will still be valid in those states for which no such objection lies.

The Madrid Agreement does not provide for a world trade mark – it merely eliminates the necessity for multiple applications to acquire a series of national trade marks. Since the UK is not a member of the Agreement, UK companies and traders cannot take advantage of it, except by means of applying through a subsidiary based in one of the signatory states. The protection afforded by the regime is however said to be weak. Clearly the regime could be very advantageous to a country with a relatively lax regime and consequently disadvantageous to those such as the UK with a very rigorous pre-registration investigation, which explains why the UK (and, incidentally, the USA) has not participated. So UK traders may not take advantage of it.

5.3.3 The Trademark Registration Treaty 1973

This too aims at acquisition of protection in a wide range of countries through a single application but in this case the regime is designed to enable the applicant to apply direct to a central office without the necessity of applying first through the national office of the trader's home state. The Treaty's central administration is the WIPO but the Treaty has just come into force since its signing in Vienna in 1973, and is to date only effective in a few (so far unimportant commercially) states, and again the UK has not signed.

5.3.4 The Community Trademark Convention

Proposals for a Community trade mark (CTM) are now well advanced. In outline the proposal is this. The trade mark will apply throughout the whole of the EEC instead of a bundle of national marks – you must choose one or the other, though if the application for a CTM fails or becomes invalid, you will be able to revert to your national mark in the UK.

Registration will be obtained for both goods and services but compared with the present regime applied in the UK the investigation procedure will be much more restricted; it will seek absolute bars to registration, i.e. the absence of certain characteristics which are required of a mark under the CTM regime or the presence of disqualifying factors connected with morality and public order. Thereafter proprietors of an existing trade mark will be able to oppose the registration on the basis of relative grounds, i.e. similarity to their own mark.

Once the registration is effected it will stand or fall throughout the whole of the EEC. Traders with a mark which has not been registered in their own national office will not be able to oppose the registration but if theirs is a well-known mark of more than merely regional importance they will be able to apply for the cancellation of the registration later.

When the CTM is in force, once the Community mark is registered it will be protected from not only any identical marks anywhere in the Community but also from similar marks even if the traders concerned are not competitors if their use of the similar mark might be detrimental to you. To remain on the register it will be necessary to renew the mark every ten years, and to renew the proprietor must be able to show use of the mark in the previous five years. The same qualification will govern the right to enforce the mark against others or to oppose new registrations, once five years of his or her own registration have elapsed. There will be a strict doctrine of exhaustion of rights but the mark will be assignable without the goodwill of the trader's business.

There will be a central office to administer the CTM. There will be a quasi-judicial body to hear appeals against decisions of the office, with appeals on points of law ultimately going to the European Court of Justice. The national courts will have jurisdiction to hear enforcement and infringement actions and counterclaims for invalidity. Judgment will be effective throughout the EEC.

5.4 WHY BOTHER WITH REGISTERING A TRADE MARK?

This in fact raises two questions. Why bother with trade mark protection at all? And why bother with the expense of trade mark registration when there is legal protection for unregistered trade marks through the common law?

5.4.1 Why bother with trade mark protection?

The answer to that question is really provided in Section 5.1 merely by stating the functions of the trade mark. Trade marks help to keep counterfeiters of your designs at bay even if you have not got very wide protection or indeed any protection for the product's design itself. The uses to which consumers put trade marks, their quality guarantee function, mean that they are often the only way in which you can make any quick impression on a consumer faced with a wide range of competing, functionally identical, technically interchangeable, similarly priced, products. They assist you in advertising the product by creating an easily memorable image in the mind of the consumer. They help to sell style and product image in products where other characteristics are taken for granted or are subordinate in the mind of the consumer. If design of a product is essential to its success in the modern consumer market, trade marks reinforce the image that the design has created.

5.4.2 Common law and registered trade marks

As for the second question, the answers lie in the additional scope of protection given by registration and the ease of proving infringements compared with the protection afforded by the common law and by the much greater scope for exploitation by licensing the use of the trade mark if it is registered. A registered trade mark is much less vulnerable than a common law trade mark. These are considered further in Section 5.15.

5.5 COMMERCIAL FACTORS IN SELECTING A TRADE MARK

In selecting a registered trade mark both legal and commercial factors have to be taken into account. This gives you less flexibility in the choice you have to make than you might have in choosing a common law mark. Never underestimate the financial implications of trade mark selection. A trade mark is potentially such a valuable asset that one should not be parsimonious in spending money on its creation. Trade marks are more than the icing on the cake of a product. They play a central role in its marketing and sale. When reviewing proposed trade marks you should keep in mind a number of questions.

If you can afford to do so, the design of a trade mark package by an advertising consultancy, checked out by your trade mark agent, is the best course of action. Whether or not you follow that course you should bear in mind the following matters in selecting a trade mark.

- Do you have an existing mark with a reputation? If so do you want to use it with the new products you are bringing to the market. The advantages of doing so are that the name is already known, and advertising has a spin-off for other products in your range. You can combine the use of an existing house mark with a new mark to give that a helping hand into the market place or to strike a new image. The new mark may eventually take over from the old one. The campaigns leading to the replacement of Datsun with Nissan by the Japanese car company of that name are an instance of success in this tactic. You may want a family of related trade marks as in the use of, for instance, the trade marks Kodak, Kodachrome, Kodacolor, etc. This may require the design of a trade mark which fits comfortably into the general house style of the portfolio of the trade marks already owned, perhaps with subtle variations designed to match the aura of the particular products to which it will be applied.

- Is the mark commercially attractive and flexible in use? Can you use it successfully visually, orally, in colour, in black and white. Is it adaptable to a wide range of products or confined to a particular technical field? What consumer reaction is there to it in trials? You may need extensive market research.
- Does the mark conflict or come close to existing registered or unregistered marks which might give rise to conflicts now or in the future, should you wish to expand or another trader wish to expand his or her range of usage into other products? It will be essential to conduct a preliminary search to determine the risks of trade mark conflict. It may be necessary to buy out any obstructive trade mark rights which smaller traders may have.
- Is the mark exportable with the products you want to use it for? For each country where you envisage trading the same questions as already posed above will have to be asked. The answers will often be different. Particularly with word trade marks, language barriers make the use of a universal trade mark more difficult. Well-known examples include the inappropriateness of use of the rather wistful English term 'mist' for a deodorant spray in German-speaking countries, where the word means manure. With each country you consider, conflicts are more likely with existing marks. At the very least some attempt should be made to have a suitable single trade mark for each of the language groups, so that the same English language trade mark may be used throughout the UK, the USA and the Commonwealth, the same French mark in Francophone countries, and so on.
- Is the trade mark registrable? Even if you do not intend to register a mark at this stage it will be wise to devise your trade mark to comply with the legal requirements of the *Trade Marks Act 1938*, so that if at some stage in the future you do wish to register the mark, there is one less obstacle in your way. In any event, there is much to be said for registration of a mark as soon as possible. It cannot prejudice your position since registration of a mark does not in any way deprive you of common law trade mark protection, but failure to register a mark may with the passing of time prevent you from doing so in the future and it would be foolish to throw away the benefits of an accrued reputation when you do want registered protection by having to abandon an existing and well-known registered mark in favour of a new and unknown registrable trading style.

5.6 LEGAL RESTRICTIONS ON TRADE MARK REGISTRATION

It is not enough to have thought up a good trade mark from a commercial point of view if you want it to be registered. This is because only certain types of mark may be registered under the UK legislation. In all cases of course the wisest action is to consult a trade mark agent to establish whether a proposed mark is of a type registrable and whether anyone else has registered it already. The following notes are a summary only of the complicated limitations on the type of mark which may be registered and they do not affect what may be used as a trade mark which is protected by the common law.

5.6.1 What types of trade mark are protected?

The statutory scope of protection was in October 1986 dramatically extended by the amendment in 1984 – the *Trade Marks (Amendment) Act 1984* – which provided for protection of service marks as well as marks for goods.

The three main legal criteria by which a trade mark application is judged are that:

(1) It must be distinctive
(2) It must not be descriptive of the product
(3) It must be different from other trade marks

Within limitations, the form of mark protectable under the Act extends to devices, names, signatures, words, letters, numerals, or any combination thereof. It appears that it is impossible to register as a trade mark the shape of the product sold so that, for instance, the shape of the Coca Cola bottle could not be registered as a trade mark under the UK Act. But it is no objection that the trade mark covers the whole surface of the product which bears it.

To be registrable, the mark must be one which is 'used or proposed to be used'. It is clear from the definition that the trade mark need not be in actual use at the time of application for it to be registrable – there is no need to acquire a reputation with it before acquiring the protection of the statute. It suffices that the mark is intended to be used. But there must be a present intention to use the mark, not merely an intention to prevent others using it. This provision is intended to prevent the monopolizing of too many marks by a trader of which he or she has no need, and to discourage the trading in marks independently of goods or services, as if these marks themselves were commodities.

In one exceptional set of circumstances it is possible to register a mark in respect of a class of goods in which you have no present intent to trade. This is known as defensive registration. Where the applicant is a registered proprietor of a trade mark made up of an invented word (and only an invented word trade mark, no other type), which has become so well known in respect of the things for which it was first registered and used that the use of the mark on other things would be likely to cause confusion, then he or she may register it in respect of goods for which he or she has no intention of using the mark. The procedure is set out in r37 of the Trade Marks Rules 1986 but the section is hardly ever invoked for the simple reason that it is easier to register in more than one of the classes whether or not you are actually using them.

Although the level of applications for defensive registration is low because of the restrictive attitude taken to the section by the courts, it is an interesting provision in theory because of its implications for the marketing concept of brand loyalty and the fear of a large manufacturer that the mark's prestige might be diluted by its use in other fields by smaller traders, whether or not the large manufacturer wishes to diversify into those fields at some later date, or thinks he or she might wish to do so.

The sort of abuse the provision is designed to prevent can probably be caught by the common law of passing off in most realistic situations imaginable, and there is probably something to be said for the view that such protection as this affords merely expands the already substantial level of dominance which a concern big enough to rely on it already enjoys. There is still to be no provision for services to benefit from defensive trade marks; only in the case of goods will these be available even after October 1986.

There must be a connection in the course of trade between the goods or services and the person seeking to register the mark. This requirement excludes from the possibility of registering as trade marks the badges or symbols of non-trading bodies. But there is no restriction on the type of connection in the course of trade which must be shown to qualify for trade mark registration. So the connection may be by way of trade manufacture or any type of dealing, such as the sale or selection (as in the case of the St Michael trade mark) of goods (but as opposed to merely manufacturers of them) and there is no need for the public to know who owns the mark.

5.7 TYPES OF REGISTERED TRADE MARK: PART A AND PART B MARKS

The foregoing discussion has encompassed the basic requirements of the definition of the trade mark. Even assuming that a mark complies with these requirements, however, it must still conform to certain defined characteristics to be acceptable. The position is complicated by the fact that the Register is divided into two parts, Part A and Part B marks, each with different requirements and affording different levels of protection.

5.7.1 Part A and Part B marks

The Part A marks must meet rather more rigorous standards but provide a better level of protection. You would always seek a Part A trade mark unless you had an existing and valuable common law mark which you wanted to gain more protection for but which for some reason was not registrable in Part A, in which case you might content yourself with a Part B registration if one was possible. To be registered under Part A, a trade mark must (unless it is a certification mark, considered below) be of a certain type and be distinctive within the meaning of the Act. Your trade mark agent will make sure that whenever possible you gain a Part A mark. The types of mark which qualify under Part A are:

(1) Names of companies, individuals or firms, represented in a special or particular manner
(2) Signature of the applicant or some predecessor in business
(3) An invented word or words
(4) A word or words having no direct reference to the character or quality of the goods and not being according to its usual meaning a geographical name or a surname: most attempted registrations fall under this head
(5) Any other distinctive mark but only with evidence of its distinctiveness: see the table of evidence (Table 9).

Even if registration under Part A of the Register proves impossible it may nevertheless be possible to register under Part B. The requirements of Part B are clearly intended to be less stringent than those of Part A.

In practice any marks which are to some degree descriptive but which have an expectation of becoming distinctive of the user are registered under Part B as if to serve an àppenticeship until the desired distinctiveness appears. Likewise for a geographical or surname mark. So the function of Part B registration is to serve as an intermediate stage of protection by statute for a period in which time the user can qualify for Part A protection by use of the mark in such a way that from being merely capable of distinguishing his or her goods it becomes actually distinctive of them. It is also a useful resort for the court or Registrar where the entitlement of the mark to Part A status is in doubt. The applicant can at a later stage apply for promotion to Part A protection when he or she has acquired sufficient evidence of actual distinctiveness. It seems in practice that about five years' use of the name in the UK to establish distinctiveness is the norm.

Other provisions may bar the way to registration of a Part A or Part B mark.

- The mark will not be registered if it would cause the public confusion.
- A narrower provision exists, designed to prevent the registration of marks which would duplicate any existing registrations. This provides that no UK trade mark can be registered in respect of any goods or description of goods that is identical with a trade mark belonging to a different proprietor and already on the register in respect of the same goods or description of goods, or that so nearly resembles such a trade mark as to be likely to deceive or cause confusion. The restriction is extremely narrow: it relates only to 'goods of the same description', unlike the first objection, where the confusion or deception might well arise out of use of a similar mark in respect of an entirely different class of goods by another trader, and it protects only registered marks from subsequent registrations, not unregistered marks. A parallel provision exists in relation to services; and closely related service marks and goods marks will also be examined by the Registrar.

Table 9 *Information required for trade mark evidence of distinctiveness*

(1) Date of first use of the trade mark and a precise and detailed list of the goods or services for which the trade mark has been used.

(2) Area throughout which the trade mark has been used, for example the whole of the UK, or England and Wales, etc.

(3) Details of gross turnover of goods or services sold under the trade mark before the date of filing of the Application.

(4) Details of annual retail turnover figures of goods sold under the trade mark for each of the seven years preceding the date of filing of your Application, or such shorter period of years as is available.

(5) Amount of money spent on making the trade mark known to the public, e.g. by way of television, newspaper, periodical or billboard advertisements, use of trade mark on letterheadings, invoices, etc., representatives' expenses, etc., or any other media wherein the trade mark is used in relation to the goods.

(6) Details of all major towns and cities in the UK where the goods or services are sold under the trade mark together with details of the retail and/or wholesale outlets in such towns and cities.

(7) Copies of any items such as listed in (5) showing the use of the trade mark used in relation to the goods or services.

The Registrar may require the applicant to disclaim exclusive rights to parts of his or her trade mark to which he or she is not entitled as a condition of registering the whole. Any subsequent infringement action brought by the applicant on the basis of his or her registration cannot be founded on the disclaimed matter. But the whole of the mark will be considered when determining whether to admit a later applicant to the Register with a similar mark.

5.7.2 Substance names etc.

No word which is the commonly used and accepted name for any simple chemical element or single chemical compound, as distinguished from a mixture, shall be registered as a trade mark in respect of a chemical substance or preparation: but this provision does not have effect in relation to a word which is used only to denote a brand or make of the element or compound as made by the proprietor or registered user of the trade mark, as distinguished from the element or compound as made by others, and in association with a suitable name or description open to the public use. It is for this reason that chemical substances developed by the pharmaceutical companies seem to acquire a multiplicity of names; the long and often unpronounceable technical chemical name; the shorter and pronounceable name by which they come to be commonly known (the so-called generic name); and the many brand names under which the various companies sell their versions of the substance.

5.7.3 Further restrictions on registration

A number of sensible restrictions on registration which would no doubt be imposed by the Registrar in the exercise of his or her discretion anyway are expressly set out in the Trade Marks Rules. In summary they are these:

- The words Patent, Patented, Registered, Registered design, Copyright Entered at Stationers Hall, and the like may not be registered as a trade mark. Nor may words Red Cross, Geneva Cross and the like, or representations of the Royal Family.
- The Royal Arms or Imperial arms or devices so similar as to be capable of being mistaken for them, or the word 'Anzac' may not be registered, nor may the words 'Royal' or 'Imperial' or any letters or devices which might lead to the belief that there was royal approval or patronage.
- Armorial bearings, insignia, etc. of a town or city borough, body corporate, society, etc. require evidence of consent from the person or body represented.
- Use of the name or representation of a person living or dead requires consent.

5.7.4 Restrictions on textile registrations

The line heading alone of a textile is not to be registrable.

5.7.5 General discretion to refuse registration

Quite apart from all these statutory exclusions the Registrar enjoys a residual discretion to refuse admission to the Register or to admit registration only on conditions or limitations. The Registrar may impose conditions on the acceptance of a trade mark including the disclaimer of exclusive rights to parts of the mark registered and the restriction of the mark to use on certain types of product.

EXAMPLE
Edwards' Application (1945)
The applicants sought to register the trade mark Jardex for disinfectants. It was found that there was another mark, Jardox, used for meat extracts. Both substances might be used in hospitals where negligence or carelessness on the part of staff might lead to the two substances being confused, with consequential harm to patients. Registration was refused.

5.8 HOW DO YOU REGISTER A TRADE MARK?

A trade mark can only be registered in respect of particular goods or services and use of the mark on goods or services not in that list is not protected. The Register is divided up into classes of goods and services. One application can only cover goods or services in the same class in the Register, though it may cover one, some or (exceptionally, if use justifies this) all of the goods or services within that class. Thus a single registration will be able to cover, for instance, petrol and oil, both of which are in Class 4 of the Register; but a separate application will have to be made in respect of the same mark for use in connection with chemicals or paints, which are in different classes of the Register (Classes 1 and 2 respectively). Protection in more than one class therefore requires more than one application. Where two or more applications are made for the same mark for similar types of goods and/or services, the Registrar will usually require the association of those marks where the two classes, though different, are commercially related – this is examined further below.

If you have done your homework properly you should be aware in advance of registered trade marks which may be cited against your application, though there may be little you can do about unregistered marks maintained on a fairly low profile. If the search or subsequent events reveal a potential conflict, you have a number of options available, depending on the nature of the existing mark.

- Abandon the attempt to register the mark and seek another one, writing off any expenditure so far incurred on development of the goodwill and design of the first. If your mark was one which had already acquired some common law reputation you may be able to persuade the opponent to buy you off, thereby cutting some of the wasted expenditure.
- Where the registration attempt is for a mark which you have already used for some time, abandon the registration and continue to rely on any common law protection which you may already have acquired in the mark, if this does not infringe the registered rights or common law rights of the opponent. At the same time develop a new mark and start using the two together so that the new one, registered early on in its life, may eventually replace the old one.
- Proceed with the registration application arguing that although the marks are similar they are not too similar and that there is no risk of public confusion. You may obtain the other owner's consent as evidence of this fact.
- Buy off opposition by purchasing an assignment of the trade mark, together with all goodwill associated with it, if this is financially viable and the owner can be persuaded to sell.

5.9 AMENDMENT OF TRADE MARK REGISTRATIONS

5.9.1 Loss of the mark by non-use

A mark may be removed from the register on the ground that it should not have been registered; or on the ground that it was registered without a bona fide intention to use it and there has been no bona fide use of the registered mark by any proprietor up to one month before the action was brought or up to the date one month before the action was brought a continuous period of five years had elapsed during which there was no bona fide use in relation to those goods by any registered owner or the mark. Although the burden of proof is initially on the applicant for removal, once he or she has established a prima facie case of none-use, the burden falls then onto the registerd owner to show sufficient use in good faith.

Where there is a registered licensee of the mark, permitted use by him or her of the mark shall be attributed to the registered proprietor for the purposes of the Act. But the Act raises great dangers for licensors who fail to keep an adequate watch over the use to which their marks are put.

Use of the mark on purely export trade business will suffice for these purposes.

There is a discretion to accept use of an associated registered trade mark or of the trade mark with additions or alterations not substantially affecting its identity, as an equivalent to use of the trade mark in question.

EXAMPLE

Re McGregor Trade Mark (1978)

The McGregor company registered a trade mark in respect of clothing including dressing gowns. They later entered into registered user agreements in which licensees subsequently registered as users of the mark were permitted to use the mark on dressing gowns. This was done for some time but then from 1973 the licensee began to use the mark on other clothing too and in breach of certain restrictions. In 1976 the licensees sought rectification of the Register to remove the McGregor registration for non-use. They succeeded. Their unauthorized use, going beyond the scope of the licence, could not be attributed to the owner for the purposes of the Act.

5.9.2 Loss by trade marks becoming generic

A best-selling product combined with slack use of the trade name under which it is sold can be the cause of the loss of the mark. Where the trade name of a product has become the generic name or description for that product in the trade (not however by the buying public, though such usage may eventually spread into the trade so it can be just as dangerous), any interested person may apply for the removal of the mark from the Register.

Furthermore, where the word registered as a trade mark has become the only practicable name or description of the product which was manufactured under a patent and a period of two years or more has elapsed from the cesser (i.e. the cessation of term) of the patent, any person interested may again seek removal from the Register of the mark. Where the name forms only a part of the trade mark, for instance, when it is combined with a device, corresponding provisions exist for the Registrar to require disclaimer of the part which has become generic. Finally, even without removal from the Register the proprietor's exclusive rights disappear.

In practice a number of practical steps are taken by companies to prevent this from happening and in the USA very elaborate campaigns of public and trade education directed at teaching the consumer and retailer to use the trade name properly and not as a generic or descriptive term are undertaken by the larger corporations with very valuable trade names such as Xerox. And in relation to pharmaceutical products, as has already been noted in relation to initial validity, the practice is to coin a convenient short trade 'brand' name as well as short generic name.

5.9.3 Loss by failure to pay fees

The trade mark registration must be renewed after seven years and thereafter every fourteen years. The trade mark may be removed from the Register if when the time comes the renewal fee is not paid. However, note that when a trade mark has been removed from the Register for non-payment of fees it is nevertheless deemed to be a trade mark for the purposes of any application to register another mark for a period of one year after its removal unless there has been no bona fide use of the trade mark that has been removed for a period of two years immediately preceding its removal or that no deception or confusion would be likely to arise from the use of the trade mark that is the subject of the application by reason of any previous use of the trade mark that has been removed.

5.9.4 Loss by failure to observe a condition

The mark may be removed for failure to observe a condition imposed on registration. Alternatively it may be varied. Any person aggrieved may apply.

5.9.5 Amendment of trade mark registrations

Apart from loss or restrictions required because of third party oppositions to your mark, you may yourself wish to have the mark amended in some way. There are two ways of achieving this without making a fresh application.

5.9.6 Correction at request of proprietor

At the request of the proprietor, the register may be amended to correct errors, strike out any class of goods for which the mark is registered or enter a disclaimer; at the request of the registered user an error may be corrected. Further, at the request of the registered proprietor the mark may have insubstantial changes to it registered.

5.10 SPECIAL TYPES OF REGISTERED TRADE MARK

Two special types of trade mark deserve a separate mention at this stage.

5.10.1 Certification marks

The Trade Marks Act (1938) provides that trade marks may be registered as guarantee marks or certification marks, which indicate that the goods to which they are applied are of a certain quality or come from a certain region rather than that they originate from a particular trader. Indeed, the applicant for such a trade mark may not trade in the goods himself or herself; the certification mark is designed for use by bodies such as the trade associations which set out standards and test products against them. The mark may certify the origin, material, mode of manufacture, quality, accuracy, or other characteristics of the goods. There are not very many such marks – well-known ones include the Woolmark and Woolblend certification marks and the British Standards Institution's Kite mark. To gain a mark the association seeking it has to gain the approval of the Department of Trade and must allow the use of the mark by any trader reaching the prescribed standards, whether or not he or she is a member of the association. There will be no certification marks for services even after October 1986, only for goods.

5.10.2 Associated trade marks

These have already been mentioned in passing. Where a trade mark that is registered or is applied for registration is identical to or so closely resembles one already on the Register in the same type of goods or services owned by the same registered owner that it is likely to deceive or cause confusion, the Registrar may at any time require them to be registered as associated marks.

Associated marks are in commercial terms marks which closely resemble each other but which provide a variety of forms which enable a trader to use what is basically the same mark in a number of styles which vary in accordance with the nature of the product and its associated image. Thus for instance a well-known trade mark used in relation to clothing at the bottom end of the fashion market might require a mark in a different style from that applied to clothes at the top end, because of the different clientele attracted to those types of clothing. If one area of activity dominates the production of the manufacturer his or her failure to produce clothes in the other for some time will not prejudice the mark because use of one of the associated marks counts as use of the others.

Trade marks registered which are parts of other registered trade marks and marks which are registered in a series will be deemed to be associated marks. The registration of associated marks will not be accepted in relation to different classes of goods. Where the registered proprietor applies and the Registrar is satisfied that it will not cause confusion or deception the association of the marks may be dissolved.

Where marks are associated they may only be assigned as a whole and not separately but they are for other purposes regarded as separate marks. This only applies where all the goods in respect of which an associated mark is registered are subject to the assignment. Where the assignment relates to some only of those goods, the assignment is governed by special rules. The same rules may apply to services.

5.11 PRACTICAL STEPS FOR SAFEGUARDING A TRADE MARK

The two principal dangers a trade mark owner faces are inadequate use of the mark and excessive use of the mark, in the sense of use of it in an incorrect fashion. These are reflected in the provisions for loss of the trade mark registration by non-use and by the mark becoming generic. There is also the problem of avoiding unnecessary conflicts and warning off invaders who might otherwise gain rights by estoppel (i.e. rights arising from failing to act in previous cases of infringement). A number of practical steps may be taken in policing the use of the mark to make it safe from such dilution or loss.

5.11.1 Avoiding loss through non-use

This has three basic elements:

- Ensuring use of the mark as registered rather than some variant which is not covered by the actual registration of the mark. This includes the regular marketing of goods under any ghost marks which you have been obliged to register in order to obtain indirect protection for another unregistrable mark or to expand the areas of protection for a trade mark which would otherwise be vulnerable to dilution through widespread use of similar but not infringing marks. You should use the mark on trade literature, letter heads, etc., in exactly the form in which they are registered and if you have cause to use a modified form of the mark that too should be registered.
- Ensuring use of the mark on the goods for which it is registered and not just on other goods for which there is no current registration. Always make sure that your registrations are up to date with your current range of manufactured products so that you are not manufacturing only goods for which there is no registration, leaving you open to the accusation that there is no use of the mark as registered. Where you have a house mark and a specific mark for each product, make sure that each is made to appear on the product.
- Ensuring the proper use of the mark by licensees in accordance with the terms of the trade mark licence and in accordance with the current registrations. The McGregor trade mark example cited above in Section 5.9.1 illustrates the importance of this. It is also considered in more detail when looking at licensing of trade marks in Chapter 12.

5.11.2 Avoiding loss through genericism

We saw in Section 5.9.2 that a trade mark may be lost through genericism, i.e. when the trade mark becomes the word normally used for the type of article concerned rather than the particular brand or source of manufacture. There are many examples of this, such as nylon and linoleum, to name but two former trade marks which became generic terms. A number of steps can be taken to avoid this fate.

- Always use the term as a proper adjective, never as a noun, and never play around with it so as to make it a verb or invent a word of which the trade mark is merely a component. You would not receive many thanks from the manufacturers of 'Elastoplast' for coining the phrase 'I'll elastoplast your knee.'
- Always expressly distinguish between the generic term for the product and the trade name for the product and keep them together in all advertising and labelling of products so that there is no temptation in the public mind to use one word as a substitute for the other. So you would use 'Elastoplast' Fabric Plasters, not Elastoplasts. Or at least 'Elastoplast' brand.

- Make it clear that the word Elastoplast is a trade mark. Although they have no significance in this country as such the symbols ®,™ are often used and seen to serve their purpose. They in fact are used in the USA to indicate that the mark is a registered mark with the US Trade Mark Office or is an unregistered mark. It is possible to use the term 'trade mark' even where the mark is not registered. If the mark is registered overseas but not in the UK any goods imported or marketed in the UK should be checked to ensure that they bear no representation that they have a UK Act registration for that is a criminal offence. It would be sufficient to make it clear that the goods have a trade mark registered overseas only.

If the mark is used in contexts other than on the product itself or if there are very pressing reasons why an indication that it is a trade mark is undesirable, then, as an alternative, in advertising copy of any length, a footnote indicating that the word is a trade mark of the company concerned may be a useful device. If you are exporting to countries where it is not a registered mark, ensure that you are not falsely representing that it is on the packets or products there too.

- Always where possible use the trade mark in a special style of lettering or at least with a capital letter to indicate that it is not an ordinary word and try not to use too many variants of the registered form of the trade mark in your literature and advertising copy.
- Make sure that staff and especially sales staff use the trade mark properly and are aware of what proper usage is. In house manuals of intellectual property practice and procedure should equally include details on the proper use of trade marks. Staff can often be effective policemen for the mark in the business world since they have the necessary degree of frequent contact with traders to be able to detect any improper use.
- Some American firms with particularly successful product trade marks feel obliged to conduct a public education campaign instructing the public or the retail trade in the use of their mark.
- Maintain a vigilant eye over usage of the trade mark by others especially in the mass media and if you spot an improper use point it out and ask for improper practice to cease. This is particularly important where the mark is used by franchisees and licensees. Details of proper usage should be included in the franchisee's manual or even in an appendix to the licence and any breaches of this code jumped on. A sample letter to this end is included in the Appendices.

5.11.3 Avoiding trade mark dilution

This requires a policy of chasing infringers and challenging similar trade marks when any attempt is made to register them, preserving your registration in as wide a range of register classes as possible consistently with the requirements of bona fide usage and seeking to maintain the widest possible scope for your mark. Design of the mark can be an important factor here. If the trade mark design is too complicated with many distinct elements it may be vulnerable to dilution through use of other marks which although reminiscent of yours nevertheless omit certain details and which render a rival's mark sufficiently different to escape liability. If the mark is too simple it may not afford you the flexibility you want from a mark. With licensees it requires control over their use of the mark and insistence that the mark is not associated with, say, one of their own marks on the same product without consent.

5.12 FOREIGN TRADE MARK PROTECTION

Although most countries have registered trade mark laws, as you might expect trade mark laws vary from place to place. About sixty nations follow the UK model fairly closely, but other countries have mere deposit regimes for trade marks. The strength of protection varies quite widely. In some places unfair competition laws provide very good levels of protection without the necessity for any registered trade mark at all, though even there trade mark protection is worth having, and countries with strong unfair competition laws may well have strong trade mark laws too. This is a question not only for commercial decisions but one that also depends on advice from your trade mark agent.

5.12.1 Preparing for foreign registrations

In seeking protection abroad the trader may either wish to make use of any unfair competition laws of the country concerned or of the trade mark registration law or both. The trader should bear in mind a number of facts when planning the campaign.

The first point to note is that some overseas countries require a certificate to the effect that you have already obtained a UK registered trade mark before admitting a registration in their own state, including in some limited cases the important markets of the USA and Canada. This is a restriction on your scope of freedom in altering a trade mark design for foreign consumption because you will have to stick fairly closely to the design used in the domestic registration. The answer is to register your foreign design trade mark in the UK alongside your domestic version and then seek a fresh registration of the new version overseas. What you do with the new version in the UK then has no effect on the subsequent life of the overseas registration. A certificate of this kind is obtainable from the trade marks registry. It is another reason for preferring registration to common law trade mark protection.

However, most countries do not require proof of an existing registration in the applicant's home register, so you will be free to devise completely different trade marks from any UK marks you may have. Whichever be the case, the point has already been made that what seems a wonderful trade mark in the UK may even if it is registrable overseas be wholly unsuitable for the purpose because of cultural or linguistic factors. So this should always be checked out first.

Which countries you seek protection in will always be a commercial decision in the same way as for patents or designs, with many of the factors already considered (especially in Chapter 4 on designs) being relevant here too. Because most countries do not require actual use of the mark before registration is accepted it is possible for many markets to register the mark well in advance to prepare the way for when you want to enter that market with advertising etc. It may be necessary in some countries to avoid the extra expense of having to buy a well-known UK trade mark from someone who has registered it in anticipation of your arrival in his or her country where the UK mark is well known through media contact such as advertisements in magazines etc. which are bought in that country.

If it is desired to indicate that the mark is registered as a trade mark and you want to export to places where this may not be the case it may be better to expand the trade mark notice to indicate that it is registered in the UK, by such a device as UK Reg TM or similar means. Your trade mark agent will always be able to advise you on the specific regulations for each country.

Some countries such as Japan require proof of usage before allowing the renewal of a mark which has been registered, so consideration should be given to the means by which such proof may be acquired and then produced in the country in question. Requirements range from mere affidavits to evidence of production and advertising – and even local production rather than import in some places. It is worth bearing in mind that steps will have to be taken to police the use of the trade mark in overseas markets in just the same way as in the UK and it may be necessary to appoint representatives overseas to do that if you do not have your own representatives there already.

5.13 REACTING TO OTHER PEOPLE'S TRADE MARKS

Most conflicts over trade marks can be resolved relatively painlessly by out-of-court settlements and simple negotiation. But litigation costs are not such a deterrent to challenges in this field as in patents. A lot of time and trouble may be solved by a successful application for an interim injunction.

How to react to the appearance of another person's new trade mark similar to yours will depend on whether the mark is a common law mark, a registered mark or an application for a registered mark, and on whether you yourself have any similar marks and whether they are registered or unregistered for the class of goods or services in question, or whether you have planned the use of that or a similar mark in the near future. But whichever the reaction, it should be put into effect as promptly as possible, to minimize any accrued common law reputation which might benefit the opponent. You should act as soon as possible, in general, though an example of where a strategic delay pending an action was used profitably is the case where the plaintiff used the time to market enough products bearing his trade mark to avoid losing it for non-use.

- Any registration application should be challenged at the earliest stage possible in that procedure.
- Any unregistered mark should be pursued on the basis of infringement of existing registered marks or passing off or both if that is possible.
- Where the other person's mark is one in which you have an interest but no tangible protection you will be in a very weak position and at this stage it may be necessary to buy out the other person's rights. This has been the case on odd occasions where a preliminary search prior to a large launch advertising campaign has not turned up any existing marks conflicting with that chosen for, say, a new motor car, but after the launch such a conflict does appear. The inconvenience of the other mark may warrant expenditure to buy it out where there has already been a large amount of expenditure on advertising etc.
- It is generally highly undesirable to attempt to design a vaguely imitative trade mark, at least in going beyond mere period 'feel' or emotional appeal. It will inevitably limit your scope for development of the trade mark into variants and associated marks if you have constantly to keep an eye on another person's mark to avoid infringement actions or challenges. Better to begin with something which is really distinctive of your own product or service.

It may require a degree of self-discipline to restrain the commercial competitive instincts of sales and publicity staff from unfair or unethical practices. It may require a lot of money in damages, litigation costs, redesigning any advertising packages and starting up a publicity campaign from square one again if self-discipline is not exercised from the beginning and you come unstuck. Indeed the costs may be more than those of the initial design of the product.

5.14 THE COST OF REGISTERED TRADE MARK PROTECTION

Professional fees from trade mark agents will be comparatively low. The costs of design and advertising advice will of course vary enormously with the prestige of the agencies concerned. UK registration and renewal fees are fairly low. They are given in Table 10.

Table 10 *Some trade mark registration fees in the UK (as at May 1987)*

	£
Application fee for registered trade mark	60
Registration fee for registered trade mark	84
Registered user registration fee	45
Renewal of trade mark fee after 7, 14 years	79

5.15 ALTERNATIVES TO REGISTERED TRADE MARKS

You may choose to rely only on English common law protection of trade marks. In summary the difference in the scope of protection is as follows:

- Whereas the common law's protection of the trade marks and names through the torts of passing off and trade libel is founded in a desire to prevent deception of the public and consequential injury to the trader's reputation, the statutory regime of registered trade mark protection goes much further than this in that it enables the conferment of a monopoly on the proprietor of a trade mark which protects him or her even where the use of that mark by another engenders no risk of confusion or deception of the public. So whereas to win an action in passing off one must prove reputation in the mark in every case, the registered trade mark is independent of any reputation and its presumed validity even becomes a conclusive validity after seven years.
- Being independent of reputation the registered trade mark requires far less activity to keep it in force than a common law trade mark and also gives you protection throughout the whole of the UK and not just in that area where you trade.
- Action against comparative advertising is easier from the position of a Part A registered trade mark than is the case with common law or Part B trade marks and even perfectly fair comparative advertising may often be prevented by a Part A trade mark owner in the UK.
- To some extent it is possible to protect marks other than those actually used and in respect of goods other than those actually used by use of registration. Common law protection extends as far as the reputation in the mark which with a successful company may well be a long way beyond the range of products actually traded in. Registered protection through defensive marks may only partially reproduce the scope of common law trade mark protection.
- The purpose behind enacting the statutory regime was the reduction of uncertainty and expense of protecting the goodwill of the trader's business through its trade marks. But although the degree of protection conferred is greater than at common law, the scope of marks protected is narrower even after the *Trade Marks (Amendment) Act 1984* gives the possibility of statutory protection for trade marks in respect of services as well as goods. There are many very valuable common law trade marks in existence which can never be registered because of their form.
- Furthermore, the failure to register one's mark in no way deprives one of the right to use it or of rights of action in respect of its use by another if the requirements of passing off or trade

libel are met. This is something of a defect in a system which seeks to eliminate uncertainty; there is no certainty that one is not infringing another's rights merely because that other has not registered them, nor is there any certainty that registration will procure a true monopoly for there may be persons with prior rights which survive registration even if they do not prevent the registered proprietor from continuing to use his or her registered mark.

Nevertheless, this should not be exaggerated. The acquisition of registered trade mark does confer a much greater degree of security and helps to ensure that for the future someone looking to see if they can use a mark will see yours and presumably seek to avoid it: so there is a limited means of enabling others to discover whether any particular trade mark is open to another trader, often supplemented by a business names search, so that even a not wholly comprehensive register is an aid in avoiding the problem of trade mark conflicts.

While small traders may feel they can do without registered protection and content themselves with their locally acquired trading reputation to fight off invaders, they may find that this decision costs them dearly if and when they later attempt to expand their geographical trading area.

EXAMPLE

Two traders used the names Eradicare and Eradicure respectively, in the same type of business, but in different parts of the country. Their businesses expanded and came into conflict. The conflict between the traders in the same field of operation expanding their marketing territory, one from the North and one from the South, meeting in the Midlands, was left by the courts to be fought out in the market place. A prudent registration of one of the trade marks by one of the traders would have assisted him in his exploitation of the trade name considerably in this case.

- Licensing of a registered trade mark is much much easier than that of an unregistered trade mark, which is an extremely hazardous venture. This is a major reason for obtaining registered trade marks, to avoid the risk of losing the trade mark as an indirect result of the licence agreement depriving you of the common law reputation in the trade mark.
- Because the essence of common law protection of trade marks and designs is protection of their trade reputation, usage is the only way in which this form of protection may be acquired. Thus prominent claims to link the trade mark and name with your own source of production and service are essential. But this form of protection should always be regarded as a back-up to other forms of protection such as copyright, registered design and trade mark protection. Unless you are constrained to do so you should never rely on it alone.

Not only is common law protection more vulnerable to attack and proof of infringement of your rights more difficult and often dependent on the court's view of the evidence of public opinion, tested out in market surveys to determine reputation, but also the exploitation of the marks and designs through licensing is virtually impossible if it is desired to retain common law protection. Useful as it is, common law protection is still a second best. It may be cheap in the short term but it can cost you dear in the long run.

5.16 FOREIGN PROTECTION OF UNREGISTERED TRADE MARKS ETC.

The major countries of the British Commonwealth all have a common law regime affording protection to trade marks and designs which is in almost all respects identical to that outlined in the preceding sections.

The USA has a tort of 'palming off' which is the direct equivalent of the UK's common law as well as the specific statutory provisions of the *Federal Trade Commission Act 1914*, which deals with questions of unfair competition, prohibiting the deceptive marketing of goods and services in any way (whether by use of a trade mark or not) which may cause injury to one's competitors.

A number of western European countries have laws of unfair competition which prohibit practices like those we have been considering as well as a good many more unethical trading practices such as false sales or other similar trading gimmicks including puffery (advertisement with false praise). In France, Italy and Holland there exist court decisions giving at least as extensive protection to trading reputation as the common law, derived from the general legal codes of those countries. In Austria, Belgium, West Germany and Switzerland, for example, specific laws have been enacted dealing with such unfair competitive practices.

6 Special protection problems

6.1 INTRODUCTION

A number of new developments in technology and in pure science with potential technological significance have stretched most existing systems of intellectual property rights to their limits by transcending many of the traditional scientific concepts and the boundaries between technical fields. Among the most significant are those in biotechnology and computer technology. Other, older fields of knowledge also cause continued problems, not least because of disagreement over how they should be treated. Pharmaceuticals and medical treatments provide very important instances of this.

Some of the very newest technologies such as the modern applications of biotechnology (in itself not a new field of scientific endeavour but one which is now unrecognizable to those who associate it with cooking, brewing and fermentation) may also impose very acute demands on professional advisers, including the patent agents, who are faced with the difficulty of advising clients on the best means of protection, and the examiners in the patent offices of the world, who must deal with these advances even though they too may have little personal experience of dealing with the new technical fields being explored and have no existing scientific base of established knowledge on which to draw in the same way as in the more traditional fields of research and development work.

It is difficult for patent offices to attract such suitably qualified people in these fields when they could be making a lot more money researching into them! The same of course is true for those courts faced with disputes over patentability etc. Bridging the wide gap in communication between the research workers and their professional advisers is a major problem.

This will make it incumbent on research workers to provide additional assistance and information over and above that which might ordinarily be required of them when their firm wants to consult a patent agent or other legal adviser to consider possible forms of protection, and the research worker in turn requires a special understanding of the problems facing the advisers in trying to procure the best possible protection for new developments. And, at a later stage, any technology transfer is likely to be dependent on considerable assistance from the licensor.

Whereas in the usual course of things a patent agent will be able to take a very positive role in exploring the precise boundaries of the invention and in improving its presentation for patentability, the obligations on the research and development worker to explain and explore on his or her own initiative will be paramount in these new fields where the patent agent's general scientific background may be insufficient to confer on him or her the necessary high level of understanding which a proper patent specification will demand, to determine whether patenting is worth attempting, and how best to argue the case in the face of opposition.

6.2 COMPUTER FIRMWARE AND SOFTWARE

Both software and firmware have proved very much susceptible to imitation despite the various and the increasingly sophisticated methods of physical security which manufacturers have devised to protect their investment. The law has been relatively slow to adapt to the demands of this new technology, but it is now beginning to do so and specific legislation has at last covered the question at least partially in most of the markets which matter.

6.2.1 Copyright protection in the UK

In the *Copyright (Computer Software) Amendment Act 1985* which came into force on 16 September 1985 the UK Parliament confirmed what the courts had already been saying, that copyright protection is available for computer programs on disks, tapes, etc. The Act nowhere defines what a computer program is or indeed what a computer is. It does provide that the unauthorized translation of a computer program from, say, COBOL to FORTRAN, is an adaptation of the work and therefore an infringement of it and that mere storage of a computer program in a computer is 'reproduction in a material form' and again therefore an infringement if it is done without consent of the copyright holder. The acts which will infringe copyright in a computer program if done without consent of the copyright owner are:

(1) Reproducing the work in any material form, i.e. copying in a backup or storing in memory, reproducing the program in written form, etc.
(2) Publishing the work, i.e. through sale of copies
(3) Performing the work in public, i.e. on a TV or on a screen
(4) Broadcasting the work, i.e. over the radio or TV
(5) Causing the work to be transmitted to subscribers of a diffusion service
(6) Making any adaptation of the work, i.e. converting it for use in another make of computer
(7) Doing in respect of any adaptation of the work any of the above acts

So far as firmware is concerned it has been possible to say that the drawings plotting the topography of a chip are artistic works and therefore protectable under the Copyright Act too, relying on the test of whether a non-expert would recognize the three-dimensional chip as reproduction of the two-dimensional plotting drawings, in the same way as the components of engineering machines can be protected by the blueprints used in their production: this was discussed in depth in an earlier chapter and will not be considered here.

The possibilities of patenting new developments in computer software and firmware in the UK are considered in the discussion of patent protection in Europe below in Section 6.2.3. There is of course no obstacle to the use of trade secret protection if it is physically possible to impose such a legal regime successfully. A use of source code retention, physical obstacles to any unauthorized reproduction and backup copies, and the usual contractual bars on unauthorized use may be quite effective in many cases, though only usually on limited distribution programs and the use of toggles and such similar devices is not universally popular because of the inconvenience it can cause to the user, which is a major deterrent to their use in a highly competitive software market. As a practical matter the use of one's copyright if only as a legal support to other forms of protection seems extremely desirable because of the scope of protection it affords and the ease with which it may be acquired and asserted.

6.2.2 Protection in the USA

Provided that care and resources are available, there is no difficulty in protecting developments in computer hardware and other electronics in the USA. The protection of patents is clearly available if the basic requirements of novelty and non-obviousness can be met by the applicant. With firmware and software there has been more doubt and there is now specific legislation dealing with the matter which differs markedly from that of the UK in what it demands of the owner to protect his rights.

Firmware too now has clear and certain protection in the US law. Under the *Semiconductor Chip Protection Act 1984*, a completely new regime for firmware has been created. The 1984 Act protects semiconductor chips by protecting the patterns of the chip. These the Act calls mask works. These may be purely functional but protection does not extend to those characteristics of the chip which are dictated by the function it has to perform. Nor does protection extend to the wider idea underlying the chip: only the mode of expression of the idea is protected under the Act.

To gain protection the chip must be original. The Act provides that the work may not consist solely of designs that are staple, commonplace or familiar in the semiconductor industry, or variations of such designs, combined in a way that, taken as a whole, is not original. To qualify the chip must be registered with the copyright office, and the owner of the mask work as it is known must be national or domiciliary of the USA or of a country party to a treaty with the USA protecting semiconductors, or stateless, or the work must be first commercially exploited in the USA, or the work comes within the scope of a presidential declaration extending the Act's protection to it on the basis that chips are in some way protected in the other country where the owner is domiciled or resident or where the work was first commercially exploited. Transitional provisions liberalize this to some extent by providing for temporary extensions to certain countries. The UK has persuaded the US Secretary of Commerce that the copyright law of the UK covers semiconductor chips.

Applications must be made within two years of the first commercial exploitation in order to be able to register. The effective date of registration is the date on which application, fee and deposit of the identifying material is made. But the Act is retroactive to 1 July 1983 if claims were made before 1 July 1985. The Copyright Office has issued regulations for registration of claims. The filing fee is $20, and registration must be sought using their form MW. Where there has been commercial exploitation of the chip concerned four examples must be deposited, but in the case of chips not yet exploited, the Copyright Office permits the waiver of this requirement. Some limited provision is also made for the protection of embryonic chips, those in an early stage of development, for which an interim registration may be sought.

Special provision is made for foreign applicants who wish to apply pending a decision of the Secretary of Commerce to make an extension order in respect of their country.

Regulations are made for the form of mask work notice which is not compulsory as the software notice is, but which is merely prima facie evidence of valid registration and protection. The form of notice is the words 'MASK WORK' the symbol *M* or the symbol M with a circle around it, plus the names of the owners of the rights in the work or an abbreviation by which the name is generally known (e.g. IBM).

The Act requires deposit for complete registration but concessions are made in respect of deposit where trade secrets are involved. The general requirement for commercially exploited chips is that there should be filed four copies of the chip as first commercially exploited, plus one full set of visually perceptible reproductions of each layer of the chip, at a level of magnification of twenty to thirty times the actual size. For works which have not been commercially exploited, the identifying material must generally be a full set of plots of each layer of the chip design.

Where in a commercially exploited work there is some trade secret the applicant may seek to protect no more than two layers of the chip in a work consisting of at least five layers by submitting four chips as first commercially exploited and certain identifying portions of the design instead of reproductions of the layers, which consist of a print-out of the data pertaining to the chip in microform with the sensitive parts blocked out or stripped off from the deposited material. If you cannot fit within this requirement for a trade secret claim you may ask for special relief which will be heard on a case by case basis.

The scope of protection afforded by the Act is *sui generis*, neither copyright nor patent, but a short-term form of protection. The term of protection is ten years from the date of registration or date of its first commercial exploitation, whichever comes first. During that time the Act confers on the semiconductor rights owner the exclusive right to reproduce the chip by any means, to import or distribute a product in which the chip is embodied, or to knowingly induce or cause another to do any of these things. However, there are limitations on this apparently rather wide scope of protection. It is thus permitted under the Act to reverse engineer the chip for the sole purposes of teaching or investigating the make-up of the chip, and the results of such investigation may be applied in producing and distributing other chips so long as they are original within the meaning of the Act.

There is also an exhaustion of rights doctrine in so far as purchasers of chips from the owner of the right may resell them, though not reproduce them without the consent of the owner of the rights in the chip. And any innocent purchasers of an unauthorized chip may resell it on paying a reasonable royalty. Infringement takes place if the chips are substantially similar, though functionally dictated patterns must be disregarded. The scope of protection from imitation is probably narrower than for software under copyright. All the normal remedies are available and it is possible for the rights owner to elect for statutory damages of up to $250 000 in lieu of actual damages.

The rights in such a chip are fully assignable or licensable on compliance with formalities prescribed in the Act for recoding the change of ownership etc. with the registry; basically a transfer in writing with the signature of the owner or his or her duly authorized agent, plus a fee dependent on the length of the document recorded. The document must be the original or a certified reproduction.

So far as software is concerned, the USA has opted for copyright as the principal means of protection. But unlike the UK law, copyright protection is dependent on registration and on certain formalities if you wish to enforce your rights. Copyright definitely subsists in software under the USA law. The 1975 Copyright Act as amended by the *Computer Software Protection Act 1980* provides that copyright protection will subsist in any 'original works of authorship fixed in any tangible medium of expression, now known or later developed, from which they may be perceived, reproduced, or otherwise communicated, either directly or with the aid of a machine or device'. This therefore eliminates any possible requirement that the work be readable by the human eye.

The 1980 amendment to the Copyright Act reinforced this view by defining a computer program for the purposes of the Act as 'a set of statements or instructions to be used directly or indirectly in a computer in order to bring out a certain result' and, secondly, by making express provision for the scope of protection afforded by copyright to computer programs, limiting the rights of the copyright owner to some extent by providing that it is not an infringement of the copyright merely to boot the program into the computer either directly or indirectly; and by providing that the owner of a copy of a computer program does not infringe the copyright in it by making a copy of it for the purpose of using it in a machine or for archival purposes. So mere backup copies are not forbidden under the USA law, unless you as the consumer have by contract expressly undertaken not to make them.

However, in Atari Inc v. JS and A Group (1983) in Illinois, the court held that this only permitted the making of copies for archival purposes of any programs which are subject to damage or destruction by some mechanical or electrical failure, and that in the instant case the copying of programs in ROM by a machine constituted an infringement since programs in ROM were not subject to this kind of damage or destruction. Furthermore the law provides that these backup copies must be destroyed when ownership of the original program disk etc. ceases and if the original is assigned the copies must pass with it. So it would not be possible to circumvent the law merely by making copies when owner and passing the original on in a ring of users in much the same way as schoolchildren do with video games.

There was some discussion of whether the 1980 Act definition excluded from copyright protection the system program and only admits the application program. It is now clear that both are protected.

As in the UK, all copyright in the USA arises automatically, but in order to gain the protection of the Act so as to avoid loss of the copyright or loss or remedies for its infringement, certain formalities have to be complied with. The general requirements have been considered in Chapter 4. But let us look at them as they are specifically applied to software.

Firstly, all published and visually perceptible copies of the copyright work must contain a notice of copyright in the prescribed form. The US Copyright Office regulations provide that for works reproduced in machine readable copies (such as magnetic tapes and disks, punched cards or the like) from which the work cannot ordinarily be perceived save with the assistance of a machine or device, the following constitute acceptable methods of affixation and position of the notice:

(1) A notice embodied in the copies in machine readable form which in visually perceivable print-outs appears either with or near the title or at the end of the work
(2) A notice that is displayed on the user's terminal at sign on
(3) A notice that is continually on the user's display
(4) A notice which is permanently legible on the label securely affixed to the copies or to the box, reel, cartridge or other container used as a receptacle

The form of the notice is as follows:

(1) The word COPYRIGHT or its abbreviation COPYR. or the UCC symbol ©
(2) Followed by the date of publication
(3) Followed by the name of the copyright holder
(4) In general copyright matters some UCC countries require the statement 'all rights reserved' (basically, the South American states). This formality should be observed too for safety's sake, particularly in view of the fact that the American legislation may be taken as a model in those countries.

If this notice is not present the copyright in the work is not necessarily lost but any person making unauthorized copies of the work without notice may be held to be an innocent infringer and therefore be held immune from an award in damages.

In practical terms it will be advisable always to place a readable copyright notice on all the printed materials, source and object codes, packs holding them, chips, etc., and a copyright notice in proper form programmed to appear each time the program is run, and even on the text of the hard copy print-out. Various UCC countries apply the requirement of the © symbol in varying degrees of strictness.

So, the first requirement is a proper notice of copyright. Secondly, published programs are required to be registered if the owner wishes to be able to sue for infringement. Registration is possible at any time during the copyright term – it is not restricted to the commencement of the right. If you are worried about loss of secrecy you could delay registration until infringement has begun, but that is something of a double-edged sword. Unless registration is effected within three months of publication of the program you lose damages and costs for the period of infringement prior to registration. If you delay more than five years after the date of first publication you lose the presumption of copyright validity which registration would otherwise afford you. To accomplish registration, a Form TX has to be filled in, and a small fee ($10) paid. The Form TX is obtainable from the US Copyright Office, the address of which is Library of Congress, Washington DC 20559. At one time you had to use their forms:

photocopies would not do, but now major budget problems have resulted in photocopies being accepted in more recent times. You can phone up for the forms if you want to do so: the Forms Hotline (202) 287–9100 will accept calls leaving a recorded request for the forms. The form is simple to fill in with a set of instructions on what to do. The form requires you to provide information about:

(1) The author (which may for these purposes be the employer of the real author if the employer takes ownership of the copyright)
(2) The date the program was finished, published and whether it was assigned to the claimant
(3) It is also possible to register the associated documentation such as manuals on this form, providing information about the printer and typesetter.

In addition to the Form TX you can also register the design of the appearance of parts of the program on the VDU, by using a Form PA obtainable from the same address. This enables you specifically to protect the format of software for video games and their artwork and story boards. This requires an additional $10 filing fee plus a videotape of those formats you wish to protect. Dealings with the copyright in the program have too to be registered, by sending another $10 filing fee and a copy of the assignment. Your patent agents should have appropriate forms in stock.

Thirdly, a 'visually perceptible' copy of an 'identifying portion' of object or source code of the copyright material to be registered is deposited with the Copyright Office. Quite what this identifying portion is we will look at below. This requirement of an identifying portion is instead of the more usual requirement of a deposit of two complete copies of the copyright work in question, a requirement designed for older media such as books. And computer programs are altogether exempt from the requirement of deposit of one 'best copy' of the work with the Library of Congress.

The minimum for an 'identifying portion' of the object or source code for these purposes is either the first or last 25 pages or equivalent units of the work together with any page containing the copyright notice if any exists. The point of this is to make proof of infringement and of copyrightable material easier. If the source code is confidential the Copyright Office will accept object code together with a statement that the material submitted is work of copyrightable authorship, which the Office will accept but which it will label as not having been examined. Those concerned with security may adopt the ruse of padding out the program with 25 pages of uninteresting material, thereby shielding the part of the program which they really want to protect. There are disadvantages to this – the difficulty of proving infringement in the absence of a presumption created through the similarity of the material deposited to that complained of. The portion supplied must be in the form of paper or microform. Deposit of the ROM chip containing the program will not be acceptable to the Copyright Office as a valid deposit. There is provision for discretionary relief from the requirement of deposit in certain circumstances. Where deposit is validly made, a certificate of registration with the Copyright Office is issued. Deposit has been said in one case not to deprive you of the legal right to rely on trade secret protection but as a matter of practice in the absence of the 25-page ruse mentioned earlier, deposit means you have, as the Americans would say, 'blown it' so far as confidentiality is concerned.

Unpublished programs are now also eligible for registration. There is some disagreement as to whether a copyright notice should be put onto these programs. One view is that such a notice implies publication and therefore denies the plaintiff any possibility of bringing an action for breach of trade secrets if the unpublished program is misused. Another view disregards this risk, and says that for safety's sake all such material should bear the notice lest accidents happen and unpublished in-house materials escape or leak to the outside world or beyond the limited user licensee. Perhaps the best way out is to put on the copyright notice together with a statement that the program has not been published by the copyright owner and has been disclosed only to third parties under a licence prohibiting unauthorized disclosure copying or use, and forbidding unauthorized copying disclosure or use by anyone.

Again, as in the UK, the protection of copyright in the USA extends only to the manner of expression and not to the ideas expressed. There is no monopoly, but merely the right to prevent copying, though of course copying may be inferred from similarity and a number of tricks of the trade exist for assisting you in proof, such as the inclusion of dead program material, salting as it is called in the USA. So if a rival can achieve the same result through a different means, no action can be taken.

There is no express exclusion from utility patent protection in the USA but the USA Act's definition of a patentable invention, as a machine process manufacture composition of matter or improvement of these, clearly excludes a computer program as such from patentability and the courts have excluded patent claims in respect of machines whose novelty resided in their program by relying on the old judicial exclusion from protection of mathematical formulae, algorithms and mental steps. As early as 1972 the Supreme Court had stated in the case of Gottschalk v. Benson that, while not all such computer programs were necessarily unpatentable, a claim to a program *per se* would be so because it would have the effect of protecting an algorithm and remove a general law of science from the public domain. Thus a code for converting binary-coded decimal numerals into pure binary numerals was held non-patentable.

Then, in Parker v. Flook in 1978, the Supreme Court indicated that a process using a scientific principle or algorithm is patentable only if the process itself and not merely the algorithm is novel and non-obvious. The application in question involved a computer program for determining the alarm limit in a process for the catalytic conversion of hydrocarbons. The process was not novel; the only novelty lay in the formula for computing the alarm limit and this was an unpatentable algorithm. In the view of the majority, all it did, or at least all that was claimed to be novel in it, was that it was a program which was to perform arithmetical calculations. This was the low point of patentability in the USA.

The minority in this case had argued that it did not matter that the novelty resided in the program where that was one step in a process and the claim was for the process. This minority view gained acceptance in the later Supreme Court case of Diamond v. Diehr in 1981. In the Diamond case the claimant had developed a process for curing synthetic rubber in which a computer program was used to determine continually the precise temperature in the mould, and thereby to determine the time for which the rubber should be kept in the mould, and when the time elapsed equalled the time determined, to instruct the machine to open up the rubber moulding press. The form of the claim was for the curing process employing the program. The formula for working out the curing time was long known and the only new element was the ability to calculate very precisely the variable temperature – which had before been impossible – and therefore to determine precisely the variable time too. Nevertheless, the patentability of the claim was confirmed by the Supreme Court.

Thus, following Diamond v. Diehr in 1981, the US Patent and Trade Mark Office guidelines were amended to provide that whereas a claim which seeks to protect merely the mathematical formula of a computer algorithm or its use is not patentable, a claim for a novel process which includes steps involving the use of steps carried out through the use of a computer program will be patentable if the other requirements of non-obviousness etc. are satisfied. This involves a two-stage examination of the claims by the examiner. If the claims either directly rely on the program or if they indirectly rely on the program, then whatever their wording, the examiner must be satisfied that the claim does not merely outline an algorithm or a method of calculating something (for if it does it will be unpatentable) and that it is used in a specific application to improve a process or define the specific operation of a machine governed by the computer program. In other words you must limit your claim to use of that program in a specific application or applications and not claim the program itself for this would be to claim an algorithm which is only a general law of science and unpatentable.

6.2.3 Protection in Europe

Throughout Europe software protection is best sought through the law of copyright. The basic concept of copyright protection for software has been conceded in France, in Italy (at least for video games as a cinematographic work, even though the player of the game has a degree of influence on the precise form taken by the display) and in Germany, where programs as such count as copyright works without the need to rely on the artificial use of VDU displays as in Italy, provided the requirements of the general law of copyright in those countries as to novelty and intellectual content are met. Video games seem to have run into trouble for no good reason, and here there has been some degree of reliance on the Italian approach of protection as cinematographic works, though again with some doubts in certain decisions. There have been some adverse decisions in Belgium and Switzerland about ROM chips.

France has passed legislation dealing specifically with the application of copyright to new technologies, designed to take account of their so-called industrial characteristics. This is the Law of July 1985. Under this law, which deals very widely with the new communications media such as cable and the like, video games and in general programs will be covered by French copyright but – taking account of their industrial character in French eyes – this copyright will not be the normal period of life of the author plus fifty years but a limited period of twenty-five years. The interesting thing is that it would appear to be the case that computer software created before 1 January 1986, the day on which this legislation comes into force, should be protected by the French law of general copyright and therefore enjoy a full life of the author plus fifty years term, whereas the new twenty-five year limit will apply to software created after that date.

The computer programs governed by this new regime will be protected from copying in the same way as normal copyright, but it makes specific provision for software on a number of matters. No copies may be made except for a single backup copy without the consent of the right owner. Enhanced powers for the seizure of counterfeits are introduced by the Act.

In the UK and indeed all the countries party to the European Patent Convention, the scope for patent protection of programs is fairly limited. Thus in the UK version of the patent legislation required under the European Patent Convention of 1973, the *Patents Act 1977*, which implements the European Patent Convention, provides that: 'a scheme, rule or method for performing a mental act, playing a game or doing business or a program for a computer' cannot be patented. Quite apart from this express exclusion, some computer programs might have been excluded simply on the basis of being mathematical methods, scientific theories, methods of doing business, presentation of information, etc., which by virtue of the Act are not patentable, so this provision at least tidies the statute up by excluding all and not just some such programs as such, which has the merit of eliminating some anomalies. The *Patents Act 1977* also excludes from patentability the mere presentation of information and computer programs might often be excluded on that basis instead. Thirdly, of course, there is the difficulty that computer software would still have to overcome the requirements of being new, inventive and capable of industrial application.

Generally, it seems that computer programs as such, even if patentable by one of the drafting devices available (e.g. seeking to draft a patent specification as an application for a machine as programmed rather than as a program itself), are really unsuitable for protection by patent and most will not be inherently novel enough anyway. The disclosure required is often undesirable, the expense considerable for very little more protection than can be achieved through the law of copyright and indeed the scope of protection may at times be less.

The computer program appears to be excluded only as such from the benefits of patent protection. The UK courts have allowed a certain amount of patenting of computer software and firmware in well drafted patent specifications which contribute some technical advance in the art. But by contrast the German courts in their equivalent legislation have been extremely restrictive. Of the English approach, it could well be argued that it leaves open what might be seen as an absurd distinction: exactly the same invention may succeed or fail depending on the manner in which the claims are drafted; if as a computer program it will fail, but as a novel method of programming or as a computer as programmed or as the means by which the program is stored, it will succeed. The European Patent Office (EPO) takes a mid-way view of things. At one time the view expressed in the *European Patent Office Guidelines for Examiners* (the manual of instructions for the officials who assess the patentability of inventions at the European Patent Office and which represents EPO policy on borderline questions) was fairly vague, stating merely that:

> 'If the contribution to the known art resides solely in the computer program then the subject matter is not patentable in whatever manner it may be presented in the claims. For example a claim to a computer characterised by having the program stored in its memory or to a process for operating a computer under control of the

program would be as objectionable as a claim to the program *per se* or the program when recorded on magnetic tape.'

But there are now a new and more precise set of guidelines under which it would appear that more EPO patents may be available in this field, for instance to patent a known form of hardware as programmed by a novel program. This confirms what had in fact become practice. Under the new guidelines it is said that

> 'If a computer program is claimed in the form of a physical record, e.g. on a conventional tape or disc, the contribution to the art is still no more than a computer program. In these circumstances, the claim relates to excluded matter as such and is therefore not allowable. If, on the other hand, a computer program in combination with a computer causes the computer to operate in a different way from a technical point of view, the combination might be patentable.'

These 1985 *European Patent Office Guidelines for Examiners* go on to say that:

> 'A computer program claimed by itself or as a record of the carrier, is not patentable irrespective of its content. The situation is not normally changed when the computer program is loaded into a known computer. If however the subject matter as claimed makes a technical contribution to the known art, patentability should not be denied merely on the ground that a computer program is involved in its implementation. This means, for example, that program controlled machines and program controlled manufacturing and control processes should normally be regarded as patentable subject matter. It follows also that, where the claimed subject matter is concerned only with the program controlled internal working of a known computer, the program could be patentable if it provides a technical effect. As an example consider the case of a known data processing system with a small fast working memory and a larger but slower further memory, in such a way that a process which needs more address space than the capacity of the fast working memory can be executed at substantially the same speed as if the process data were loaded entirely in that fast memory. The effect of the program in virtually extending the working memory is of a technical character and might therefore support patentability.'

This is drawing the European position closer to that in the USA and there is therefore much greater scope for patenting machines which are controlled or made more efficient by the use of chips of software run on a microcomputer than many people imagine. Indeed, the larger firms such as IBM have always adopted a very high profile in protecting developments in the software field by patent and the absence of patent protection is largely a myth derived from popular belief and a degree of superficial awareness of the new patent laws.

In the UK, for programs which are not subject to very wide distribution, protection by trade secret is still the most common form of protection. The scope of protection available for trade secrets has already been considered in general and the general principles apply equally here. It is of little use to wide distribution programs, such as video games and the like. A note of warning should be sounded about licences in European states. The civilian legal systems give rather less protection to trade secrets than the UK, since they lack a general doctrine of confidentiality and one has to rely more on contractual terms and the law of unfair competition. Always seek local professional advice on trade secret licences in continental Europe.

6.2.4 Protection in the British Commonwealth

Most of the British Commonwealth has a copyright law based on the UK Statute and so in the absence of amending legislation we may expect the position there to be the same as it was in the UK before the 1985 Act.

As in the UK, Australia seems to have taken conventional copyright as the best mode of protection for software. Indeed when there was a suggestion at first instance – later overturned on appeal – that no such protection was available in Australia, the Australian government at once introduced urgent legislative proposals to rectify the situation, which it withdrew on the appeal decision being reached. But there has now been Australian legislation on the matter, amending the Australian Copyright Act. The *Copyright Amendment Act 1984* removed some of the uncertainties created by some difficult litigation.

The Act makes it clear that computer software, at least in the source code, is protected by copyright, and makes some provisions of specific relevance to copyright, including a legal presumption that the owner of an authorized copy of software is entitled to make a single backup copy for use in the event of loss or destruction of the original. Criminal offences are introduced in respect of distribution by telephone or radio of unauthorized copies of programs, and knowingly advertising the supply of infringing copies. Where appropriate the use of confidentiality to protect the software through trade secrets is equally available and in that respect the law is identical to that in the UK.

In Canada, despite a generally favourable view taken towards copyright in computer programs, the legacy of doubt left by judicial decisions on Canadian copyright law even more antiquated than that of the UK led the Canadian Government's White Paper on copyright, entitled *From Gutenburg to Telidon*, to recommend a detailed legislative proposal.

Under this proposal, if it is accepted and enacted in present form – and it has met with a mixed reception so far – high level language i.e. a human readable program would retain copyright protection including the right to restrain machine code programs based upon the high level material, enduring for a period of five years from the end of the year of creation of the high level version. But machine code programs by contrast would acquire only a 'computer program copyright', giving the right to publish (by sale, lease, offers for sale or lease or other trading transactions), copy, adapt or convert into high level language, but excluding from the regime rights relating to performance, broadcast, transmission, use, importation and leasing – all of which are associated with conventional copyright. Moreover the duration of the right will be only five years from creation (if unpublished) or five years from publication if published and a failure to publish in five years from creation will deprive the machine code of all protection. Dealing in an infringing copy will infringe if the dealer knows or has reasonable cause to suspect that it is one. Protection is extended to all foreigners whose countries are party to the UCC or Berne Convention except those whose national law expressly excludes machine readable programs from the scope of copyright protection.

Canadian patent law would appear to be at the same stage as American patent law in this regard before the latest Supreme Court decision. It is no objection that there is a computer program in the machine it is sought to patent but the Guidelines for the Canadian patent examiners are rather more restrictive than those of the USA.

6.2.5 Protection in South Africa

South Africa again has a different, almost unique legal system, based on the Roman Dutch law, a sort of mix between common law and civilian legal systems. The use of trade secret licensing through the law of confidential information is possible as in the UK.

The *South African Patents Act 1978* like the European patent legislation excludes computer programs from any patent protection as such. But South African copyright law differs from that of the UK. The 1978 Copyright Act, despite its enactment in the computer age, rather extraordinarily makes no mention of copyright and computer software at all. The leading court decision is in the case of Northern Office Microcomputers (Pty) Ltd v. Rosenstein 1981 in which the judge held that a source code program which is in compliance with the requirements of the Act, written down, recorded or otherwise reduced to material form, by for instance being printed out, would qualify for protection. In other words copyright can subsist in a source code recorded on a computer print-out. It was not necessary that the print-out should have any meaning in language. Copyright protection would arise automatically. The position of object code was left open.

In a subsequent case, Apple Computer v. Julian Rosy 1984, in an interlocutory proceeding, the court held that object code is an adaptation of or translation of source code which may likewise be protected through the law of copyright. If this is followed it will resolve a doubt created by the text of the South African copyright legislation. Although the English text of that legislation refers to adaptation as including 'translation', the Afrikaans text, which must be read with and control the meaning of the English text, uses the term 'vertaling', which apparently means the translation of material from one human language to another. Does this therefore restrict the scope of protection and deny copyright to object code? Only time will tell.

6.2.6 Protection in Japan

Hardware protection in Japan is attainable by the use of patents in much the same way as in the UK. So far as software is concerned, although Japan has not been a large importer of software it is a potentially important market and with the opening up of IBM compatible hardware markets, both pre-existing and overseas software designed for such machines may find a larger market than before.

Apart from patent protection, Japan has fairly precise legal regulation of computer technology. Indeed, some recent legal developments have given Japan arguably the most sophisticated laws on computer technology of all. Unfortunately, the complex provisions of these new laws cannot be discussed here for lack of space, but their broad essentials may be mentioned.

Copyright protection is expressly conferred on computer programs and its operation fully worked out in a 1985 amendment to the basic Japanese Copyright Law. The law grants the standard period of protection of fifty years plus the life of the human author or if the author is expressed to be an entity then fifty years from first publication or creation – whichever comes first – is adopted. No formalities are necessary though there is provision for voluntary registration of the work but no real advantages in practice seem to lie in registration of the work.

Some of the most interesting features of the 1985 law include the provision that use of an unauthorized copy in the course of business makes the user liable to the owner of the copyright, even where that is for purely internal use and there is no reproduction for sale or hire etc. Microcode does not seem to be protected under the law by copyright nor are program languages as such so that a program language may be used to the extent necessary to enable writers to produce compatible add on programs and the like. Likewise the algorithms of the program seem to remain unprotected.

The second new Japanese Act, called prosaically the Act Concerning the Layout of a Semiconductor Chip of 1985, closely models the USA's Semiconductor Chip Protection Act. The Japanese Act differs from the USA law in a number of respects, however. Firstly the Act extends its new protection equally to all nationals whereas the USA law is confined to Americans and those countries giving the same or equal protection to Americans. Secondly, date of registration alone is the date which governs the period of protection, and not also the first commerical exploitation as in the USA, but as in the USA provision registration must take place within two years of the first commercial exploitation. Registration is in fact essential if it is desired to license or assign the rights in the chip. Thirdly, criminal penalties exist for infringement of the Act as well as civil remedies.

The term of the new Act's protection is ten years. The legal nature of the protection lies somewhere between basic copyright and patent protection with what is in principle absolute monopoly protection being tempered by a number of defences for independent creators and innocent infringers. There are very similar provisions of secrecy to the US law.

It remains to be seen what effect the adoption of a *sui generis* regime for semiconductor chips by two of the most important producers will have on Europe and other countries where the automatic protection of copyright has so far been thought to be the better approach. Much pressure will clearly be brought to bear through the use of the USA's discretion to grant protection to only those whom it believes to have enacted or enforced some equivalent level of protection and the initial periods of protection granted to a number of countries soon run out and require further temporary extension. For it was clearly important to Japan to gain permanent reciprocal treatment from the USA and this explains the speed with which the Japanese chip legislation was enacted. EEC legislation is expected soon.

6.3 BIOTECHNOLOGY AND GENETIC ENGINEERING METHODS

Biotechnology, microbiology, genetic engineering, biological engineering are all aspects of basically the same scientific field. Biotechnology is the generalized term for the application in industry of some biological phenomena to a practical end. At a more specific level, genetic engineering is the manipulation of the genetic material in cells to determine the production by cells of new metabolic products or the functioning of the cells in new ways.

Many microbiological processes are very old – the classic examples being those of baking and brewing – and an American patent for disease-free yeasts was granted to Louis Pasteur as far back as 1873. But in modern times biotechnology is one of the fastest growing fields of scientific development into which vast investment is being made. It has superseded microelectronics as the most exciting field of research in Western Europe. It has implications for many non-polluting energy efficient processes in important human needs such as hygiene, foods and environmental control. Its commercial implications are that it can be used to make things more cheaply by reprogramming the genetic material which makes up the substances themselves, to convert harmful bacteria to useful drugs, increase the yield of crops, shorten the reproductive cycle of plant and animal life, and so on. It has proved potentially very valuable in plastics and mineral extraction.

More specifically, the commercial aspects of this 'new' scientific field include, firstly, cellular organisms, which are used in animal feeds and as the subject matter of processes to produce drugs, convert harmful substances into neutral ones; secondly, the products of microbes, the enzymes, which are large molecules which are present in nature and used in many known processes but which can now be created artificially in known or improved form, more cheaply, and which can be used instead of natural organisms in a number of processes; and thirdly, compounds of these enzymes which can produce a vast range of useful new results. The products include hormones, vaccines and drugs, new fuels, foods, and environmental controls such as a substance designed to 'eat up' oil slicks such as those left by accidents at sea. Also by combining sophisticated computational techniques with biotechnology – the so-called science of bioinformatics – a whole new range of highly efficient compounds for pharmaceutical, veterinary and food processing use can be designed by computer thereby opening up a new dimension to existing practices.

There is a marked need for legal protection of the results of research and development of biotechnology. Comparison is often made between the development of the microelectronics industry in the 1970s and that of the biotechnology industry in the 1980s to argue why the latter needs patent protection far more than the former and why the fact of the former's survival and phenomenal growth despite certain inadequacies in the level of legal protection afforded to it is not a good indicator for the latter.

Whereas relatively low levels of research investment produced relatively high levels of return in the microelectronics industry, an industry whose growth was guaranteed by the rapid obsolescence of its product and an almost entirely open commercial environment using applied technology from the start, the biotechnology industry is heavily capital intensive by comparison, relies more heavily on pure fundamental research, with longer periods for the returns (five to ten years, not one to three), more durable end-products and considerably more restrictions on the commercial exploitation of the results, especially with food and drug administrations: a good example is the necessity for the government approval of Humulin, the artificially produced insulin which had been produced through the use of recombinant DNA technology, before it could be administered to patients, a long process.

Biotechnology thus poses the classic situation in which good patent protection is said to be essential to promote adequate investment. Such protection does exist but it is hedged about with qualifications.

Following the European Patent Convention, the 1977 Patent Act excludes from patentability any variety of animal or plant or any essentially biological process for the production of animals or plants, not being a microbiological process or the product of such a process. Thus there is no protection at all for animal varieties. This – it is said by the EPO – is because other means of acquiring protection exist in many EEC countries, considered below in Section 6.4, but it poses real problems, to be discussed below. As for the biological processes, which at first sight seem to exclude many biotechnological processes, this requires a little more consideration. The *European Patent Office Guidelines to Examiners* explain that:

> 'The question of whether a process is essentially biological is one of degree depending on the extent to which there is technical intervention by man in the process; if such intervention plays a significant part in determining or controlling the result it is desired to achieve the process would not be excluded. To take some examples a method of crossing, inter-breeding, or selectively breeding, say, horses involving merely selecting for breeding and bringing together those animals having certain characteristics, would be essentially biological and therefore unpatentable. On the other hand, a process of treating a plant or animal to improve its properties or yield or to promote or suppress its growth, for example, a method of pruning a tree, would not be essentially biological since, although a biological process is involved, the essence of the invention is technical; the same could apply to a method of treatment of a plant characterized by the application of a growth stimulating substance or radiation. The treatment of soil by technical means to suppress or promote the growth of plants is also not excluded from patentability.'

As for microbiological processes, which are not excluded from patentability, the *European Patent Office Guidelines to Examiners* explains that this is to be interpreted as meaning:

> '. . . Not only industrial processes using micro-organisms, but also processes for producing new micro organisms, for example, by genetic engineering. The product of a microbiological process may also be patentable *per se* (product claim). Propagation of the micro-organism itself is to be construed as a microbiological process . . . consequently the micro-organism can be protected *per se* as it is a product obtained by a microbiological process. The term micro-organism covers plasmids and viruses also.'

This is all rather vague. What does it mean in real terms? It is important to recognize the real scope of this exclusion which does not exclude from patent protection many examples of innovation in the processes and products of genetic engineering. This is of immense importance.

It appears that propagated micro-organisms *per se* will be patentable whether or not they occur naturally. For the exclusion does not apply to any microbiological processes or the products of such processes. Thus EPO patents can be obtained not only for processes involving micro-organisms but also for inanimate products and presumably also micro-organisms themselves when produced by a microbiological process. The extension of industrial to agricultural use had as one of its objects the desirability of permitting patents to be granted for such things as fertilizers as opposed to the production of new types of animal and plant life.

The best way of illustrating this is to consider the sorts of claim which might be made and to see if any of them can be used in connection with an invention in this field. Several possible claims might be made and of course where the unity of invention concept applies they can be combined in the same patent application. Typical types of claim might include:

(a) Process claim for producing a new micro-organism
(b) Product claim for micro-organism produced by a new defined method
(c) Process for using a defined micro-organism to produce a defined product
(d) Product *per se* claim for new substance produced by use of a defined micro-organism in a defined way
(e) Product by process claim for a substance produced in a defined way using a defined micro-organism
(f) Use of combinations of micro-organisms to create special effects or exploit or suppress certain characteristics

In the context of some typical biotechnological inventions, these might produce the following results.

(1) Discovering or isolating for the first time a micro-organism found in nature: these face the major objection that they are a mere discovery and therefore not patentable. In the Australian Rank Hovis MacDougall case a patent application for the isolation in a soil sample of a new strain of micro-organism called *Fusarium graminearum* was rejected on the basis that no invention was involved in the mere discovery or the mere identification or the mere isolation by an unspecified method of something that occurred in nature. This was despite the fact that the discovery was of itself extremely useful. There are moral and ethical arguments against recognition of such an invention as patentable.
(2) Creating artificially a micro-organism found in nature: this appears to be quite susceptible to a claim of type (b), assuming that the other requirements of reproducibility and inventiveness were met. If the culture of a micro-organism which is purer than that found in nature is extracted, that in itself might make a claim under (b) possible. This was a view taken in the USA decision in Re Bergy: a pure version of a newly isolated strain was patentable.
(3) Creating a micro-organism not found in nature, e.g. by genetically manipulating a natural organism to produce a new enzyme: a claim of type (a) or (b) would appear to be possible, assuming again that the other conditions are met. This was the ground upon which the famous American decision in the Supreme Court in the Chakrabarty case was reached: a genetically engineered micro-organism was patentable.
(4) Making a new or old product from the use of micro-organisms for the first time, will in the former case be patentable under claim type (d) or (e) and in the latter case a claim under (c) or (e).
(5) Making new uses of micro-organisms, for example in the fermentation or related food industries, may also be framed in a claim type (f) in proper cases, as may some medical applications, subject to the comments made below about legal restrictions on the patenting of methods of treatment. Likewise use of micro-organisms as reagents etc.

(6) Fermentation processes: fermentation of bacteria, fungi, cultivation of viruses, etc. which produce useful products such as drugs and foodstuffs, have always been regarded as clearly patentable as processes or product by process or even product claims under (c), (d), (e).

6.3.1 Special deposit requirements

The definition of the micro-organism requires some thought with patent offices becoming increasingly demanding in their requirements of adequate data when characterizing the micro-organism in the specification. A mere reference to the culture collection in which the micro-organism has been deposited may be enough, but its precise scope of protection is not yet clear. This leads on to the next major problem for those in this field, that of the deposit requirements for such micro-organisms.

Because of the difficulties and indeed on occasion virtual impossibility of reproducing a micro-organism from a description of it in a patent specification, most of the mature patent systems require deposit of a strain in a culture collection at the patent office and its availability under controlled conditions for testing and examination by others.

In the case of the UK, one is required to make available to the public the micro-organism used in the performance of the invention including the deposit of a strain in the culture collection not later than the date of filing the application and supplying the name of such collection, the date of deposit and the accession number of the deposit, within two months of filing the application. Applicants in the specification must give as much information as is available to them on the characteristics of the micro-organism. Where this has been done the culture will be made available to anyone who after the publication of the patent application makes a valid request, certified as such by the Comptroller of Patents. This requires the person seeking access to undertake not to make the culture available to anyone else until the patent either expires or is withdrawn, and to use it for experimental purposes only.

In the case of the European Patent Convention, the rules state that deposit must be made where the micro-organism is not available to the public and cannot be described in the patent application in such manner as to enable the invention to be carried out by a person skilled in the art. But the rules only require that the strain be made available to an independent expert after the first publication of the application and only on the second publication need it be handed over to others. The EPO has a list of officially recognized independent experts to examine it on behalf of those requesting such access, without making the strain itself available to them.

Supply of strains to the patent office of all the jurisdictions in which one sought protection might become very burdensome and inhibit the reduction in duplication which the rest of the international patent mechanism is seeking to achieve. Under the Budapest Treaty on the International Recognition of the Deposit of Micro-organisms for the Purposes of Patent Procedure 1973 ratifying States accept a deposit of a strain with any of the Treaty 'international depository authority' culture collections as sufficient deposit in their own. A culture collection can become such an international authority by undertaking to keep to the requirements of the Treaty on acceptance, storage, documentation and on supply of samples, which is only made in response to a request certified by a patent office to which a patent application involving the micro-organism has been made as valid under its national law.

6.4 NEW PLANT VARIETIES AND SEEDS

The *Patents Act 1977* provides that any variety of plant is excluded from patentability. This requires more consideration. There is no suggestion that it is not possible to patent the product or process of genetic manipulation of plant material to produce, for instance, some characteristic such as enhanced colour, flavour or perfume. They appear to be a pure process or product by process claim which has always been in principle accepted by Patent Office examiners.

If protection for the new plant variety itself is desired, in the UK it is possible to acquire monopoly rights in new plant varieties outside the UK's patent system altogether, subject to a requirement of official testing of the new variety by the Plant Variety Rights Office to establish that the subject matter of the application meets certain legal requirements.

The *Plant Varieties and Seeds Act 1964*, amended in 1983, gives this legal protection which is available only in respect of those species and genera for which a statutory scheme has been drawn up by the relevant government minister, and both the scope and duration of protection may differ according to the scheme and genus concerned, ranging from twenty to thirty years. To take an example, a new variety of fruit or forest or ornamental tree must be protected for at least eighteen years, and this protection may be extended for up to a maximum of twenty-five years in exceptional cases, but on the other hand protection for certain other varieties endures for only fifteen years.

On the assumption that a scheme exists for the species or genus concerned, to be protectable under the Act, the new variety must be shown to be novel, to be distinctive and to be uniform. These require some explanation.

Unlike patent law, in this context novelty is confined to a requirement that the new seed or new variety for which protection is sought should not have been the subject of prior commercialization anywhere in the world before the introduction of the scheme for the relevant genus or species, or by the applicant after the introduction of the scheme ignoring previous sales by him or her abroad in the preceding four years. This distinctiveness requires that the variety must differ from others in one or more of its morphological, its physiological or its other characteristics. Uniformity demands that a variety must retain such characteristics after each propagation or reproduction cycle, i.e. its genetic characteristics should remain stable with each generation of plant.

These legal requirements must be tested for by the Plant Varieties Office which may take some time, but in the interim an applicant may be granted a direction by the Comptroller of Plant Varieties by which he or she is entitled to prevent others from doing anything which would be an infringement of his or her rights if a grant had already been made, on the condition that he or she makes an undertaking not to market the variety himself or herself during that period.

Even if protection has been given, the right may be withdrawn if the variety is shown to have lost its distinctiveness by testing in the Plant Variety Rights Office, but not for subsequent lack of uniformity or stability on reproduction or propagation. The right can also be revoked for failure to supply reproductive material for the variety, for failure to maintain stocks, or for breach of requirements to grant compulsory licences, etc. There are only very limited rights for third parties to make representations over the matters.

The rights conferred on the person who registers such a new variety are rather more limited than those available under the Patent Act for an invention. He or she has the exclusive right over marketing of the reproductive material intended for reproduction for sale (as opposed to consumption as food etc.) or reproduced for use as seed by the owner of the material (such as a farmer wishing to use seed from his or her own crop of the material) although the scheme may extend his or her rights to cover production or propagation for consumption if the Minister thinks it appropriate.

Abroad, the UK's 1964 Plant Seeds and Varieties Act was followed by the 1968 International Convention for the Protection of New Varieties of Plants to which are now party the UK, USA, Belgium, Denmark, Eire, France, West Germany, Israel, Italy, The Netherlands, New Zealand, South Africa, Spain, Sweden, Switzerland, all of whom have basically similar provisions, with some local variations. But some other countries have in fact preferred to make their plant and seeds regimes rather close to patents with any necessary modifications rather than creating a separate and somewhat complex regime such as that in the UK. In the USA, there are now three routes to plant protection to be considered.

(1) It is possible to acquire a plant patent upon inventing or discovering and asexually reproducing any distinct and new variety of plant, including cultivated sports, mutants, hybrids, and newly found seedlings, other than a tuber propagated plant or a plant found in an uncultivated state. The term plant seems to cover conventional trees, flowers and fungi, but not bacterial products. The same requirements as for general patents are made.
(2) It is possible to acquire a plant variety protection certificate under the *Plant Variety Protection Act 1970* upon breeding and reproducing any novel variety of sexually reproduced plant (other than fungi, bacteria or first generation hybrids). The legal regime and the rights granted are somewhat different from those of the plant patent, being dealt with by the Department of Agriculture, and is broadly similar to that in the UK *Plant Varieties and Seeds Act 1964*.
(3) It is possible to acquire a utility patent in some cases, where one can claim for a novel method of production of the plant rather than for a new variety of plant itself, for example, where the result is a more rapid reproduction cycle or increase in a plant's reproductive capacity.

6.5 PHARMACEUTICALS AND NEW CHEMICAL SUBSTANCES

6.5.1 Novel chemical substances and discoveries

(a) New things not in nature

Where one can claim to have made an entirely new substance never before known, one may make both product claims and product by process claims to protect the substance and the way it was made. A claim to a new thing as opposed to a way of making that thing is very broad indeed; it effectively reserves to the claimant any means of making using or dealing with the thing claimed.

Claims to a substance which does not form part of the prior art are permissible under the UK law and the laws of most of the 1973 European Patent Convention States and the *European Patent Office Guidelines for Examiners* permit the claiming of a whole range of compounds where it can be predicted with a fair measure of certainty that all will share the characteristics of those representatives tested even though the applicant has not made or tested them. It does not matter that the substance is made from known ingredients and by conventional methods but the properties of the new substance must be special for the inventive step to be found.

This may cause a real problem where the compound one seeks to protect is very close structurally to existing known compounds and the chemical distinction between them is relatively unimportant, giving rise to no unexpected difference in performance etc. or characteristic properties.

(b) New things in nature

However, the mere discovery of something which was already in existence in nature is less easy to patent. The 1977 Patents Act, implementing the European Patent Convention, provides that a discovery, a scientific theory, or mathematical method as such cannot be patented. In other words, what we need for a patent is not merely a discovery that something theoretically or indeed actually exists but the application of that discovery to some end. This has implications for the discoveries of naturally occurring chemical substances which have at the time no apparent use, and which may seem difficult to patent either on this ground or on the ground of being incapable of industrial application. The same difficulties afflict those who discover some new properties of known substances or for those who discover that theoretically something ought to exist but cannot yet be isolated by known methods.

The effect of current practice in Europe is that merely to discover a new property in a known material or article is mere discovery and unpatentable. If one can put that property to practical use one has made an invention which may be patentable. To take an example from the *European Patent Office Guidelines for Examiners*, the discovery that a known substance is able to withstand mechanical shock would not be patentable but a railway sleeper made from that material could well be patentable.

Alternatively, even though the substance itself is not new it may be possible to patent it in combination with some other known substance when together they are used to achieve some novel process or result, because the result utilizing the previously unknown property of the one in combination with the known properties of the other is such that no one would ever have thought of combining the two substances in that way before, as where the substance with the newly discovered property is introduced as an active ingredient to some other known substance to produce an insecticide or fertilizer or drug.

(c) Things of unknown structure

A further difficulty lies in the case where one has discovered or created a new substance but its precise structure is not certain or even wholly unknown to the inventor. The problem here is one of defining what it is one is seeking to patent. This is a feature of some microbiological patents, where in the past resort has been had to description of the properties of the substance and the method of creating it if that is readily reproducible.

(d) New compounds and synergism

Combinations of existing substances into useful compounds may often be the subject of a valid patent application. A mere placing side by side of the two integers so that each performs its own proper function independently of any of the others is not a patentable combination, but when the old integers placed together have some working interrelation producing a new or improved result then there is patentable subject matter in the idea of the working interrelation brought about by the collection of the integers.

If a combination claim of this type is accepted for grant it confers only a very limited range of protection. Any use by others of individual parts of the combination is beyond the monopoly of the patentee. The patent protects only the particular combination of components as arranged and interacting.

An example of the combination of known elements which will demonstrate sufficient inventive step to be patented is the combination of two substances known in themselves but whose effect in combination is one of synergism or alternatively which produce the same effect but with reduced side effects.

(e) Methods of producing a known substance

Where the invention comprises the discovery of the composition of the substance and how to produce it synthetically the inventor cannot of course claim the substance for that forms part of the prior art. But he may well be able to claim a product by process patent. Where the composition of the naturally occurring substance was already known the claimant will be restricted to a claim for its artificial creation.

(f) Patents for groups of substances

On occasion it may be the case that one has come across a new group of substances but it is found that there has been a disclosure of just one of that group. It would be tiresome to have to spend much more time and effort identifying individually the remaining members of the group for the sake of one known member. Where it appears, for instance, that one member of a class which one would like to patent does form part of the state of the art, and that class shares common characterisics not obviously different from the one member, as it almost inevitably will, it may still nevertheless be possible to gain protection for the class less that one member (a $n-1$ claim) by disclaiming it. Whether it will be possible will depend on the reason for that member being known to the prior art.

As a very great generalization, if the member was known of generally the claim for $n-1$ will be bad. But if the member was, perhaps, only a 'paper' discovery in a prior specification and was put forward in that specification to solve some entirely unrelated end, where its application to the new end is not obvious, then, all other things being equal, the '*n*-1' claim may succeed. *N*-1 claims may often be encountered in specifications which have been amended as a result of some anticipation unexpectedly turning up in the Patent Office search. Nevertheless the burden of showing that discovery of the class *n*-1 was not obvious in the light of the known existence of '1' is a heavy one. '*N*-*m*' claims, where more than one member of the class are already known are extremely rare and, in the light of the less generous attitude to applications widely expected since 1978, instances of such grants may be expected to become even more rare.

(g) Selection patents

In some cases a whole group of substances is known to exist, but someone later discovers that of the group of known substances one or more of its members happen to display certain known characteristics of the group more usefully or in greater or lesser degree than the others. In such cases, notwithstanding that the whole group is already known, it may be possible to claim a so-called selection patent in respect of the substance or sub-group of substances displaying these advantages. Where another has patented a whole range of substances (the question rarely arises in respect of mechanical patents, being confined in practice to claims in respect of chemicals), in accordance with the EPO's Guidelines, it is open to others to claim selections from the class where all or most of the members of the sub-class selected share some advantage over the generality of members of the wider class sufficient to furnish evidence of a further inventive step and the special sub-class was not the subject of specific mention in the original claim.

However, although it is possible to claim the substances themselves it is safer to claim the use to which these members of the sub-class may advantageously be put. This is because the applicant often faces a particular difficulty because of the requirement that his or her advantage must be special to the group for which he or she claims. He or she may have a fair idea of the size and of the membership of the group but he or she would have to test every single member to guarantee one hundred per cent accuracy which would be an incredibly wasteful use of research resources. If others outside the sub-group share these advantages the claim might well be invalid, but a claim to, for example '*X* plus all those members of the *X* group which possess the following special characteristics namely *a*, *b*, and *c*' would seem to be of doubtful validity because it is insufficiently precise: it identifies merely the characteristic rather than those substances in the group actually possessing it.

(h) Use of known things for new purposes

In general the application of known substances for new uses as opposed to mere advantages may be claimed, provided that they are inventive, for example, the new use of a known compound as a wood preservative. Such claims often are made where a selection patent may be thought vulnerable. One cannot, obviously, claim a full monopoly over the known substance used but a patent for the method of using it may be claimed. However, this type of claim causes particular problems in the case of pharmaceutical products used in new methods of medical treatment, discussed below.

Equally, although to find a previously unknown substance in nature is also a mere discovery and therefore it is unpatentable, if a novel and inventive process for obtaining it is developed, that process will be patentable as will the substance itself as produced by that process, a so-called product by process claim. Where the substance is already known in nature and the invention comprises the discovery of the composition of the substance and how to produce it synthetically the inventor cannot of course claim the substance for that forms part of the prior art. But he or she may well be able to claim a product by process patent. Where the composition of the naturally occurring substance was already known the claimant will be restricted to a claim for a method of manufacture.

(i) Process improvements patents

One example of an improvement patent in this field is the use of a known chemical to improve a known process, either a completely new method of synthesizing a product or a mere improvement in the efficiency of an existing method by altering the conditions in which it is undertaken, poses no real difficulty, provided that it is suitably inventive. Of course, if the use of this new method or improvement is not easily discoverable in examining the end-product, it may often be quite pointless to patent it unless one believes that others will soon come across it themselves independently. In fact, particularly where the improvement claim is one by which a process for making a particular type of chemical product is made significantly cheaper by some modification to (say) the catalyst by which it is produced, the inventor may well find that preservation of the trade secret is better than publication of the patent whose infringement it may be difficult to prove.

A final obstacle which it is conceivable one might run into in extreme cases is the objection that the substance invented is an immoral invention. The 1977 Act provides that a patent will not be granted for an invention, the publication or exploitation of which would be generally expected to encourage offensive, immoral or antisocial behaviour; it is not enough for that behaviour to be prohibited by the law. This way of putting things was designed to overcome differences in the national laws of the signatory states to the EPC, but the moral opinions of the national courts are hardly likely to be less diverse.

Under the old legislation the same exclusion was available either through reliance on the Statute of Monopolies 1623, with its proviso to s6 excluding inventions mischievous to the State, or by exercise of the Royal prerogative. This prerogative was for many years used to deny letters patent to inventions relating to contraceptives though such patents are now thought to be acceptable and have been for some time: see abortificeants, as in Upjohn Co (Kirton's) Application (1976), where apart from an objection (sustained) that a method of inducing abortion was a method of medical treatment a disclaimer had to be inserted as to any method that was contrary to law. An example suggested by the Patent Comptroller of an invention which might be subject to the provision is one that claims a substance use of which infringes the *Carcinogenic Substances Regulations 1967*.

6.5.2 Pharmaceuticals and methods of treatment

The precise drafting of claims for patents in the field of the pharmaceuticals is extremely difficult. Under the European Patent Convention as applied in the UK Patents Act an invention of a method of treatment of the human or animal body shall not be taken to be capable of industrial application and is thus not patentable even though a product consisting of a substance or composition is not to be prevented from being patented merely because it is intended to be used in any such method. The text of the Convention in fact states a somewhat wider express exception for such products, allowing in particular (but not restricting it to) substances or compositions, there is unlikely to be any difference in practice on that basis.

This appears to deny the possibility of patenting methods of treatment as such and also claims for methods of use of known substances for treatment or for prevention of a disease or medical condition. Other countries do permit the patenting of methods of treatment as such and the presumably policy basis for the exclusion is not wholly clear. Three practical examples may illustrate this rule.

EXAMPLES
ICI (Richardson's) Application (1981)
An application claiming a method of producing anti-oestrogenic effect in warm blooded animals, including man, but excluding any method of treatment of the animal or human body by therapy, was rejected, on the ground that the qualification deprived the invention of any content at all, the specification describing only the application of the compound in the treatment of breast cancer and infertility.

Stafford-Miller's Application (1984)
An application to patent a method of controlling lice on humans was allowed because, although this was 'at the frontier' on the balance of evidence this was not a method of medical treatment, but verged on mere hygiene.

A clearer case of patentable treatments are those treatments to strengthen nails and condition hair which are primarily cosmetic in effect.

EXAMPLE
Oral Health Products Inc.'s Application (1977)
A claim to a method of cleaning teeth by a certain process was refused on the basis that it was a method of treatment, but claims amended to extend only to the cosmetic not curative effect of the process were allowed.

However, the comment of the *European Patent Office Guidelines for Examiners* should be noted, that:

> 'Surgery is not limited to healing treatments, being more indicative of the nature of the treatment; methods of cosmetic surgery are thus excluded from patentability.'

For this reason, where a development is of a known substance for use in some medical treatment it is common instead to draft claims for the known substance as used in that method of treatment rather than for the method itself. However, again following the European Patent Convention the UK *Patents Act 1977* also provides that in the case of an invention consisting of a substance or composition for use in a method of treatment of the human or animal body by surgery or therapy or of diagnosis practised on the human or animal body, the fact that the substance or composition forms part of the state of the art shall not prevent the invention from being taken to be new if the use of the substance or composition in any such method does not form part of the state of the art.

This is a crucial provision for pharmaceutical companies. It is generally interpreted as meaning that the first but only the first medical application of a known substance is patentable. This again is a very controversial provision, for it is difficult to see the distinction in principle between an inventive first and and inventive second, third, etc. medical application of a known substance. Some commentators challenge the view that only first applications are admitted and various attempts have been made by applicants to circumvent this prohibition on the patenting of second medical use. Some examples follow to illustrate some of the means employed.

EXAMPLES

Organon's Application (1970)

The patentees discovered that the administration of certain known drugs in a certain order produced a highly effective contraceptive. They patented a package containing these known pills, packed in that order together with instructions to take them in that order. This patent claim succeeded. The court overrode any objection that it was merely an arrangement of known material, and held that the novelty and the inventive step lay in the packaging in that particular order, which no one would have done had they not known of the discovery.

The pack claims device is, nevertheless, a very small loophole. In the absence of an inventive step in the arrangement, the mere packaging of the drugs with instructions will not suffice: there must be something new in the method of administration as well as the method of using the drug in that way.

EXAMPLE

Re Nolan's Application (1977)

A method of producing a surgical dressing by combining two substances at the point of the injury in such a way that they became a dressing. It was held that this amounted not to a method of treatment but a claim to surgical dressings as produced by a particular method.

However, generally attempts to rely on very narrow constructions of the exclusions of methods of treatment have also failed.

EXAMPLES

Unilever Ltd (Davis's) Application (1984)

An application for a method of vaccination and immunization of poultry by mixing a chemical in with the poultry's food at subclinical doses was held to be within the scope of the concept of a method of treatment of the animal body by therapy and thus not capable of industrial application. The word therapy was held to be wider than simply curative treatment and covers prevention of illness.

Re Sopharma's Application (1983)

The applicant argued that a substance could still be taken as new under the terms of the Act if its use was unknown in that method of therapy even if it was known in another. This argument was rejected. The Act rendered unpatentable use of the substance once it had been used by surgery therapy or diagnosis in any method of treatment.

Where the new pharmaceutical product is to be used not for treatment but for diagnosis or tests there is no difficulty in drafting a suitably worded claim for patent protection, assuming that it is novel and has an inventive step. So, for instance, a claim for a method of testing involving the treating of a blood sample with a substance to determine, perhaps, the presence of the Aids virus or the level of cholesterol in the blood is a patentable claim, all other things being equal.

In some recent applications there does seem to be conceded a wider scope for claims for in effect second medical applications and for methods of treatment. The EPO has held that although claims directed to use may not be granted for the use of a substance or composition for the treatment of the human or animal body by therapy, nevertheless, a patent may be granted which has claims for the use of a substance or composition for the manufacture of a medicament for a specific new and inventive therapeutic application. In other words, claims for use for the purposes of treatment are not acceptable, but claims for use in the preparation of medicaments to be used in a new and inventive medical application are.

The trouble with this is that unless the medicament so prepared is presented in some novel form it may be difficult to establish novelty in such a claim. Where the use is the first pharmaceutical use that in itself may be novel even though the substance is known. But where it is the second pharmaceutical use, what can be done?

The EPO also rejected the suggestion that the European Patent Convention text necessarily excludes second medical applications of known substances from patentability. For, on their interpretation of the text, it was possible to argue that it was justifiable by analogy to derive the novelty for the process which forms the subject matter of the type of use claim now being considered from the new therapeutic use of the medicament and this irrespective of the fact whether any pharmaceutical use of the medicament was already known or not. The legal position thus remains up in the air.

Outside the field of medicine, or course, the ordinary claim to a novel method of applying a known substance to a new use is perfectly acceptable, novel methods as such being wholly unobjectionable save when methods of treatment of the human or animal body.

Finally, one passage from the *European Patent Office Guidelines for Examiners* should be quoted in this context:

> 'If an application discloses for the first time a number of distinct surgical, therapeutic or diagnostic uses for a known substance or composition, normally in the one application independent claims each directed to the substance or composition for one of the various uses should be allowed. . . '

7

Ownership of inventions and designs

7.1 INTRODUCTION: THE SOURCE OF OWNERSHIP DISPUTES

This chapter considers two problems:

(1) The rights to ownership and rewards of employees and independent contractors such as consultants in any intellectual property which may result from their work
(2) The duties of such employees and consultants to their employers: what knowledge may they take with them when they leave the job and what must they not

These questions can give rise to three basic types of conflict:

(1) Between employer and employee
(2) Between employee and fellow employee
(3) Between new employer and old employer

Most of them can be avoided by adequate management procedures, and even where they cannot be avoided, most of their consequences can be markedly mitigated if the same procedures are followed. As in many of the issues considered in this book, the question requires planning and policy to be formulated in advance and not when the first dispute arises.

In determining these questions of ownership there is a conflict between the interests of an employer in receiving the benefit of his or her employees' ideas and any inventions, as part of their labour for which he or she has paid, and the interests of the employees in being able to sell their labour freely to whichever employer will pay him or her the most. There is also a much wider economic interest in employee mobility and equitable bargains as well as a public interest in ensuring that those best placed to exploit inventions can do so.

Conventionally there has been little argument that the employer is usually best placed to exploit, and the law favours ownership by the employer over the employee in most cases, although a Government Green Paper of 1983 suggested that in many cases the inventor may be the best promoter of his or her invention and said that in respect of inventions which a company had failed to exploit the employee might be given more rights than the present law allows. The subsequent White Paper in 1986 was non-committal and a further inquiry was to be undertaken.

We will look at the present law before considering the management implications of the law. This chapter is broken down into three areas for discussion. The first deals with the position on legal ownership of those inventions, designs, drawings, etc. made by employees and by independent consultants. The second deals with the duty of such persons to keep employer's information secret and not to use it for the benefit of themselves or others, even if the employer does not himself or herself exploit the information. The third deals with questions of personnel management, internal procedure and the limits on the contractual arrangements which may be made by the employer so as to avoid questions over these rights and duties becoming a problem.

Unfortunately, there is no uniform UK law on the ownership of intellectual property rights. For although there are similarities between the position for each of the rights, there are also very important differences. Accordingly, it is necessary to look at each in turn, before making any general comments about dealing with inventions etc. so as to minimize problems. Equally, the law differs from country to country. So here we will assume that the invention has been made or the drawing or design etc. drawn in the UK by an English employee.

Ownership and employee regimes are basically similar in the rest of the British Commonwealth and in the USA to the general common law principles discussed below. In many European countries however very different regimes apply in which special awards may be payable by law to employee inventors over and above their salaries in certain cases, some countries rewarding not only their general employees but also research staff in this way.

7.2 THE EMPLOYEE'S POSITION AT COMMON LAW

At common law the employee is very unfavourably treated. It is an implied term in the contract of employment of an employee that anything he or she produces in the course of his or her employment with the employer belongs to that employer. An invention or other discovery is made in the course of employment:

(1) When the employee concerned is employed to invent, in other words, when he or she is either expressly or impliedly employed to undertake work likely to lead to inventions being made (and not for instance where he or she is employed merely to sweep the floor and one day comes up with a device for improving the efficiency of the machines in the building he or she sweeps – such an invention would under the common law be the inventor's to do with what he or she wishes), or

(2) Where because of his or her position of seniority in the company, the employee owes a duty

of trust and good faith towards the company, such as a director, or other senior employee.

This obligation to hold an invention for the benefit of the employer coincides with a further obligation to preserve its confidentiality pending any patent or other publication by the employer. A few short illustrations of how the two rules coalesce may help to explain how they work in practice at common law.

EXAMPLES

Triplex Safety Glass v. Scorah (1938)

The employee worked as a research chemist for the plaintiff and when acting on the express instructions of his employer he devised a method of making acrylic acid used in glass processes. The company failed to take up the invention. When he patented it himself they changed their mind and succeeded in a claim that he held the resulting patent for them.

Worthington Pump Engine Co v. Moore (1903)

The employee was the manager of the UK branch of the plaintiff's business. He was paid a high salary and was the main representative of the company in the UK. He developed improvements to pumps, which were the technical field in which the company worked. It was held inconsistent with his duty of fidelity to the company to seek to keep the patents himself.

Re Selz's Application (1954)

This is a borderline case. The employee was manager of a factory making lampshades. He was not a director and not specifically employed to undertake research of any kind. He devised a method of coating metal in plastic, suitable for use on lampshades but also on other products too. It was held that this was outside the scope of his normal duties and therefore the invention belonged to him and not to his employers.

A point which should be made here is that the English law does not have the concept adopted in American law of so-called shop rights, though commonly business people assume that such a 'right' does exist in the UK too. Shop rights were sometimes conferred on employers by American courts as a half-way house between giving the full rights of exploitation to the employee and to the employer: where the invention is on the borderline of the employee's duties or peripheral to his or her normal activities, the courts could award the employer a free licence to use the invention himself or herself but provide that the employer cannot prevent the employee from licensing the invention to outsiders if he or she so chooses. There are no equivalent provisions for such orders to be made in the UK. Either the invention belongs wholly to the employer or he or she has no rights over it at all. There is no half-way house solution. Nowadays the American courts tend to favour the employer more and award full title to him or her.

This implied position at common law can always be modified by express contractual clauses depriving one party or the other of the rights which the common law's position would dictate. The agreement may be express or implied by the courts from the facts. It can be made in or subsequent to the main contract of employment. In practice of course if such agreements are made then usually it is the employee whose position is worsened and subject to a possible statutory restriction the only limitation on this is the common law's doctrine of restraint of trade, where the employee's position is rendered unfairly poor.

EXAMPLE

Electrolux v. Hudson (1977)

A storekeeper working for the Electrolux company invented at home an adaptor to fit vacuum cleaner bags. His contract of employment provided that the storekeeper was to assign all rights in inventions to his employer. It was held that though the invention related to the employer's business the clause was void as being in unreasonable restraint of trade.

7.3 EMPLOYEE INVENTIONS AND THE PATENTS ACT

The 1977 Patents Act to a degree merely codifies the common law rules just discussed but it refines them and makes certain provision for the consequences of the employee losing ownership rights in an invention which goes beyond anything in the common law. The 1977 Act provides that an invention shall be taken as belonging to the employee as between the employee and his or her employer, with only two exceptions. However, these exceptions are very important in practice and appear to reflect the old common law position which gave most inventions to the employer. The common law rules on when the employer takes the invention remain relevant for inventions made before 1 June 1978 and in a few cases to inventions made after that date, but the case law will not be considered in detail here. It is arguable that this law applies to all inventions made after 1 June 1978 even if not patentable, though the present author's view is that it relates only to inventions patentable and/or patented.

The great change from the common law position is not so much when who takes the invention in the absence of any contrary agreement but a prohibition on certain contractual terms diminishing the employee's rights of ownership and provisions for proper compensation when he or she does not hold the invention beneficially. Before the 1977 Act the only means by which the court could strike down such agreements depriving an employee of what would but for a term in their contract of employment have been theirs was by invoking the common law rule of restraint of trade. If the 1977 Act applies to all inventions and not just patentable inventions, then its effect will be very wide and extend to invalidate many employee invention schemes.

The Act provides that the patentable invention will belong to the employer if it was made:

(a) In the course of the normal duties of the employee or in the course of duties falling outside his or her normal duties but specifically assigned to him or her and the circumstances in either case were such that an invention might reasonably be expected to result from the carrying out of his or her duties; or

(b) In the course of the duties of the employee and at the time of making the invention, because of the nature of his or her duties and the particular responsibilities arising from the nature of his or her duties, he or she had a special obligation to further the interests of the employer's undertaking.

With regard to (a) it will be clear that research staff will almost inevitably lose their inventions to their employers if these inventions are even remotely connected with the employee's field of activity. This will be so even if the invention was actually made when working at home if it was connected with the research they were undertaking at work, or an extension of that research; but if it was in a completely different field the invention will probably not have been made in the course of their employment, and will therefore belong to the employee beneficially.

As to other employees, their job description and the terms of employment under which they contract may be crucial, in determining whether the invention was in the course of normal duties. Are they employed to invent? Were they specifically told to work on the project in question in such a capacity even though not normally told to do so? However, employers cannot go too far in imposing artificial obligations to invent, for there are dangers in doing so and the court may well simply ignore such clauses in looking at the true position. This was considered in the first case on the law of ownership under the *Patents Act 1977*, the case of Re Harris' Patent, Reiss Engineering v. Harris (1985), discussed below.

With regard to (b) it will be seen that this is specifically designed with the director or other senior management employee in mind whose activities and duties may be very poorly defined if at all in their contract of employment. Such senior people owe a strict duty of fidelity to their employers, an obligation to do more than the letter of the contract and to advance more generally the commercial interests of their employer.

EXAMPLE

British Syphon Co. Ltd v. Homewood (1956)

The head of the company's design department patented an improved syphon design. Although he had not been asked to do this or to give any advice or otherwise to participate in any such project, it was held that he must hold the patent for the benefit of his employer.

Nevertheless, the provision may have an effect wider than that strictly intended, for whilst such a duty has always been clearly established for directors and other senior employees such as scientific heads of departments and the like, English law has been willing to impose the status of at least a quasi-fiduciary on employees quite a long way down the ladder, albeit that a proportionately diminishing level of fidelity has been required the further down one goes.

EXAMPLE

Hivac Ltd v. Park Royal Scientific Instruments (1946)

A fairly junior skilled craftsman was restrained from working in his spare time for a rival instrument maker because this was inconsistent with his duty of loyalty to his employer.

The common law also implies into the contract of employment the obligation to refrain from publishing any invention which one has made and which one holds for one's employer, without his or her consent, so as not to deprive him or her of the chance of applying for a patent or keeping a trade secret, and an obligation arising out of the fiduciary relationship will in many cases be an equally strong ground on which to found the obligation. Nothing in the Act changes this.

In Re Harris' Patent (1985), a middle management salesman was employed to sell valves. During his employment with the plaintiffs he invented a valve which solved a specific design problem encountered by his employers. Who owned the new patentable design?

It was held that the provisions of the Act might not be declaratory of the old common law, although the old cases were useful guides, but it was not necessary for it to decide the point in this case. The obligation of the employee to his employer depended on his normal duties which were the actual duties he performed and his general obligation of fidelity was not helpful in formulating what those normal duties were. In the case in question he was not employed to invent and his only task was to sell the valves and to ensure after-sales service and report back any problems which arose, not to solve those problems. If he had a special obligation to his employer under paragraph (b) it was to sell and report back and beyond that there was no special obligation to further the employer's interests. He had never been taken on himself to design valves, and indeed there was evidence that when he tried to make a suggestion to his employer the employer was not interested and told him to get on with his real job of selling.

As a result of these findings of fact, therefore, the invention he made was not within the course of his normal duties nor was it made in circumstances where an invention might reasonably be expected to result from his carrying out his normal duties. The circumstances which had to be taken into account were those in which the actual invention in dispute had been made and not those in which any invention whatsoever might have been made.

Nevertheless, the court was prepared to accept that in examining the scope of any duty of fidelity it was permissible to examine the old cases under the common law and that guidance could be had from these in determining the responsibilities and status of the employee.

The significance of the Reiss Engineering decision for our purposes is the attention of the court on what happened in the actual employment of the salesman and not what his contract of employment might say or what any duty of fidelity might otherwise imply. This is a welcome stance and it also indicates the wisdom of a regular review of contracts of employment to see if they accurately reflect the current activity of the employee. There are in any event dangers, as we shall see, in trying to cover the Act by providing a wide job description for the purposes of listing normal duties. The decision to some extent counters the fear expressed on the enactment of the statute that the employer could turn the provisions of the Patents Act against his employee in this way.

Nevertheless nothing in the 1977 Act seems to be calculated to make any easier the task of the courts at common law of determining what is in the course of his employment or in the course of normal duties, concepts which have created a vast body of case law in those other fields in which the term has arisen.

There is thus a change of stance inasmuch as one does not start with the contract of employment but with the Act, for the contract of employment can no longer diminish the rights of the employee as we shall see later. But the factual problems may be even more acute than in the past when the contract would be the final arbiter in most cases subject only to the common law doctrine of restraint of trade.

7.4 REGISTERED DESIGNS AND EMPLOYEE DESIGNERS

The *Registered Designs Act 1949* provides that the author of a design shall be treated for the purposes of the Act as the proprietor of the design, provided that where the design is executed by the author for another person for good consideration, that other person shall be treated for the purposes of the Act as the design's proprietor.

This ensures that the employee who produces work in the course of his or her employment or the independent contractor who works on a commission will generally lose the ownership of the registered design to his or her employer because the salary of the employee or the fee paid to the consultant or other independent contractor will be good consideration. Note that an assignee or licensee must have acquired the right to make, not merely sell, the design if he or she is to be able to apply, and attention must be paid to the risk that there has been a prior publication if dealings in the design have already taken place before the date of the application.

7.5 EMPLOYEE DRAWINGS, DESIGNS AND THE COPYRIGHT ACT

Whereas patents and registered designs rely on a grant for their validity copyright arises automatically and accordingly could be an entirely different matter from the employer–employee point of view. The general rule is that the copyright vests in the author, and the author is the person who contributes the skill, labour and judgement which make the actual work, not the mere contributor of the basic idea. So where more than one person collaborates in the work it will be a question of fact who is the author or whether there are in fact joint authors, the answer being dependent on the amount of skill and labour put into the work by those who were connected with its production.

There are four exceptional cases.

(1) In the case of a photograph, generally it is the person who owned the materials on which the photograph was taken who owns the copyright in that photograph, in the absence of any agreement to the contrary.
(2) Where a person commissions the taking of a photograph or making of an engraving of any subject matter, or the painting or drawing of a portrait, and he or she pays or agrees to pay for it in money or money's worth and the work is made in pursuance of this commission, that person is the copyright owner in the absence of agreement to the contrary.
(3) Where a work is made by an author in the course of his or her employment by a newspaper or magazine or similar periodical for publication in a newspaper, the employer takes the copyright for publication in any newspaper; but the employee takes the copyright for all other purposes, again in the absence of an agreement to the contrary.
(4) Copyright in works produced by employees in the course of a contract of employment or apprenticeship vests in the employer in the absence of any agreement to the contrary and subject to the newspaper provisions just outlined.

7.6 EMPLOYEE COMPENSATION SCHEMES

No compensation schemes exist under English law except for the inventor of an invention which has been patented in the UK. Even the inventor of a patentable invention will not be able to claim a reward under the UK Patents Act if the invention was never actually patented by his or her employer.

All the provisions in the Patents Act concerning employee inventors are 'home grown' and the result of a Government White Paper of 1975, though there are some European precedents – in particular the German Law of Employees' Inventions which dates from 1957 and which provides a detailed legal regulatory regime. The UK's present statutory provisions are not wholly without precedent even in the UK.

In the old 1949 Patents Act, an attempt was made to provide for the possibility of an employee winning compensation from his or her employer, aimed basically not at the paid research worker but the junior employee whose invention was quite unrelated to his or her duties but who for some reason appeared likely to lose ownership of the invention to his or her employer, perhaps as a result of a term in his or her contract of employment. But harsh judicial interpretation of the Act effectively rendered the 1949 provision redundant.

Now, in the case of a patent, where the employee has lost his or her invention to the employer under these UK legal provisions but also where the employee originally owned the invention and transferred exclusive rights in it to the employer for a payment which was inadequate in relation to the benefit obtained by the employer, special statutory compensation provisions may come into play.

An initial point which should be noted is that the effect of the Act is that references to the making of an invention by the employee extend to his or her making it solely or jointly with any other person; but not to merely contributing advice or assistance in the making of an invention by another employee. This can cost the court faced with a claim much difficulty in the context of inventions which are the product of team research supported by technicians, by draftsmen, draftswomen and the like who contribute some technical know-how to the fundamental idea developed by the main scientific staff.

7.6.1 Invention owned by employer

Where the invention was owned by the employer right from the start the statute requires three conditions to be satisfied before an award may be made:

(1) The invention must have been made the subject of a patent grant. Thus the compensation provisions have no effect in relation to inventions which for any reason are not made the subject of a patent grant.

(2) The patent must be of outstanding benefit to the employer. It is important to remember that it is from the patent itself and not from the invention that the benefit must arise. The Act does not seem to require consideration of the validity of the patent granted except in so far as that question of validity has any bearing on the value of the benefits to the employer. Thus a patent which earns licence fees for many years before being held invalid may still qualify the employee for compensation.

It is necessary to consider the size and nature of the employer's undertakings, among other things, in determining the level of benefit, and quite how this test is to be applied is not made clear. Perhaps a patent which arose to an employer who already enjoyed a dominant market position so that sales of his products were little affected might well be found to have conferred no outstanding benefit even though it was revolutionary in concept. This might seem to be somewhat harsh on the employee. By the same token, a smaller company, for whom the invention may represent a greater benefit in relation to its turnover even though the gross profits are the same as those which would have been made by a large company, will probably be in a more difficult position to meet any award.

Whichever approach is taken, it is important to pay careful attention to just what makes a particular product a success – the patented part, or a clever advertising campaign, attractive design, etc. A further point is that consideration must not be restricted to the profits directly made by the employer out of licensing or exploiting the patent, himself or herself, but to the benefit he or she has derived from use of the patent to keep competitors out of the market in that invention.

Indeed, in a sense it is this ability to exclude rather than the ability to exploit which is the central criterion (as we saw when looking at the nature of the monopoly rights conferred by a patent grant) and as such the court or Comptroller would be quite justified in making a very large award in respect of an invention which had not been exploited profitably by the employer at all (perhaps because he or she lacked the financial resources to exploit it) but which had been used as a blocking patent to prevent others from coming onto the market with a highly competitive product to the employer's detriment.

(3) It must be just that compensation should be awarded, by reason of the factors mentioned in (1) and (2). This is of course a matter of fact for the court to decide in every case.

7.6.2 Invention owned by employee

Where the invention belonged to the employee and was patented but he or she assigned the patent or the invention before patent to the employer, or granted the employer an exclusive licence in it, then the award provisions will operate where:

(1) The benefit derived by the employee from the assignment, grant or any other ancillary contract is inadequate in relation to the benefit derived by the employer.

(2) By reason of these facts it is just that the employee should be awarded compensation by the employer in addition to the benefit derived from the contract between the employer and employee.

These provisions only apply to inventions made on or after 1 June 1978 but otherwise had immediate effect on all contracts of employment notwithstanding anything in such contracts, and apply in relation to patents granted not only by the Patent Office in the UK but also to the European Patent Office and foreign patent offices.

The compensation provisions will apply to all employees who are mainly employed within the UK and also to employees who though not mainly employed in the UK or having no definite place of employment are employed by persons having a place of business within the UK to which they are attached, whether or not also attached to somewhere else. If, because the employee is not a 'UK' employee, i.e. one within the categories described above, the Act has no application, then the employee's rights will be governed by the law governing his or her contract, for instance by German patent law if the employee worked for a German company mainly in Germany. If by chance the actual law of the contract happens to be English law, the Act will still not apply, but the old common law rules will: as, for example, an English employee employed by an English company under a contract of employment governed by English law, but employed mainly in Saudi Arabia.

An employee includes government employees and employees regardless of age or nature of employment including part-time employees and former employees where they made the relevant invention while still an employee, and members of the armed forces. This will of course raise the time-honoured problem of deciding who is and who is not an employee, and in particular the distinction between an employee and an independent contractor. This is not the place for a discussion of the considerable body of case law but it is mentioned briefly below when considering the position of an independent contractor.

Finally, the compensation provisions provided by statute will not apply to an employee covered by any other compensation scheme operative under a collective agreement made by or on behalf of his or her trade union with his or her employer or employer's own association, where the employee belonged to the union at the time he or she made the invention. There is no requirement that the collective agreement be a better scheme from the employee's point of view. A trade union for these purposes need only be a temporary, even *ad hoc* body got together by the firm employees for the purpose of making such a collective agreement: a research team might, for instance, appoint one of its number as their delegate to negotiate an agreement on their behalf.

Assuming that the UK Patents Act provisions apply the statute lays down guidelines by which a court or the Comptroller General of Patents must assess what compensation is payable. The employee may apply to either for a decision.

Where the invention belongs to the employer the employee must be given a fair share (having regard to all the circumstances) of the benefit which the employer has derived or might reasonably be expected to derive from the invention: in other words it is assumed for these purposes that the employer makes reasonable efforts to exploit or defend it from infringement once he or she has it. Regard must be had to:

(1) The nature of the employee's duties, his or her remuneration and other advantages (e.g. promotion) derived from his or her employment or in relation to the invention under the Act.
(2) The effort and skill he or she devoted to the invention
(3) The advice and assistance he or she received from the firm's other employees
(4) The contribution by the employer in the way of finance, facilities, managerial and commercial skill, etc.

Different guidelines apply to the employee who owned but has assigned or licensed the invention in determining what is his or her fair share. The court or the Comptroller must take into account, among other (unspecified) things:

(1) Any conditions in a licence or licences granted under the Act or otherwise in respect of the invention or patent
(2) The extent to which the invention was made jointly by the employee with any other person, and
(3) The contribution made by the employer to the making, developing and working of the invention, by the provision of advice, facilities, and other assistance, opportunities and by his or her managerial and commercial skill and activities.

In both cases, where the employer assigns a patent to a connected company, then the benefit to the employer for these purposes will be the amount which could reasonably have been expected if the assignee company had not been connected. This eliminates the possibility of circumventing the Act's provisions by an assignment of the patent between connected companies at an undervalue.

However, it does appear arguable that the employee may only claim compensation from his or her employer at the time of the invention; thus, if the business for which he or she worked is sold or goes into receivership he or she may find himself or herself without remedy. If the employer dies, benefits which vest in his or her personal representatives are included in the assessment, and the personal representatives of a deceased employee may claim any award to which the employee would have been entitled.

In determining benefit, the Act provides that any benefit attained under UK and foreign patents and 'other protection' are to be taken into account. The meaning of this phrase has been put in doubt by some who suggest that it could be used to frustrate the attempts of an employer to circumvent the Act by failing to take out a patent.

At first sight, the simplest way of defeating the employee's right to compensation is not to patent the invention belonging to the employer. To avoid this conclusion, it has been pointed out that since the Act refers to 'other protection' as well as patents, one might hold this to protect the employee's rights under the statute even if an idea is not patented but remains simply confidential information which a court would protect or through a UK registered design or copyright so long as it was also patentable.

But this seems most unlikely. The argument would be a very bold one to adopt and it would be difficult to apply it to the other provisions in this part of the Act; its real purpose was to ensure that benefits from such rights abroad as the German law's Gebrauchmuster, which are often taken out instead of full patents for some procedural reasons, would be taken into account. In any case, what employer with a very promising patentable idea is going to risk it by not taking out a patent, all other things being equal, simply to frustrate an employee compensation claim?

7.7 INDEPENDENT CONSULTANTS' PROPERTY RIGHTS

The general rule is that unlike an employee an independent contractor such as a consultant will enjoy the full ownership of any rights which may arise as a result of his or her working for another, in the absence of any agreement to the contrary.

This right can still be surrendered by contract, whereby such a consultant undertakes to assign any rights which may result, and the only restriction on such assignments or contracts to assign would be the common law doctrine of restraint of trade.

Too often this is left up in the air, however, and many contracts assume that the employer takes the full rights in certain contexts such as consultancy work on the premises, and many assume that in a case, say, of a computer software consultant creating a bespoke system program, that the software house retains copyright in the program. Neither assumption is necessarily wholly correct. Two factors complicate the basic premise that an independent contractor owns what he or she invents or draws.

Firstly, certain people who are apparent so-called consultants may in fact turn out to be employees. For example, a consulting engineer in fact employed full-time on a particular project relating to hydraulic presses who makes a number of inventions in the course of the project has been held to lose the inventions to the company in a case in 1952, Re Loewy's Application.

The borderline of who is an employee and who an independent contractor, and what is in the course of employment, may be very difficult to draw. It is a question of fact in every case, one of impression and one difficult to define with any precise guidelines.

Secondly, there is the possibility of arguing that notwithstanding the common law and statutory provisions a copyright work may nevertheless vest in the person commissioning the work in certain circumstances, as in the case of Massine v. De Basil (1938). A choreographer agreed to compose and arrange a series of dances for a ballet at Covent Garden. He was in the event held to be an employee so that the work vested in the employer, but the court held that the position would have been the same had he been an independent contractor because in that event the court would have held him to be a trustee of the copyright in the work since he was commissioned and paid to create the work for Covent Garden.

This may be of considerable practical importance to independent contractors. When a legal and equitable title are split in this way it can be inconvenient. To be able to sue to restrain an infringement, unless he or she is seeking purely equitable relief, the equitable owner will have to join the legal owner as a party to the action.

For convenience's sake it is better to take an express assignment of the legal title to the copyright, together with the right to sue others for any past infringement. But if the contract is silent, the courts may be willing to use equity or to imply a term to the same effect in favour of the person who pays the money.

This is clearly covered in the *Registered Designs Act 1949* where even the independent contractors lose out to the person commissioning the work. The reticence of the courts in fields such as copyright probably relates to the fact that usually the cases coming before them are of an artistic or literary character rather than a purely industrial character. In other countries such as the USA the courts are far more willing to imply that title vests in the person commissioning the design or other piece of work produced by the consultant.

Thirdly, the consultant may despite his or her apparent ownership of the rights be unable to use them because of an implied obligation of confidentiality owed to the person commissioning his or her work. This is considered in Section 7.8.6.

7.8 CONTRACTING OUT OF EMPLOYEE RIGHTS

In many contracts of employment employers take the sensible step of eliminating any doubt about the legal position with regard to ownership of rights by making express provision for it. The provisions are usually very unfavourable to the employee, though in some institutions such as the universities the reverse is true and employers tend to be exceedingly generous with commercially viable ideas devised by their employees.

In the bulk of cases where express provision is made the employer claims all inventions, drawings etc. made by the employee during his or her employment with the employer, and insists on confidentiality for any trade information acquired during employment even after the employee leaves. How far can he or she go?

7.8.1 Contracting out at common law

So far as questions of ownership were concerned, the employee could do little about such claims except to argue that they were in unreasonable restraint of trade. This argument would work in extreme cases only. An example of such a case has already been given in Section 7.2 in the case of Electrolux v. Hudson.

7.8.2 Contracting out of the Patents Act 1977

The Patents Act makes provision for attempts to contract out of its ownership and compensation regimes. Contracts seeking to avoid the statutory provisions might be of one of two types: those requiring a transfer of the employee's rights in the invention itself; and those providing an agreed formula for compensation. Both types of agreement are dealt with by the Act. No corresponding provisions exist for the other intellectual property rights, though the common law of restraint of trade may be of some assistance, and this is considered below, after a discussion of the Patents Act provisions.

Under the Patents Act, any term in a contract with the employer or with some other person as well as the employer or with some other person made at the request of the employer or under the contract of employment with the employer, which diminishes the employee's rights in an invention which belongs to him or her made after the date of the contract or in or under a patent or application for a patent for such an invention is unenforceable.

This is a somewhat curious provision, for it only expressly affects what might be termed pre-invention contracts; it does not appear to affect contracts made after the invention. Perhaps the Act's reasoning is to exclude the possibility of going back on settlements made out of court. What the statute does not say is what precisely is meant by 'diminishing' the rights of the employee. Does an agreement to assign for a proper valuation reached by means of an arbitration clause diminish rights? A clause requiring notification of any invention made by the employee, even if it belongs to him or her? A clause requiring the employer's consent to the employee's exploitation of an invention which will compete with the employer's products?

The Act provides that nothing in the Act overrides the employee's basic obligation of confidentiality owed to his or her employer. This can be used to mitigate the position within the overall limitations of the law of confidence and restraint of trade, which are many, as we have seen in Chapter 2, and in particular when looking at the decision in Faccenda Chickens v. Fowler.

A further alleged anomaly is that the Act does nothing to exclude the possibility of an employee agreeing, before the time when an invention has been made which belongs to his or her employer, to forgo his or her rights to compensation under the Act, since the subsection refers only to rights in inventions or in or under patents or applications for them, not to the statutory right to compensation provided for by the Act. It would be surprising, in the context of these provisions as a whole, if the Act were held not to protect the employee against such a clause, and his or her entitlement might fairly be called a right under a patent.

Under the 1977 Act a contract cannot exclude the employee's rights to compensation under the statute in respect of an invention which belonged to him or her and which he or she assigned or granted an exclusive licence for to his or her employer and for which the remuneration was inadequate. It will be seen that this says nothing about excluding the employee's rights to compensation in respect of an invention which belonged to the employer.

7.8.3 Contracting out of entitlement to copyright

Contracting out is possible under the Copyright Act, subject only to the restraints imposed by the common law. Any agreement contrary to the presumptions would have to comply with the requirements of s37 of the Act, on formalities. Unlike the presumptions the exceptions can be varied by the parties without recourse to s37 so long as there is consideration or a deed under seal.

It is quite possible to conceive of cases where the copyrights and the entitlement to the patent vest in different persons. If an employer wants to ensure in advance that copyright in any materials produced by an employee outside the scope of his or her employment would vest in the employer he or she would have to obtain an undertaking in compliance with the formalities laid down in s37 of the Copyright Act for it to operate as an automatic assignment as opposed to a mere contract to assign.

Many firms employ outside draftsmen and draftswomen *ad hoc* for drawing new designs and plans. Generally in such cases a s37 assignment will be necessary, since independent contractors take the copyright even if commissioned unless the work is a photograph, portrait or engraving. A lawyer should draft a model assignment for use in all such ,cases.

7.8.4 Contracting out of the Registered Designs Act (1949)

This appears to be possible in the same way as for copyright, subject only to the restrictions of the old common law doctrine of restraint of trade, discussed above. The same precautions should be taken.

7.8.5 Using obligations of confidentiality

If an employer wants to ensure that new ideas do not leave with his or her employee-inventor or consultant, then apart from the question of claiming ownership of the invention, he or she may want to restrain the employee from disseminating any information about it.

In this respect more scope is left by the law of confidence for even the Patents Act provides that its provisions do not apply to contract clauses which impose obligations of confidentiality on the employee, although any such contractual confidentiality clauses too can be subject to the restraint of trade doctrine.

The more recent case law indicates a willingness on the part of courts to confine employees' obligations of confidence to cases where there is a substantial conflict of interest and to reject clauses which seem likely to prevent the employee from seeking work anywhere else within the same trade. Furthermore, the courts take a much harsher view of confidentiality clauses the lower down the ladder the employee is to be found. Hence a director of research or a professor will be in a much weaker position than a junior technician. The scope to which an employer could restrain use of information through the obligation of confidence was considered in the case of Faccenda Chickens Ltd v. Fowler (1983) discussed already in Section 2.5. An earlier example of the law of restraint of trade being used to cut down a restrictive clause may be useful:

EXAMPLE

Commercial Plastics v. Vincent (1965)

The employers had an 80% share of the market in thin PVC calendered sheeting for adhesive tape. They took many precautions to preserve the secrecy of their new discoveries in this field. When the employee who had been working in this area of research left he challenged a clause in his contract of employment which had forbidden him from working in the field for any competitor for a period of one year. The clause was struck down: it was worldwide, covered the whole PVC field and not just that of production of sheeting for adhesive tape, and was thus too wide, even if the period in itself was acceptable.

7.8.6 Management obligations of fidelity

Quite apart from these rules the general fiduciary obligations of senior staff and management apply too in this field, requiring the advancement of the employer's interests even against those of the employee. What may be unreasonable in the context of a junior employee may be expected of a senior employee. This is why inventions which would not be held by the company if made by a junior employee might fairly often be lost by a senior director. This has been discussed in part already in looking at the ownership provisions of the 1977 Act.

7.8.7 Drafting employment contracts on ownership

As has been seen the position of the employer is relatively strong when it comes to ownership of most rights, only those concerning patents giving rise to any great problem. An employer may wish to ask his or her lawyer to devise means of evading the legislature's prohibition on contracting out from the statutory regime as to ownership of patent rights. Whether the lawyer is advising the employer or the employee he or she must bear in mind a number of factors.

Clearly the wider the job description under which an employee works, the more likely it will be that anything he or she invents will belong to the employer. Thus the normal duties of an employee perhaps with express obligations to invent, might be drawn up very widely by an employer with inventions in mind. But there are difficulties with this.

(1) Firstly, a very wide duties clause might be construed as diminishing the employee's rights in his or her inventions and under the Patents Act such a contract clause will be unenforceable.
(2) Secondly, the court might well have regard not to what the contract clause says are the duties of the employee but to what the employee actually does, especially if the employee has, since joining the employer, been promoted or his or her duties have otherwise changed. In seeking out the scope of normal duties one must look to items such as the original recruitment advertisement, the correspondence leading up to the hiring of the employee, particulars given under the *Employment Protection (Consolidation) Act 1978*, the specific contract of employment and any collective agreements on grading and demarcation, etc. It is also necessary to consider the implied duties of the employee.
(3) Thirdly, there are other dangers in formulating a wide job description: how easy is it to make an employee redundant or avoid offering him or her suitable alternative employment if his or her job description is very wide? If the employee is a technician and he or she is obliged under the contract to undertake research, will this furnish the employee with an argument that he or she should be regraded onto a research staff salary grade?
(4) Fourthly, similarly, it has been seen that over-wide provisions concerning any of the intellectual property rights or the obligations surrounding confidential information can be struck down by the court under the common law doctrine of restraint of trade.

Nevertheless, despite these dangers, there are at least two arguments for including very wide clauses in such contracts when it comes to providing for the case of employees wishing to use information or competing with their former employer.

(1) Firstly, it is arguable that they can do you no harm because even if the contractual obligation is struck down the implied obligations of confidentiality will still survive. This may deprive you of the chance of imposing as wide a clause as you might legitimately obtain by contract, but the

clause could always be split up into a number of severable obligations of gradually increasing severity so that while some fall others survive, and in any case if all fall there is still some backup in the common law obligation implied.

(2) Secondly, many employers regard these clauses not so much as weapons which they intend to use in court but as either frighteners to wave in front of employees or employers or as a tactical weapon in negotiation of terms. The morality of this is not something for any discussion here; the fact that it is a possible tactic should however be mentioned.

7.9 HANDLING EMPLOYEE INVENTIONS INTERNALLY

Hopefully, many of the problems which might arise out of employee inventions may be avoided by proper internal procedures. Part of the problem arises in the contract of employment itself, in its definition of employee duties. This has been looked at above. The rest of the problem can be dealt with by a proper procedure for handling employee inventions and employee research work. Let us look at steps to be considered.

- Firstly, make employees aware of what is and is not an invention. One part of the job of the personnel manager should be to ensure that he or she and all other employees are adequately informed about matters such as the law of intellectual property, practical steps in protecting it, following the company's regime and getting proper information from patent specifications, etc.
- Make the position as to the employee's duties and his or her rights clear in the contract of employment, perhaps by making express references to the company statements of policy on employee inventions, reporting procedures, etc. The contract of employment should also make some provision for home-based work, the use of company microcomputers at home in the evenings by company computer programmers being one very good example of a situation in which a private project may be carried out very profitably by an employee at the company's expense.
- Review regularly the contracts of employment of the company employees to ensure that they conform to their actual duties within the company. In medium to large firms this can give rise to grading and demarcation disputes. For example in some research institutions technician staff are by their contract of employment expressly prohibited from ever undertaking research in employer's time or on their employer's premises: for to recognize their capacity to do so would serve only to strengthen their union's argument for pay parity with research staff whose functions at a junior level may overlap considerably with those of the much lower paid technicians.
- Establish clear and obligatory procedures for fully reporting staff inventions. This may involve a system of forms to be filled in *ad hoc* as well as of regular reports on the research activities conducted by each research team. A sample staff invention record form is included in the Appendices together with the explanatory letter which accompanies it. Make it an obligation to fill in such invention forms wherever the invention was made. Make sure that the subsidiary participants' roles are clear and not just the team leader's name when they report inventions. This may avoid much trouble, when it comes to filing the applications for patents and it may ensure that compensation claims are less complicated by lack of evidence. It also helps staff morale and gives a better idea of who is doing what.
- Operate an updating system, so that an employee is kept informed of the progress of any invention of his or hers which has been taken up by the company as a patent application etc. and expect similar updating from him or her on similar inventions, improvements, etc.
- Do not impose artificial obligations on non-research staff to invent.
- For non-research, non-technical staff, operate a full employee suggestion sheme, with a

staff reward system to back it up. This will involve the provision of some confidential disclosure form, which can be channelled through a central point for consideration and action. The procedure can be assimilated with that proposed for dealing with outsider suggestions, discussed at length in Section 7.12 below. The scheme should be for a voluntary notification or suggestion scheme for ideas even if they are outside the normal field of research and development and treat all such suggestions with equal seriousness. Make the response mechanism quick to respond to any such report of an employee invention to ensure that it is apparent that a company is genuinely interested in the ideas of its staff and to ensure that the employee always knows he or she will not have to wait for a response indefinitely. It may he worth considering a rejection mechanism, whereby, if an employee suggestion is declined, and it is one which ordinarily would have belonged to the employer, the employee should be given the go-ahead to attempt to exploit it himself or herself if he or she so chooses, on condition that he or she notifies the employer if he or she is seeking patent or other legal protection and also grants the employer a first option on a licence under such rights at a reasonable royalty. Unless there is a real chance of the research results prejudicing company research, in which case this factor should be clearly explained to the employee personally (and not in a pro forma letter), there is no reason to frustrate his or her own entrepreneurial desires by declining either to take up the idea or allow him or her to, but merely to leave it around gathering dust on a shelf simply because, for instance, it is too small for the scale of activities on which the company operates to tackle it itself. But all such schemes do require very careful drafting if they are not to threaten the existing rights of the company.

- Operate an employee invention and full research paper vetting procedure which ensures that responses to any proposed publications are quick and fair. Extend this to conference addresses and impress on the employees the need to avoid premature off-the-cuff disclosures in oral discussions, question and answer sessions, etc. Act reasonably. Do not impose unreasonable obligations of confidentiality in respect of some mere hypothetically commercially valuable information. Mere publication will stop anyone else gaining a dubious monopoly and will encourage an interchange of ideas which may well be beneficial to the company. Again, explain why permission for publication is being refused. And respond quickly.
- Consider introducing a collective agreement providing a more precise regime for company employee compensation schemes. This may court more applications but has some advantages for company morale and may encourage other staff. At the very least a company statement on policy may be beneficially introduced along the lines of that suggested in the Appendices. And in either case employee compensation and award schemes of a mere discretionary nature might be formally introduced to persuade such employees to seek that route before going to court.
- Debrief employees on their departure from the firm. There are two ways of going about this. One is a formal leter sent by the personnel manager to the departing employee. A sample letter is set out in the Appendices. But a much better way is to sit down with the employee in informal debriefing session with the personnel manager and the technical director to discuss what the employee has done with the company, who the employee has worked for and with, any inventions, any published articles and any invention disclosures he or she has made or should have been made but has omitted and to ensure that all documents are returned to the company. This information can then be recorded and agreed by the parties, with a signed copy being sent to the employee.
- Consider whether to write to the new employer of any research staff, the company informing the employer of the work which the employee did (in a broad sense) for the company, and letting the new employer know that the company will be looking carefully at any new products over the next year or two to determine whether or not the new employer has apparently used any of the company proprietary information which should not then have been revealed by

the employee. Some people question the real ethics of this and its implications for employee morale and clearly it is not a tactic which should be used with the intent or result of harming the employee's chances of obtaining employment elsewhere by labelling him or her as a potential legal liability.

7.10 HANDLING EMPLOYEE COMPENSATION CLAIMS

If the matter cannot be resolved amicably, then compensation contests between employer and employee over either ownership or compensation can be heard either by the court or by the Comptroller General of Patents. The factors the courts take into account in such claims have already been outlined. But in handling such rather delicate claims a number of other factors should be borne in mind by the company advisers.

- An order may be made for compensation by way of lump sum or in periodic payments: there must be borne in mind the tax implications of what is sought on behalf of an employee. It is also possible for the court or Comptroller to vary or suspend an order already made and also to hear applications which have in the past failed or to revive suspended orders. If the UK patent compensation claim should reach either the court or the Comptroller the Comptroller has a complete discretion as to costs, and the court if it hears the claim shall in determining the question of costs have regard to all the relevant circumstances including the financial position of the parties. This with the relatively uncertain scope of the jurisdiction and uncertainty in the manner in which the court will exercise it will make assessment of one's chances of success somewhat difficult.
- It is worth bearing in mind that in almost all of the cases it will be former employees who are seeking such remuneration. Current employees naturally may be very unwilling to take action and may even be bought off in practice fairly cheaply. Costs may deter even a former employee unless backed by a new employer in any later ownership dispute.
- On the other hand some factors may weigh in on the employee's side which may make a reasonably generous settlement attainable: firstly, the threat of trade union intervention, though of course in the newer and more highly technical fields in which one might expect more claims, union penetration into research and development is not high; secondly, the effect on the employee morale and the concern over the precedent value within the company of a first pay-out may have some effect, though this may be a double-edged sword.
- A patent the ownership of which is in dispute is not as valuable as one whose ownership has been resolved. It may be desirable to settle to ensure that the patent is marketable.
- It should be remembered that claims may well be raised some two decades after the events which give rise to the dispute. If proper laboratory and similar working procedures are adhered to then many of the evidential difficulties which may otherwise arise may be overcome but it should not be forgotten to keep accurate records too of the subsequent exploitation and development of the invention. All of this not only makes life easier when a dispute arises but of course reduces the chances of a dispute arising in the first place. One of the major problems in the employer–employee field is which employee invented what or who contributed to the team effort. Elementary precautions followed in most larger companies and university departments are equally applicable in the smaller company.
- Clearly in an ideal world the employee too would keep records and it should be part of the training of all research staff to do so. The employee should keep a day book in which his or her activities and ideas are noted. Any written work or drawings he or she produces should be dated and signed and it should be made clear whether the work was done at home and in his or her own time or at the employer's premises, using that employer's facilities (for example using one of the

employer's microcomputers which the employee has taken home) or in the employer's time. Of course this is too often not done. Because of the likelihood that the employer will have kept any relevant documents concerning the employee's contract of employment, the profitability of the invention and so on, the employer certainly does often enjoy an initial tactical advantage in determining whether to fight and how much to settle for. It may be very difficult for the employee's advisers to estimate at all accurately these figures without the expense of discovery. Some short cuts are available in industry via company accounts, publicity in company staff and professional magazines, etc., but they are of limited help. If the invention has been licensed or otherwise exploited abroad the employee's problems will be acute.

7.11 OTHER POTENTIAL OWNERSHIP DISPUTES

So far we have looked at the position of employees of a company, working for the company. Problem cases may occur outside the scope of what we have so far considered. The inventor may not be an employee but still be in a close relationship to another person providing the financial support for his or her research.

Prima facie the non-employee will be the person entitled to the grant of the patent. However, there is nothing to stop him or her assigning his or her rights by express contractual provision and if he or she does so the only way of going back on that agreement is likely to be to invoke the common law doctrine of restraint of trade. The non-employee may also find that the law of confidence restrains him or her from making any profitable use of his or her discoveries. In practice, non-employees in relationships likely to lead to such discoveries are often found to have made express arrangements. Some particular cases merit a mention.

Firstly, the invention may be made by someone working under a research grant from the Science Research Council (SRC) or a similar body. Usually the terms of the grant will make specific provision for the ownership and administration of any inventions which may result from the research work to be funded. This is always the case with SRC grants. Where the inventor is an employee of the Crown or of a research council, he or she might be prejudiced in his or her claim for a reward if the public body disposes of the invention which has vested in it for less than the full commercial value, in just the same way as if a company disposed of the patent to a connected company.

Accordingly, the *Patents Act 1977* provides that where the Crown or a research council in its capacity as an employer assigns or grants in a patent or patent application to a body having among its functions the development and exploitation of such inventions deriving from public research (i.e. the British Technology Group (BTG)) and does so for no consideration or for only a nominal consideration, then benefit derived from the invention, patent or application by that body shall be treated as a benefit to the Crown or research council. Note that here, any benefit from the invention and not merely from the patent or patent application appears to be included.

Secondly, the work which leads to the invention may be the result of research undertaken by an employee involved in consultancy work for an outside firm being done by the company. The position here will depend on the status of the employee under the contract between the company and the outside firm. The company will often have no rights in the invention under such a regime because as a measure of safety the employee is often made an employee of the outside firm for the purposes of the project so as to relieve the company of any risk of being held liable for the acts of the employee while undertaking that research.

Thirdly, the invention may be made by a student, either by a postgraduate research student or even by an undergraduate completing a course project. Such persons will usually (not always: they may be sponsored student employees of the company on leave of absence) not be employees and accordingly the normal compensation provisions of the Patents Act and the ownership provisions of the various statutes will be wholly inapplicable.

There is here much more scope for *ad hoc* arrangements between the student and the university. Some universities as a matter of practice require students to sign a form at the beginning of the course by which they agree to be treated as employees for these purposes should they invent anything in circumstances which would have led to the university being the owner of the invention if they had been employees. Other universities offer a student who has invented something the chance to opt in to such a regime after the event, as and when the event occurs. Most students are happy to join such a scheme and get the financial, legal and technical expertise of the university behind them. The split of the profits when a patent is licensed is often 33% to the student, 33% to his or her department and 33% to the university, though the figures may vary more beneficially to the student.

There are guidelines for the universities from the Committee of Vice-Chancellors and Principals (CVCP) on all these matters. In summary, the guidelines are these.

(1) The CVCP recommend a detailed set of procedures to be promulgated for all categories of staff and students governing the patenting and commercial exploitation of research results.
(2) They recommend that all academic staff contracts include an obligation to undertake research, express, not merely implied. This should be the case in most universities already for full-time staff, but probably not for the part-time staff.
(3) They also recommend that universities ensure that all agreements for research under the sponsorship of research councils and other outside bodies be subject to prior approval by the institution, which should have special regard to detailed and mutually acceptable provision for the exploitation of results.
(4) They recommend that private consultancy work should be forbidden without prior consent, given on condition that the university is under no risk of liability.

These last two recommendations are common practice.

(5) University regimes for students should be drawn up on similar lines to those of the research councils, but where the invention is made by a team of students and staff, the students should be treated in exactly the same way as the staff on the team. Steps should always be taken to safeguard the patentability of the invention by preventing premature disclosure. It is doubtful whether this is practicable.
(6) Applications for patents should be made jointly by the university and the employee inventor, and revenue distribution shares should be postponed until a good estimate of the income is possible, with provision for regular review. This may not be practicable either.
(7) Finally, the CVCP recommend that the same procedures should apply equally to non-patented and indeed non-patentable inventions, for as they rightly point out there is no reason in principle to distinguish them.

It should not be forgotten that it is equally possible to exploit a non-patented or non-patentable invention commercially as a patented one, although the means of protecting the invention are somewhat different, through the common law of trade secrets and confidential information. Indeed the successful commercial exploitation of that type of invention will need even more carefully drawn up legal agreements and personnel management than for a patent, since the scope of regime possible is completely at the discretion of the parties concerned, unlike the patent regime.

7.12 UNSOLICITED IDEAS FROM THIRD PARTIES

A related problem which companies face is that of receiving an unsolicited idea from an outsider, a third party who clearly has no connection with the company. This can raise problems which grow wholly out of proportion to the commercial value of the ideas or information submitted. An example will indicate the sort of situation which can arise and which must be dealt with quickly to nip potential dispute in the bud.

EXAMPLE
Seager v. Copydex (1963)
Mr Seager first appeared on a television programme called 'Get Ahead', demonstrating an invention, a carpet gripper. He was later invited to discuss it with the defendant company. In the course of this discussion he revealed another idea of which the company had known nothing and had not solicited. It was found that without Mr Seager's consent the company later used the idea he had revealed to them in a new product, having rejected his original idea. The company employees had been in good faith and they had forgotten about the source of the new idea which they had used. They were nevertheless ordered to pay compensation to Mr Seager for their use of his new idea.

All communications of ideas from third parties must be treated seriously. Many claims in the USA are speculative and frivolous attempts to extract money from companies by inducing them to buy off potential suitors. American companies go to great lengths to deal with this sort of problem. Equally, many ideas will be of interest to the company and related to its existing research of duplicating it. That is why the submitter of the idea chose that company. These are if anything even more difficult to deal with to avoid any allegations of misuse of information submitted.

The law dealing with the consideration of and use of unsolicited ideas of a confidential nature was considered in Chapter 2. This section is concerned only with establishing a practical management regime to deal with the issue.

The basic rule of practice is that no suggestion from outside the company should be examined by any of the technically qualified staff before it has been processed through the company's procedures for dealing with such outside submissions. All the technical staff should be instructed not to accept from a third party any oral suggestion without signed authority from the head of the R and D department concerned.

The response should always be to require the person to make a full written submission through the firm's normal procedures established for dealing with such suggestions. If by duress or accident a member of the technical staff does receive such a submission it should at once be reported to the R and D manager or perhaps the patent department official concerned with these suggestions so that the irregular procedure can be brought into line with the proper procedure at as early a stage as possible. The example of Seager v. Copydex given above is one case where this might have saved a lot of difficulty later on in the day.

When looking at written submissions of ideas from outsiders, there are two possible stances which may be taken: one negative, the other positive. These possible procedures may be looked at in turn.

7.12.1 The closed door policy

The first possible approach is to refuse to have anything to do with outsider suggestions altogether. This has the advantage of being less costly in staff time and simpler to operate.

Any suggestion is received in the mail room. All mail is opened by non-specialist staff even if it was addressed to the Head of R and D: this might be a secretary, for instance. On realizing that the letter is a suggestion it is returned to the sender without having been passed on to the specialist staff. This fact is recorded. A letter rejecting a suggestion might follow the pattern of that reproduced in the Appendices.

Of course, a consequence of the closed door policy is that even the good ideas go begging. It is thus not recommended. It is not a good image to have.

7.12.2 The open door policy

This involves slightly more organization, but not much more. It is a healthier stance from the company's point of view in terms of public relations and ideas.

Under an open door policy, in dealing with such unsolicited suggestions, these should be brought within the normal employee suggestion mechanism as soon as is possible, with some preliminary precautionary steps to safeguard the position of the company from claims which may cause disruption and loss of management and legal advisers' time.

The opening procedure is very much the same as for the closed door policy. Any suggestion is received in the mail room. All mail is opened by non-specialist staff even if addresed to the Head of R and D: this might be a secretary, for instance. Again, on realizing that the letter is a suggestion, it is then returned to the sender without having been passed on to the specialist staff.

But this time, two things are different. The letter sent with the suggestion is not a mere rejection, but something more. With it is sent an invitation to make a submission on the standard terms on which the company accepts such submissions. What these terms are varies with the company concerned.

- Some companies adopt the attitude that only a non-confidential submission, i.e. one which places them under no responsibility to the outsider and puts the outsider into a discretionary award scheme run by the company, will be acceptable to them. In other words the submission will be accepted only on condition that it imposes no burdens on the company at all, reflecting the relative bargaining strengths of the two parties. Although this may be in the short term advantageous to the company it stores up trouble for later and does not bode well for co-operation from the submitter if the idea is found to be worth having and development co-operation is sought.
- Some companies adopt a scheme of accepting only their properly regulated confidential disclosure agreements as the basis of a submission. This is the better course of action. It protects

the company from unwarranted claims and protects the inventor, by setting out clearly their respective positions and deals fairly between them, which the first option does not; it achieves a proper degree of record keeping of the suggestion; and it provides a sound basis from which any negotiations may commence if the company is later interested in taking up the idea, in an atmosphere of trust and openness. A model submission invitation is to be found in the Appendices.

Once a signed submission form is received together with the submission itself it can be forwarded to the appropriate person. A number of points of procedure must be carefully adhered to at this stage.

- Strict procedures to ensure that the information is not lost, accidentally disclosed, etc. must be taken. It is confidential in the same way as internally generated research data which have not been passed for publication.
- An initial decision must be taken as to who is to assess the material submitted. Clear records of who sees the information and his or her response to it should be kept in writing. It may be a good idea to name these persons from the start in the disclosure agreement.
- These persons delegated should assess and report on the information in writing. Patentability etc. should be considered, as should the relationship of the idea submitted to existing technology and design in the portfolio and any current embyronic R and D which might later be taken as having been derived from the outside suggestion in the absence of proof that it had been arrived at earlier.
- Respond quickly and simply to the submitter with a clear yes or no to further discussion. And preferably commit the company to a response within a short period of time in the letter sent with the confidential ideas disclosure form, and stick to that time limit. If the answer is 'no' send back all the documents which the submitter sent to the company together with any copies. Again record this. If the answer is 'yes' we then move onto the development or licence agreement stage in the normal way.
- This procedure should be uniformly followed and should be set down in writing. A copy of this statement of procedure should always accompany the confidential disclosure form sent to each submitter. If it can be established that this procedure is always followed, it is useful evidence in defending any later breach of confidence case which a dissatisfied submitter may bring.

8
Aspects of managing research and development

8.1 INTRODUCTION

So far this book has concentrated mainly on the legal background to and procedures for the acquisition of intellectual property rights as a way of protecting new technology and design and on some of the specific management decisions which have to be taken on several questions such as ownership of rights and deciding whether to take out legal protection in any given case. In the light of these earlier matters, it is now possible to make some rather more generalized points about certain aspects of the research and development department of a company. We will look at these in turn, though of course they interact markedly.

8.2 RESEARCH AND DEVELOPMENT DEPARTMENTS

If a company has an in-house R and D department and does not rely entirely on outside consultants who may be hired and fired with greater flexibility then before any management decisions can be taken inside the R and D department of course its place in the wider context of the company must be defined. The basic organization of research and development in a company must of course be resolved. This raises a number of questions the precise answers to which depend on the size of the firm, on the level of R and D activity it must undertake and so on.

(1) Who is to be in charge of the R and D department?
(2) What level of staffing is there to be?
(3) What basis of funding is there to be?
(4) What is its relationship with other departments?
(5) What is the role of the R and D department to be?

Only when these general matters, which form part of the general strategy of the company in deciding what its research profile, budget, etc. are going to be, can the more detailed matters of organization and R and D management be resolved. The essential point to make is to give some recognition to the specific needs of R and D particularly in the financial organization of its work which is not susceptible to detailed budgetary control in quite the same way as other departments. This point is followed up below in Section 8.2.5.

Aspects of managing research and development

The principal management tasks in all research and development departments can be classified into the nine following areas:

(1) New technology and design auditing and assessment
(2) Planning a rational R and D programme
(3) R and D personnel management and supervision
(4) Liaison with legal adviser and gaining protection
(5) Finance and budgetary control
(6) Liaison with licensing departments
(7) Liaison with production departments
(8) Monitoring of outside affairs in R and D
(9) Responsibility for full record keeping procedures

These must be examined in turn. This is not to say that these are hard and fast classifications of R and D departmental responsibility; rather they identify tasks that have to be performed in relation to any R and D and which must therefore be allocated somewhere, even if a member of the R and D departmental staff is not the one actually designated to assume the final responsibility. A number may well be delegated to, for instance, the company patent agent, if there is one, or to an outside patent agent if there is not, and certainly some people would identify a number of these tasks as clearly being within the functions of a patents department. In some smaller companies without departments of this kind a senior member of staff will be given overall general responsibility for such functions.

8.2.1 Technology/design auditing and assessment

One of the first and most important tasks for the firm R and D manager is to conduct a thorough technology and design audit and keep it up to date. This will look not merely for company technology and design which is a part of an existing patent portfolio but also for those technologies and designs which might be lying untapped or under-exploited on the shelf, or even unrecognized.

Firstly, it is important to appreciate the many sources of potential intellectual property rights which have to be examined. The checklist below sets out those areas which should be investigated when compiling the list which should of course be kept constantly updated.

(1) Patents department/agent/other document keeper
  (i) Existing patent, trade mark, design registrations
  (ii) Copyright material of which copies have been kept
  (iii) Existing patent applications etc.
(2) Research and development departments
  (i) Results of fundamental research activities
  (ii) Results of applied research activities
  (iii) Results of development projects
  (iv) Results of testing and screening activities
  (v) Software devised for specific tasks
  (vi) Know-how and manuals for instruction
(3) Drawing office and publicity department
  (i) Original drawings, plans
  (ii) Draftsman/draftswoman contributions to design and functional asapects of products

(iii) Unregistered trade marks, packaging design, and catalogues, brochures, advertising copy, etc.

(4) Computer department
- (i) Software developed for internal use
- (ii) Applications know-how

(5) Shop floor and technical departments
- (i) Know-how on operation of processes
- (ii) Design modifications introduced on shop floor, perhaps not reflected in drawings
- (iii) Employee suggestions schemes
- (iv) Staff training manuals and guides

(6) Sales staff
- (i) Know-how, reports on performance and uses to which products are put by customers
- (ii) Ideas for improvements on products already sold

(7) Outsiders
- (i) Company's unsolicited suggestion schemes
- (ii) Licensee improvements which are dependent on licensed rights to function
- (iii) Other technology dependent on company rights to function at all, such as compatible accessories

Once the available technology has been audited the management can begin to sift out any of the potentially exploitable material, thought to be exploitable either on its own or with outside technology which has not yet been exploited as much as it might. Companies within the same group may well have technology which is of little significance in itself but which can be combined with another piece of technology held by an associated company and which together produce a viable product. The assessment will be carried out jointly with the company licensing staff, production staff and patent agent, and this task then slides into that of ensuring that legal protection is available (discussed below in Section 8.3.3) and taken when required; and that of liaising with the company licensing department and the production management to make sure that any marketing opportunities are not missed. A checklist of matter by which assessment may be pursued follows:

- Who invented it and why – what project were they working on at the time? How was the invention made and on the basis of what pre-existing work which might invalidate it? All factual background should be explained. What problem were they trying to solve? Are all of these matters properly recorded? Have they published any information about it? This relates to the question of record keeping discussed in Section 8.2.9. below.
- What are the principal features of the new design or technology and what information and testing results etc. does the company have about it, what drawings and what experimental results? Again record keeping matters are essential.
- What is the technical field, the state of the art, the existing technology or design in the market? What obstructing patents etc. are there which might hinder its development? Here assessment overlaps with the existing monitoring procedures which may be very informative if the invention is related to a field in which the company is usually active and therefore usually monitors others. See Section 8.2.8.
- What lead time over others does the technology have? What advantages over other technology does it have in terms of performance, cost and reliability, etc?
- What investment of a capital nature does it require to set it up in production? What further R and D expenditure would be required to render it viable? What are the projected returns on the investment, in terms of any savings of production costs of a new process etc?

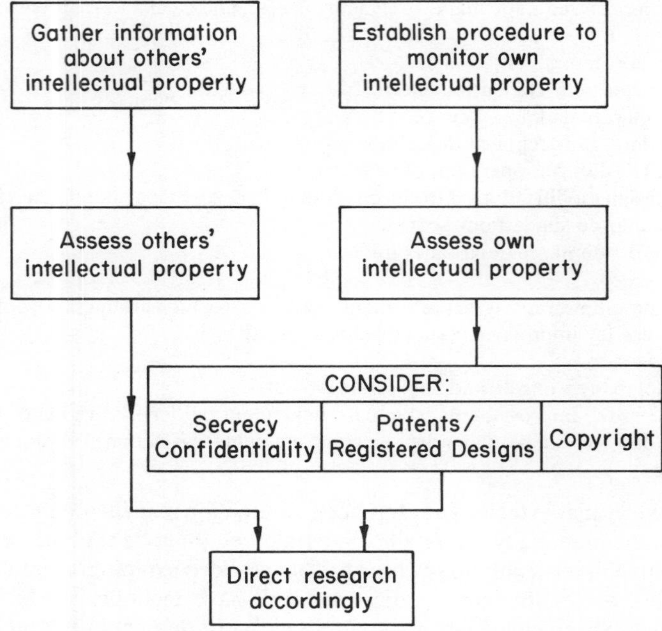

Figure 5 Planning a research and develpment programme

- How flexible is the technology: is there scope for the development of the invention in other technical areas, other materials, other processes, other design conditions, scales of operation, etc?
- How does it fit in with the existing portfolio of intellectual property rights and its current and past projects? What related expertise and know-how does the company have to back it up? If it is not now readily exploitable through production are alternative means of exploitation e.g. through licensing a possibility?
- Last and very definitely not least: will anyone buy it? The main failing of so many R and D assessments is the customer orientation of the project. What is technically very good may be wholly unappealing to the customer because of the implications of servicing, maintenance costs etc. of a sophisticated system which may be better but simply may not save any money in the long run. What will customer reaction be? Some products are too advanced for the market they are aimed at.

8.2.2 Planning a rational R and D programme (Fig. 5)

To what ends are your R and D efforts going to be directed? This is not such a silly question as it may at first sight seem. A number of considerations may be relevant to deciding how R and D policy and planning is going to proceed. In the light of the existing audit of technology and production it will be possible to sit down and plan a rational programme of R and D to fit the needs of the company and its overall market strategy.

The first stage is assessment of existing and past work, to give you a firm factual basis upon which to make the decisions about R and D planning. This will thus require you to answer questions which include:

(1) What fields of research has the department entered? What has this activity produced?
(2) What does the department do well? What badly?
(3) Has the department in the past overestimated time and budget or underestimated time and budget requirements? Is its performance equally good in all stages of pure, applied research and product development?
(4) Are all existing projects going in the right direction? Should some be dropped? Are they all suitable for in-house production or are they going to lead to licensing only?
(5) What are its competitors doing? How does their performance compare in the fields which it has been involved in?
(6) What levels of expenditure have been committed? (What return they have produced, if it is possible to judge this at all, is an unreliable guide to performance.)

This enables planning decisions to be taken with the benefit of a reasonably objective assessment of past success, failure and present capacities. In the task of constructing a R and D plan, this information will enable you to identify what is most likely to go wrong with any given project and eliminate those where more things are more likely to go wrong than is the case with others.

- Does the plan fit in with existing strengths of the company and its staff and its own production facilities? It is important to be able to undertake work into which it is possible to put existing expertise, for that will be a means of reducing risk and improving the accuracy of forecasting. Do not look round for projects to which you can adapt the company. Look for projects which will fit in with the company's strengths, in the R and D, the sales force and production.
- Does the plan fit within existing budgets and if not is it possible to make out a satisfactory case for increased expenditure which the company can finance?
- Is the plan customer oriented? There should be some likelihood that the plan is to produce a viable commercial return in the reasonably short term. This means planning something people will buy, looking seriously at market opportunities and not just the scientifically interesting or fulfilling. One of the most important factors here is simply whether what is produced will be widely affordable and workable in the market to which it is addressed. Will the performance be that much better that people will pay for it? Will it be affordable even if performance is markedly better? How much staff training will be required to use it in the customer's premises? What will maintenance costs etc. be?
- Does the plan involve refinements of existing technology? If so is that technology nearing the end of its life cycle? If so it may be better to think of licensing out the existing product before its decline becomes imminent and concentrating on developing products which have not yet reached the peak of their cycle. For example, it may not be wise to concentrate development resources on some marginal improvements to technology which will soon be replaced by new technology whose early achievements show that it has far from maximized its potential and will even in a relatively immature state improve on cost and performance characteristics of existing products sufficiently to attract a market.
- Does the plan involve complementary projects, all building on each other and producing a flow of new follow up products in a particular field? Do not fall prey to the sin of over-diversifying. Do not have too many irons in the fire or spread the research investment too thinly. And do not waste existing expertise. Innovation which steers a company right out of its existing area of high technical expertise is rarely a success, even if you get the technology side of things right. It is placing a great stress on the production and sales teams to make the most of an

area which they have no experience in. There is no point in having a piece of technology which your sales staff cannot sell, cannot penetrate the market with, because they do not know that market. There is little point in demanding production techniques of a shop floor labour force untrained in that field using a new plant which has had to be brought in of a kind wholly unrelated to past work and which may not be easily maintained.

- Plan research and development for today's market. It may be very impressive to say that you are designing products for ten years' time, but in most companies that will be five years after they went bust. The full potential of the computer has still not been attained and development is a long-term one. But there were immediate applications for the computer right from the start.
- Is the plan a high risk, high return plan? If so, throw it away and start again. It is difficult to think of many long-term successful businesses which exposed themselves to a high level of risk and succeeded. They define and seek to minimize that risk. It is quite easy to think of several successful short-term businesses which took risks but how many survive? The essence of long-term success is confining risk not exploiting it. Most high technology ventures that have succeeded are in any case not high return ventures – their returns are generally less on investment than in the traditional industries.

8.2.3 Personnel management and supervision

This encompasses four types of task:

- At a general supervisory level those of directing the staff into projects which fit into the general research plan constructed by the company's management, help in motivating and assessing research staff, filling gaps in expertise by recruitment of suitably qualified staff and co-ordinating activities involving liaison with the other departments, etc. as well as simply representing the interests of the R and D staff within the company at board level.
- Dealing on a personal level with those employees who have made some invention, written a paper or wish to attend some conference to discuss some new designs or technology, or who wish to leave the company after having had contact with confidential information. This matter was considered in much more depth in Chapter 2.
- Ensuring that an adequate security system is always maintained to preclude unauthorised leaks, loss of data and records, industrial espionage, and similar loss of confidential information, and ensuring that all R and D employees follow the system so devised. This will also involve the design of physical security measures, of adequate record keeping facilities and forms, etc. These matters were considered in much more depth in Chapter 2.
- Two purely administrative tasks which require careful thought are the establishment of an employee suggestion scheme if there is to be one for company non-research staff (research staff are of course under an obligation to report inventions and there will already be a proper invention record system subject to weekly review by the team leaders and monthly review by the head of the R and D department) and the unsolicited suggestion scheme for outsiders. These have been discussed to some extent already. Particular care must be taken when drafting the employee scheme so as to avoid the loss of rights which might otherwise belong to the company, for unless a company is careful it may be giving up rights in material to which it was in any case entitled under the law. Equally non-collective agreement schemes which operate to the disadvantage of the employees may fall foul of the law. This is considered in Chapter 7. Some model forms and letters for both of these regimes appear in the Appendices.

8.2.4 Gaining intellectual property protection

An important function of the R and D manager is to ensure that whenever possible legal protection is at least available to the company for whatever new pieces of technology and design may be developed in house, even if in any given case the decision is taken not to take up the opportunity but to publish instead without the benefit of legal protection. Ultimately, this may lead to the development of a large patent etc. portfolio for a particular line of research, either for its in-house exploitation or for licensing as an investment patent portfolio. This involves maintaining an adequate system of invention recording and reporting, adequate security and paper vetting, etc., and proper consultation with the patent agents and other legal advisers with the benefit of whose advice decisions whether or not to take out protection will have to be made and priorities reached.

Assessment for protection will be achieved by the review of the invention records submitted by staff and in collaboration with the patent agent. The commercial appeal of the technology will have been examined to some extent already, but of course an important part of its commercial appeal may be the ability to protect it from imitation by competitors. The agent will be asked to review any commercially attractive idea for UK and overseas patent and other protection and will produce a patentability or – more generally – a full intellectual property report indicating not only the prospects of gaining patent protection but also any alternative means such as copyright, registered design and so on. It may well be that a set list of those countries in which patents etc. will normally be sought in the absence of a decision to the contrary can be readily established, dictated by the markets in which the firm operates etc. In such a case a fairly standardized procedure can be developed.

Assuming that a decision is taken to spend money on protection – which will of course be determined by any decision taken on the available finances and the priorities of the company with respect to certain projects in the particular R and D field in which it is working, then the patent agent will be instructed to make a much more detailed search and report along the lines discussed in Chapter 3 for the acquisition of a patent or other right.

If this initial search is favourable then the full application procedures will be initiated. The patent agent will have to liaise closely with staff and gain a knowledge of a number of matters which the invention record will have touched on but which will have to be gone into in more detail at this stage. These questions should not be regarded as relevant to the applications alone. The answers will give a better insight into the potential markets for the invention, changes which might be made for better commercial exploitation and so on.

As we have stressed before the decision to seek protection is a business one, based on the factors we examined when assessing the technology or design listed in Section 8.2.1 above. The availability of legal protection is of course a prerequisite to this but just because it is available it does not mean that you will always want to take it.

Of course, it is important to stress that the role of the R and D department does not end with the papers going into the company patent agent's room. Any such invention will still need more development or amendment to avoid objections and this may later have to be incorporated into the patent agent's applications either as original or improvement patents etc. The patent agent will have put follow-up questions about the technology or design on his or her own initiative or as a result of objections from the Patent Office examiners, which will all have to be answered very promptly and accurately as a matter of the highest priority.

A further aspect of R and D–patent agent liaison is that of keeping the patent department up to date with what is happening or is likely to happen in the research department's work. For a patent agent will obviously be better able to assist if he or she is familiar with the general technological background of the company, not merely so as to be able to advise on patent matters but so as to be able to warn the company of relevant developments elsewhere with competitors and simply to understand what it is he or she is dealing with. This may then require regular visits to the R and D department if the patent agent is an outsider and regular meetings to review internal R and D work if there is an in-house patent department. A well-informed agent may even be able to suggest channels of investigation which have become worth following up in the light of failures or even successes in other published patents, in the light of what he or she is told by the company's R and D team of their own work. Such information as they give him or her is of course confidential and he or she will respect it as such.

8.2.5 Financial and budgetary control of projects

All such functions are of course circumscribed by their financial considerations. Apart from the central question of controlling all expenditure on the research and development projects as such this function extends to a number of other matters including decisions about expenditure on patent office fees, legal expenses to be incurred in connection with the defence and challenge of intellectual property rights, insurance, etc. In fact these other miscellaneous matters are generally easier to make judgements about than the question of proper budgeting and targeting of R and D work, which is notoriously difficult to get right. Many of the great financial disasters involve R and D projects getting a firm into the red. It is a characteristic of R and D that after the initial creation of an undeveloped idea of some great apparent commercial interest, which may be relatively low cost as a paper discovery, development expenditure increases very markedly proportionately to subsequent improvements and developments needed for its launch on the market place. Detailed consideration of this question is however outside the scope of this book. But one comment may be made here.

A comment should be made about patents and other intellectual property bills. Even in the very active R and D oriented companies it may be expected that the costs of obtaining intellectual property protection and paying professional fees will be insignificant in comparison with R and D work and overall turnover, only running into mere tenths of one per cent of annual turnover, but they should still not be seen as a burden which the R and D department budget must accept, as a matter of principle: to say otherwise is tantamount to penalizing the R and D department for its own success in creating legally protectable technology. Such rights are, after all, assets (albeit of unpredictable value) and assets which cost money to maintain, which enable the other parts of the company such as the licensing department and sales force to do their job much more effectively, and the costs of acquiring them should be borne as such. They are thus not matters to which money to be spent on research should be diverted. Accordingly there should be a quite separate budget for patent and other rights, kept apart from the general R and D budget, to which it does not belong.

Budgeting for intellectual property rights and items such as litigation and fee disbursements may be dealt with as an individual item in administrative costs of the company, or perhaps allocated to the patents department if there is one. A notional sum may be allocated for these purposes and further expenditure required to be justified. This is one way of seeking to impose a degree of cost consciousness on the patent department and eliminate the risk of purely speculative applications for protection without adequate initial consideration of the prospects of success or the merits of seeking protection in the case in question.

Proper budgeting for such costs can be extremely difficult, since patents and other rights tend to come in waves and not necessarily with a steady flow, except in very large operations. Equally, whereas plant expenditure and the like is time flexible to a limited degree, this is simply not the case with intellectual property registrations which have to be renewed at a fixed time. Thus there has to be some degree of budget flexibility in this regard. Dropping a certain patent registration one year because you have gone over-budget may be a very expensive false economy.

Similarly in looking at licence royalty fees paid for the importation of intellectual property rights and other technology, these should be regarded as part of the costs of sale on the balance sheet and preferably represented as such and not as a separate entry at some place in the account far removed from this. The proper place for these licence fees must be where other costs such as raw materials and assembly costs are placed for the profit-making product could not be marketed but for this initial cost.

8.2.6 Liaison with licensing department

This of course will be crucial to any successful licensing venture in or out, since the R and D department will be making a major contribution to the technical assessment of any proposed venture and at a later stage to performance of obligations in respect of back-up to licensed-out technology or maximizing the exploitation of licensed-in technology. Conversely there may be good cause to consult the licensing department at an early stage when assessing a new invention or design to gain a preliminary view of its commercial potential for such licensing.

When providing assistance in reviewing licensing-out opportunities, which is discussed in depth in a later chapter, the R and D team advisers will have to provide advice on a number of issues concerning the suitability of the licensee proposed for the agreement, the obligations of support and training which will be involved and the capacity of the company to supply these without overstretching the R and D department and its normal and on-going research work, and questions of the adaptability of the technology to the specification of the licensee.

Although the licensee will be doing this too, it is important for the company to have a good understanding of what it involved not merely to price the licence agreement but in order to be able to draft realistically the obligations which it may be called upon to fulfil under any contract. There is no point in undertaking support obligations which cannot be met and which give rise to legal liability. The negotiators are dependent on accurate estimates of just what technical support is needed and can be provided in order to do their job properly, even if the end result is that the deal does not go ahead.

Assessment of outside suggestions as licensing-in opportunities will be an important task not merely as a means of providing proper technical back-up to the other departments of the company but as a primary task in itself. The R and D department should be ready to make suggestions on its own initiative for licensing-in new technology and design and not merely respond to other people's suggestions. Is there any gap in the company's technical expertise which it would be desirable to fill and which could be filled more quickly and more cheaply by taking a licence in some piece of technology from another company than by development of some duplicate piece of technology in house?

This requires a positive approach to outside technology and one which may call for swallowing of false pride. To license-in technology is not an admission of failure or of an internal R and D impotence *per se*. You cannot do everything. To decide to license-in a piece of technology may simply be to apply tests of commercial common sense to R and D as they are applied elsewhere in the company whenever there is a choice between direct labour and contracting out work.

Many of the same qualities required for the internal technology audit have to be demonstrated when examining outside proposals which might complement – or supersede – your existing internal technology or design. One needs the objectivity to escape 'Not Invented Here' syndrome and the ability and flexibility to see ways and means of adapting opportunities to fit into the company's own existing expertise and experience. For if this outside technology provides a short cut to a solution which will be cheaper than using your own internal research facilities, licensing-in makes sense. It saves the company money overall and liberates funds for further internal research and development work which might have been tied up in the wasteful duplication of some other company's already proven success.

8.2.7 Liaison with production departments

This provides a valuable two-way supply of useful information which should be encouraged and developed as a matter of course even where the production and R and D sites of the company are geographically remote – which is the most common cause of a breakdown in adequate communications. Regular review and assessment of any shop floor improvements and refinements are essential as is examination of any difficulties leading to loss of production efficiency and reports of needs which have been identified and which might be met by further R and D work.

8.2.8 Monitoring of outside affairs in R and D

Part of the assessment of outside technology as potential licensing-in material is done already simply by monitoring outside technology to survey competitors' work for new rights, infringements of the company's own rights, and so on. Often the task is given to a patent agent or patent department. This acts in essence as the company's technical information gathering service; it is conducting legitimate intelligence work about other people's R and D and intellectual property activity, as well as monitoring the infringement of the company's own intellectual property rights. But the available sources of information go beyond what the patent agent or department customarily reviews and it should be a task in a company which can afford to do so to look to such other sources.

The sources of this information are considered in depth in a later chapter. The only point to make here is that monitoring implies not just mere observation but also analysis and doing something about the information reviewed. These are tasks which must be undertaken by the R and D department even when a patent department is doing the collecting and identifying of relevant information. In some companies copies of this information about recent developments in the field in which the company has an interest are automatically passed on to the researchers in that field so that, for instance, abstracts of new patents may be circulated in the R and D department with an initialling slip and a names circulation list to ensure that everyone who should see it has an opportunity to see it. In the larger R and D companies you may find librarians notifying staff of titles of recent articles in journals, and so on. And details of future conferences sent to the company will be circulated in the same way.

It will sometimes be necessary to cast a careful eye on the later submission of 'patentable' ideas by research staff so circulated, for either subconsciously or deliberately some staff may make use of this material in apparently coming up with 'new' ideas without fully and frankly disclosing their true source, which can waste the company time and effort in pursuing what ultimately becomes a dead end. But this is, on balance, a price well worth paying for the information value of the material circulated.

If in reviewing the press and the other sources of information available such as company newspapers, a new catalogue picked up by sales people at exhibitions etc., any infringement should appear, this should be notified to the company patent agent at once to determine if any of the company rights have been infringed. In the light of the advice taken the initial procedure for dealing with the matter may then be instituted. These initial action procedures are considered in a later chapter.

8.2.9 Responsibility for full record keeping

Communication has been a recurrent theme in the book. Let everyone know what is going on. An essential physical back-up to communication is adequate record keeping: it smooths the flow of information, provides the necessary back-up in any litigation, gives you full information on which to base decisions as to future conduct. This task overspans all the others. It is thus essential in guaranteeing the adequate communications between the various elements which go to make up a properly functioning R and D department. A number of such sample records are reproduced in the Appendices. The record-keeping duties fall in alongside those of proper security procedures. Again many of these will have been undertaken in practice by the patent agent or patent department.

In larger companies there may well be a general document control officer to whom this task may be allocated. Smaller companies may have a librarian or information officer or a member of staff in the R and D department itself to whom the task is given and this person will also be responsible for maintaining all the security measures for keeping the documentation (which in a small company may amount only to use of a safety deposit box in a bank or with a solicitior or even the company secretary) secure from natural loss and damage and from uncontrolled access, with records made of such authorized access and the use to which the documents so accessed were put.

8.3 PATENT AND TRADE MARK DEPARTMENTS

This section is concerned with three matters:

(1) What are patent and trade mark agents?
(2) What functions do they perform for a company?
(3) How is an in-house department organized?

8.3.1 What are patent and trade mark agents?

(a) Patent agents in the UK

A patent agent is a qualified professional person. He or she is qualified under examinations and most are governed by the rules of the UK's Chartered Institute of Patent Agents. Apart from solicitors, who with a few exceptions steer clear of advice on patentability and drafting of patents, only someone on the register of Patent Agents may practise as a patent agent for gain, that is represent a client in seeking a patent grant in the UK. At the time of writing this monopoly situation was under investigation by the Office of Fair Trading, concerned with its effect on small firms which of necessity have to employ outside assistance. Employees may act for a firm without qualification but not for others.

No such legal obstacles prevent anyone from charging for similar conduct with respect to registered designs or registered trade marks, though there is an Institute of Trade Mark Agents with its own qualifying examinations: many patent agents carry out trade mark work for which they are well trained and qualified by the Chartered Institute of Patent Agents without also necessarily belonging to the Institute of Trade Mark Agents and many are members of both institutes. Members of the institutes may not practice with limited liability.

The training of a UK patent agent requires a good general knowledge of a broad range of scientific and of technical subjects and a very thorough knowledge of the relevant law of intellectual property not just in the UK but also of foreign countries.

Patent agents may instruct counsel, i.e. barristers, directly or through a solicitor, to represent their client in the Patents Court and may appear personally in the Patents Court on appeals from UK Patent Office hearings. There are something over a 1 000 UK patent agents now on the Chartered Institute's register. All have passed their range of professional examinations and are equally qualified to work alone in private practice, or industry. At present few seek to advise clients on more than the questions of whether it is possible and advisable to seek a patent or other such right and on questions of infringement: though some give advice on wider issues of licensing and on drafting contracts to license rights, which is still a business more often left to the solicitor and often the barrister. But this role may increase in the future as the pattern of patent agent work changes, particularly if their monopoly does disappear. There is a growing body of general technology brokers and consultants many of whom have a background in the patent agent's world, often coming from posts employed in large company patent departments. The role of such technology brokers is considered in a later chapter.

(b) Patent agents and the European Patent Office

For practice before the European Patent Office, it is necessary to be on the EPO register of professional representatives. Although concessions were made to UK existing practitioners to be put on without any further examination, since 1979 new members of the profession must also pass a European qualifying examination to get onto it unless he or she is an employee of the patent applicant and even then that person may not conduct any proceedings in relation to European patents (UK) before the UK Patent Office unless he or she is on the EPO's European list.

(c) Patent agents in other countries

In the USA the patent profession is split into two halves. The patent agent has passed examinations set by the US Patent Office and has a science degree. He or she may practice and represent the client in the Patent Office hearings. The patent attorney has in addition a law degree and has been admitted to a state bar and can combine the role of patent agent with that of lawyer in advising and drafting contracts and representing the client in court in the same way as a solicitor and barrister respectively in this country. Discipline on the profession is imposed by the US Patent Office not by their own professional associations.

Many countries have no formal qualifications or regulations for persons representing others in patent or trade mark matters. In most European countries there is a division between patent representatives in private practice and those in industry, the individual having to choose which profession he or she wishes to follow and then being given a different professional title accordingly, so that for example in West Germany one might be a Patentassessor in industry able to represent only the full-time employer or a Patentanwalt in private practice able to represent anyone, but not both.

8.3.2 The role of the patent agent in the UK

Whether employed in a firm or in private practice, the basic role of the UK patent agent (and the design and trade mark agent: most patent agents take on this role too) is more or less similar. It combines a number of functions, which may be conveniently summarized as:

(1) Opinions on validity, infringement, patentability
(2) Drafting patent specifications and claims
(3) Prosecuting patents and other registrable rights
(4) Invention and publication vetting and clearance
(5) Monitoring patent publications etc.
(6) Information search services
(7) Renewals of rights, fees-reminder services, etc.
(8) Litigation work
(9) Liaising with overseas patent agents

These may be looked at in turn.

(a) Opinions on validity, infringement, patentability

The function of the patent department is not to take the final decision on whether protection is to be sought. It is to provide advice on that issue backed by a concise summary of the factors to be taken into account, prospects of success, alternative courses of action available and so on. Equally the decision to bring or defend infringement actions is a business decision to be made on the advice of the company's legal advisers and patent department, not one to be taken by those advisers. Foreign patenting decisions again will be made from above, though instructions to foreign patent agents and further correspondence will be direct from the patents department.

(b) Drafting patent specifications and claims;
Prosecuting patents and other registrable rights

These are the sole responsibility of the patent department in execution though the ultimate decision of whether to press on in the face of objections must again be made by others on their advice. Reports of grants of rights made to the company should be sent regularly to the director on the board responsible and sanction sought for any extraordinary expense which must be incurred on appeals against refusal to grant rights or in opposition proceedings.

(c) Invention and publication vetting and clearance;
Monitoring patent publications etc.;
Information search services

These may be tasks better performed by the patents department than by the R and D department or librarian if the company has such a department. Invention records and employee suggestions may be perused regularly not for their commercial appeal as such to the company – they may in themselves be of little use to existing plans – which is a task for the commercial employees of the company, but from the point of view of UK and overseas patentability etc. Even if not attractive for the size of company in question because its commercial significance does not justify a diversion of the main production resources to something which is marginally profitable or merely peripheral, such otherwise dormant technology may be exploitable by others and as such may still have a commercial value in excess of the cost of gaining protection.

Recently some very big companies have begun to allow technology patented in this way to be brought down off the shelf and licensed or assigned to smaller companies, often to companies run or advised by the original inventor or by a senior manager nearing retirement age who has left the licensor either on early retirement or on secondment as a consultant on loan to exploit the idea himself or herself in a commercial unit with sufficiently low overhead costs to make such a project viable; or licensed to companies which lacked their own R and D facilities but with adequate technical expertise to exploit a new technical development under licence without too much dependence on the back-up of the licensor. Even some local authorities are now encouraging this with grants to such small start-up ventures, or staff appointed to co-ordinate activity.

In regard to trade marks and pure design matters, advertising copy and the like both of the company and its rivals may have to be vetted for any trade mark, passing off and trade libel issues, though this might be done by the legal department if there is no in-house patent agent, assuming it has the material resources in house such as the trade marks journal. Again, outside patent agents will do this work too.

So far as vetting any outside patent or design or trade mark applications is concerned the patent agent will go beyond mere reportage to considering advice on whether challenges should be made to any such applications which appear to be prejudicial to the company, with the assistance of those in the R and D department expert in the field. Where necessary it may be decided to refer the matter to counsel for further opinion.

Vetting will be conducted not only in the UK but also in other markets overseas of interest to the company and with particular interest being directed to all the applications of the company's main technical and marketing rivals. This not only safeguards the company's legal position but gives a good idea of what the trade rivals are doing. Vetting will go beyond patent and other documents to journals and market reporting services within reasonable limits of time and cost to the company.

A further service which a patent department may be able to provide to the company is one of information searches for R and D workers. A patent agent may prepare summaries of updates on certain fields of technical interest to the company on its own initiative and circulate such a collection of material or respond to a request for a specific survey to be undertaken for a research team about to start on a new project, to avoid unnecessary duplication of work which may be read about elsewhere already.

The results of searches undertaken to establish the patentability of a company or rival may in appropriate cases be sent to the research team involved in the field if security or matters of legal privilege are not prejudiced. This makes more widely available information which it has cost money to acquire – why waste it by not making best use of it?

(d) Renewals of rights, fees-reminder services, etc.

These are basic administrative tasks which many outside patent agents operate as a service for clients. Any in-house service would be expected to perform this role and ensure that payments were made properly on time by the finance division of the company.

(e) Litigation work

While litigation is the concern of the legal department and the company's solicitors will run any case, the patent department or agent will provide a good deal of support not least in acting as the inside technical adviser to the solicitors and often counsel too in place of the R and D man. This is a good practice because as a scientist-lawyer, patent agents are often very good at explaining the law to the scientist and the science to the lawyer!

The decision to litigate is taken by a senior employee such as a director or more usually by the board itself, given the costs and value of most intellectual property litigation – it is not a mere matter of debt collecting, even though most cases are settled out of court at a relatively early stage.

(f) Liaising with overseas patent agents

This has already been mentioned. It is a very important function of the patent agent or department to gain sufficient knowledge of which agents to employ overseas for particular specialisms and to establish a working rapport with those agents to the benefit of the company.

8.3.3 In-house or outside patent agents?

This is principally a function of size and levels of activity rather than anything else. At what point does it become financially sensible to employ full-time staff with the overheads of personnel and physical costs such as patent library holdings, office space, etc. rather than paying the profit-cost based fees of an outside patent firm? Some factors for consideration may be mentioned.

- Does the industry have a high patent profile? If it does the task of monitoring other people's patent applications and responding to them if they appear to obstruct or infringe the company's rights or products will be a major task.
- Is the rate of technical progress very high so that a high level of patents is sought? Is speed an important factor, requiring a rapid in-house response? What price are you prepared to pay for the convenience of in-house advice?
- Does the level of output from the company's own R and D department justify an in-house patent service?
- What is the range of technical output of the company? If it is in a fairly narrow band it may be possible to employ a sole agent with special expertise. If a wide range of matters are protected then it is possible that you will want to have a number of agents in house, which increases the cost of the service (although of course there are certain fixed costs such as maintaining a patent library which do not double when you double the staffing) or instead you may choose to be free to deal with a variety of patent agents who specialize in these different fields. If you have to go outside the company too often, this makes an in-house agent poor value for the money spent.
- How do the costs compare? Calculate the rate of fees paid to the outsiders and examine the costs of setting up and maintaining an in-house service. This will often be the deciding factor.
- Does the work require such specialized knowledge of the technical field that you want a single agent to develop that expertise in house, perhaps with the benefit of in-house training in the field itself, who will use that expertise for your benefit and not for others? This may be a reason for employment rather than use of an outside agent. On the other hand an agent's over-specialization can become counter productive.

It must be said that in recent years there has if anything been a trend against in-house agents and some former in-house services have now gone into private practice, dealing with the same company's work but also taking on other work. This has enabled the company to retain a close relationship with a particular firm which knows the background to the company's technology but which does not create unnecessary overheads in a time when fixed costs are being reduced as much as possible. There is also an increasing element of price competitiveness in the profession which may work to the advantage of the company in bargaining over fees and reduce the advantages of cost control which formerly accompanied the use of an in-house patent agent.

A patent agent employed principally for patent and other technical matters will be able to tackle any trade mark business too. In some cases a company with a large trade mark portfolio and a merchandising bias may seriously consider a full-time employee for these matters even if there is insufficient patent business. Similar criteria apply to this decision too, where the main task of the agent will be monitoring trade mark applications of relevance to his or her company's trade mark portfolio and opposing any which appear too close for comfort.

8.3.4 Organization of a patents department

If the decision is taken to create an in-house service to perform the functions described above, then it may be made autonomous or part of the legal services department in the company if there is one, under the general responsibility of the company secretary or the head of research and development. If there is to be no in-house department then the company must still appoint a specific number of staff with whom the patent agents outside the company hired to do work may always liaise, so that there is within the company a clearly defined officer with responsibility for taking decisions and answering for them to the Board.

8.3.5 Liaison with licensing departments

This is a matter to be discussed in more depth in a later chapter. As has already been said many patent agents seem not to play a very significant role in the licensing process, or at least an overt role, though this is changing and they are becoming more active. There is potential for patent agents to take a greater part in technology brokerage and licensing activity without compromising their traditional professional standards. For now their role may be summarized as this.

In some cases a company negotiating team may need specific back-up in grasping the intellectual property background to a proposed deal whether as licensors or licensees, to examine the scope of protection of the patent right in the light of other patents, decisions on validity, etc., which may be a very significant factor in pricing a technology package.

In cases where a large team cannot be taken it will often be possible for a patent agent, properly instructed at base, to combine the roles of the technical and legal advisers in a team going to negotiate in a turnkey or other licence arrangement, if only in the early stages of the discussion. Many larger practices now undertake licence drafting and related work.

8.4 USING SOLICITOR/BARRISTER ADVISERS

This section is concerned with three questions

(1) What are barristers and solicitors?
(2) What roles do they perform?
(3) How is a legal department in a company organized?

8.4.1 What are barristers and solicitors?

Unlike many countries, in England and Wales the legal profession is divided into two distinct branches. There is no monopoly on giving legal advice as such, though solicitors do have a monopoly on preparing deeds for money and barristers a monopoly in representing clients for money in certain but not in all courts in much the same way as the patent agent has a monopoly in his or her field of work. Their work differs but overlaps.

Both sides of the profession have very stringent professional examinations and on-the-job training. But it should be stressed that there is no compulsory legal training in intellectual property law for any lawyer on either side of the profession: therefore solicitors and barristers may only learn the subject when taking a law degree, although there are only a few law courses that teach the subject, or on the job, or solicitors may now study it at optional continuing education courses being set up by the Law Society. This partly explains why there are a few specialists in the field who are highly trained within a large solicitor's practice or in some specialist chambers in the case of a barrister.

It is important in choosing a legal representative to ensure that you have selected one who is competent in the field. A firm which has been a loyal and efficient firm for the company in other fields in the past may be wholly unused to this kind of work and unsuitable for the purpose.

So far as the Bar is concerned, since all barristers in practice at the Bar of England and Wales are hired *ad hoc* and not employed on a salary, and most specialize in a given field rather than take on all kinds of work, so as to develop a specific expertise, one's choice is always made according to the nature of the case and it is a less delicate problem.

(a) The solicitor

The solicitor is in direct contact with the person he or she is advising and is often a general adviser on a wide range of legal and business matters, though more and more solicitors are specializing in certain areas of the law in practice, even though their training is of a more general nature, so that in a very large firm of solicitors there will be different departments employing solicitors entirely devoting themselves to specific areas such as tax, commercial contracts, or trusts and probate, conveyancing and planning, insolvency, etc.

In cases which require an expert opinion on a more specialist matter or a point of particular difficulty which requires an expert in the field to assess the matter, he or she will refer the case to a barrister for a second opinion. In the field of patents design or copyright or trade marks, a patent agent may sometimes be used for a second opinion instead. A solicitor may choose to work in a private practice either as a sole practitioner or in a partnership and may employ other solicitors (he or she cannot form a law practising company) or, like a patent agent, may be employed by a company and represent that company in its legal affairs. There are about 30 000 solicitors in practice or employed as solicitors by companies.

(b) The barrister

A barrister gives advice to solicitors or patent agents and their clients but has no direct relationship with those clients and may accept work only from such a solicitor or patent agent. A barrister may not accept work from a lay client. A barrister must work alone – he or she cannot form even a partnership with another barrister – and each piece of work he or she does is for a fee which must be negotiated for him or her by their clerk, who is a sort of office manager for a number of barristers who practice in the same chambers, and who negotiates with the instructing solicitor the fees appropriate to the standing of the barrister, the difficulty of the case and so on. The clerk may also push for payment after the event, since there is no legal entitlement for a barrister to be paid for his or her work and a barrister cannot sue for his or her fees, and some clients are extremely slow to pay. Gaps of two to three years between completing the work and receiving payment are common. A barrister cannot work for a salary or on a retainer.

There are some exceptions to this general rule. A barrister who wishes to be employed in a company may ask to have his or her name entered on a list of those who are employed to give advice or legal services to employers. These barristers are then prohibited from representing their employers or indeed anyone else in court, but may be approached direct by the employer for advice and not just through a solicitor. There is quite a large number of such barristers now working alongside solicitors in or as heads of legal departments in companies where the attractions of a guaranteed income, no rents and other office expenses to pay, and above all a pension appear more desirable than the freedom and insecurity of the private practice, who number about 5000.

About one in ten of the practising Bar in England and Wales is a Queen's Counsel (QC). These are senior barristers of great experience (usually about fifteen to twenty years) and ability appointed as such by the Queen on the advice of the Lord Chancellor as a formal recognition of their skill and experience who are hired for especially difficult or important legal cases. The QC will concentrate on the advocacy in court cases and on the difficult legal issues involved, leaving the procedural aspects of the case to a junior barrister who will assist him or her. This makes the use of a QC an extremely expensive business, and generally a solicitor will advise the use of a QC only in cases where there is a lot at stake.

8.4.2 What roles do they perform?

Use of solicitors and barristers is generally not found outside the following areas in the context of the sort of work we are discussing:

(1) Opinion work on interpretation and infringement

(2) Infringement litigation work
(3) Licence drafting and negotiation advice
(4) Negotiating settlements of disputes
(5) Liaison with patents, licensing departments
(6) Liaison with overseas law firms

(a) Opinion work on interpretation and infringement

This will be done either by outside solicitors or barristers or by internal employees from either side of the profession. It will form the heart of their day to day work. They will spend this time surveying proposals for legal difficulties, assessing the prospects of success in an action and so on. Where matters such as interpretation of a patent specification are concerned generally a patent agent will act, with a barrister specializing in patents used for a second opinion, and solicitors will not be much involved.

(b) Infringement litigation work

The ground work of collecting evidence together will be undertaken by the solicitor or patent agent or in-house legal adviser, as will instructing counsel to act in court for the firm. Pleadings, i.e. those formal documents needed to institute the case in court will be sent to an outside barrister who may or may not be the same barrister as eventually represents the client in court if the case gets that far.

(c) Licence drafting and negotiation advice

In the context of intellectual property licensing this will often be the most important role of the solicitor or other in-house legal adviser. The company legal adviser will often accompany the rest of a licence negotiation team to help in reaching agreement and in drafting documents to reflect the terms agreed, safeguarding the position of the company by ensuring that the agreement does not contain legal risks or liabilities which the lay team does not recognize and agree to. He or she will examine licence agreements for their compliance with competition laws. He or she may set up joint subsidiary companies for joint licence ventures which limit the liability of the companies involved. He or she may also work on checking that other laws such as food and drug regulations are made known to the R and D team so that they can work to comply with them and so that the company's product to be exported to a particular place is not found to be in breach of local laws. This role is considered in more depth in a later chapter.

(d) Negotiating settlements of disputes

An important role of the patent agent, trade mark agent, solicitor or other adviser will be helping to negotiate settlements of disputes out-of-court. This will involve a mixture of mature legal and business judgment. A lawyer who on being confronted with a dispute involving a long and complex agreement (which he or she may well have drafted or which may have been amended by non-legal staff in such a way as to be ambiguous or inadequate to cover a point) at once recommends litigation does his or her employer no good at all. He or she spends money unnecessarily unless it is clear that no acceptable compromise or settlement of a dispute can be obtained.

All litigation is expensive, because it involves a heavy investment in manpower. Intellectual property litigation can be most expensive of all because it involves in many cases a number of countries, multiplying the legal manpower requirements, because the issue can be very complex, and because the specialism means that the few legal specialists are in great demand and charge high fees. People who have spent this amount want the best and most qualified advisers and are prepared to pay to win. Often it is better to reach a compromise. It is cheaper and quicker and cuts risk. It requires an assessment of the chances of winning the case and the chances of ever recovering the expenditure, the publicity effects of the case continuing, its effect on other potential or existing disputes, business relations, etc.

(e) Liaison with the patents, licensing departments

This will be required whenever there is activity involved with litigation or licensing, to ensure that all the material connected with intellectual property rights in question is at hand and up to date. As part of the negotiating team full communication with other departments in planning the negotiations in advance and preparing drafts suitable to the transaction it is sought to complete so that the team has a document to bargain over will be crucial. This will be considered further in a later chapter.

(f) Liaison with overseas law firms

The lawyer will handle associated problems like the legal contracts for shipping products overseas, and will liaise where necessary with those foreign law firms handling the company's business overseas if patent or other infringement actions have to be brought there or it is necessary to negotiate locally import licences, maintain a local representative (often the case) or check on local laws. As with the patent agent the in-house legal adviser will be expected to work out who to hire overseas and may even have to negotiate with them the appropriate legal fees.

8.4.3 How is a legal department organized?

There is nothing special about the organization of an internal legal department involving intellectual property business; it follows the same rules as any other legal department. The only difference will be that the staff will be expected to be familiar with this area of the law. Management of the department will usually be in the hands ultimately of the company secretary as will the management of a patent/trade mark department if there is one, though in some companies where R and D is the principal activity the patent department might well be under the general supervision of the technical director if that post is sufficiently senior in the company whereas the legal department would never be so placed.

9 Licensing of intellectual property

9.1 INTRODUCTION

This chapter is about what a licence is in law and what from a commercial point of view its advantages and disadvantages are in comparison with other common forms of the exploitation of intellectual property right. It looks at licensing both from the point of view of the potential licensor and from the point of view of the potential licensee. It is not about the terms of a licence, about the peculiarities of the various types of licence which are commonly contracted, about the preparation and negotiation of a licence, selection of the right licensing partner and so on. These matters will be considered separately in later chapters.

9.2 WHAT IS A LICENCE?

A licence is a contractual agreement between the licensor (the owner of the right being licensed) and the licensee (the person who wants permission to do something which would be an infringement of that right if it were done without the permission of the right owner). It is simply an agreement giving one party permission to exploit the other party's rights. The basic form of an intellectual property licence is usually much the same whatever intellectual property right is being licensed, but the details of the licence will differ as between rights and will depend on the sort of right being dealt in (commercial, technical or artistic) as well as the scope of permission being granted and the identity of the licensee.

9.2.1 Licences and assignments

It is important to distinguish a licence from an assignment of rights and to remember that there are different types of licence, each with very different implications for the parties. The assignment of an intellectual property right is a transfer of all the rights that the seller has in something outright to the buyer. When *X* assigns a patent to *Y*, *Y* acquires the right to stop everyone, including *X*, from using that invention. Thereafter *X*, even though he or she once owned the patent, can only exploit it with the consent of *Y*. When you assign a right you lose whatever interest you ever had in it for ever. The assignee becomes the outright owner of it. He or she steps into the shoes of the assignor as it were and becomes fully entitled to deal with the patent as owner just as the original grantee was. To use an analogy, it represents the difference between the sale and purchase of a car outright and the hire of that car from the owner.

By contrast a licence of an intellectual property right allows others to use it but the licensor – the original owner – retains the right itself. The licence is a mere contract which permits the licensee to use the right in some specified way for a specified time, either a limited period or the whole life of the right, depending on what is agreed between the parties. It permits him or her to do things which would without the benefit of the agreement be an infringement of the right. The licensee acquires no property interest as such in the right itself: in other words he or she does not acquire a bit of the patent or trade mark. He or she acquires the right to do what he or she does not through ownership of any intellectual property rights but by virtue of his or her contractual rights against the licensor personally.

In some types of licence the rights of a licensee under the contract may be so great that they approach those of an assignee – but he or she will still only be entitled to them by virtue of the on-going contract, whereas an assignee has them by virtue of being the new owner of the property right, conveyed to him or her as a result of a contract which is now completed and a piece of history.

This gives the licensee less comprehensive rights than the assignee in law. But in practical commercial terms, of course, the licensee's position may be just as strong as that of the assignee and he or she will certainly have as great a commercial interest in the strength of the patent or other intellectual right as if he or she were the legal owner, whereas often the true legal owner who has granted a licence may have little genuine interest in it at all except in gaining the royalties.

Because of this, in some of the statutes, the exclusive licensee is given special procedural advantages in the event of infringement of the right by a third party, which we will look at later in the book. And in some countries the exclusive licensee for the life of the patent or other right is treated for tax purposes as if he or she had bought the entire right and was its owner – because for all practical purposes he or she is.

9.2.2 Types of licence

There are three basic types of licence common in the exploitation of intellectual property rights. The names given to them below are not terms defined in any Act of Parliament, but just convenient labels by which they are commonly known.

(a) Plain (non-exclusive) licence

This is simply a permission to use the invention, trade mark, design, or whatever, which confers no right on the licensee to be the only producer in the field, and which allows the licensor to license the very same rights to as many other people as he or she chooses; in the absence of contrary agreement the licensor may license it to another on whatever terms he or she chooses, or to go into production himself or herself. So the licensee may be in a weak position commercially as well as legally. On the other hand a non-exclusive licence may be more suitable for the licensee because the obligations of performance associated with a sole or exclusive licence may effectively negate the commercial advantages they appear to confer.

Legally, the licence confers on the licensee no rights against anyone other than the licensor – contracts bind only the parties to them and not third parties – and so the licensee will be heavily dependent on the licensor to protect him or her from the adverse effects of third parties infringing the patent or other right which is the subject matter of the licence. Because the law imposes no obligation on the licensor to enforce his or her own rights, the licensee will have to procure agreement in the contract that the licensor will make reasonable efforts to police the right effectively when required to do so by the licensee.

Of course, in practice most non-exclusive licensees are well advised to ensure that express provision is made obliging the owner to take steps to act against third party infringements and to defend the patent or other right from attack, either by eliminating the requirement of royalty payments in case of failure or imposing some financial relief or penalty for failure to take action upon request.

(b) Sole licence

This confers slightly greater rights on the licensee. A sole licence is a contract in which the licensor undertakes to grant no other licences in respect of that patent or other right in the territory in which the present licensee is to be permitted to operate for the duration of that licence, or sometimes in the field in which the licensee is to operate – this may be important where the technology licensed has a wide range of applications. It still permits the licensor to produce and to license outside that field or territory. Again the licensee is very heavily dependent on the co-operation of the licensor in protecting him or her from infringers.

(c) Exclusive licence

This is an agreement under which even the licensor is forbidden by the contract to produce or otherwise to use the patent or other right, let alone to license any other licensee, in the territory and time specified by the agreement. The *Patents Act 1977* in fact does define an exclusive licence of a patent or patent application, as 'a licence from the proprietor of or the applicant for a patent conferring on the licensee, or on him and persons authorised by him, to the exclusion of all other persons (including the proprietor or applicant), any right in respect of the invention to which the patent or application relates'. Thus there may be more than one exclusive licensee, for instance a number of exclusive licensees for a number of separate countries, territories or fields of exploitation.

This of course places the licensee in a position considerably stronger than other types of licensee. Indeed his or her position is substantially similar in many ways to that of the assignee, and the law recognizes this in many cases by giving him or her a right to sue infringers of the right even without the co-operation of the licensor.

It is quite common to find that the lack of involvement of the licensor in a product is such that they might just as well have assigned the right *in toto*, but did not do so for reasons of pride or a desire to be associated with the product in some way, vicariously through the licensee, for reasons of prestige. This may be a good reason for adopting that practice but it should be remembered that the licensor is more likely to be involved in litigation than the assignor. The assignor, who has parted with all his or her interest in the technology, will usually be under no obligations at all and will often not have warranted the validity of the intellectual property right assigned. The licensor may be under obligations which he or she would have left behind as an assignor.

If there is to be no real benefit from the continued ownership of the bare intellectual property right it will often be worth considering a total assignment to leave behind the liabilities and the vulnerability to litigation which ownership implies. On the other hand there will often be cases where the old research and development work has not reached a dead end and where it is worth not assigning the rights because of the potential involved in the on-going R and D programme which continues after the licence has been granted, which may interest the licensor and which may lead him or her to require disclosure of possible improvements with a view to seeking further protection. It is also true that publicly funded R and D work is rarely assigned because of the deemed 'public interest' in the State's own funding bodies retaining ultimate control over the technology licensed.

9.2.3 The licence agreement package

So far we have talked about the licence agreement as if it is just a contract permitting the use of a single right, say, a patent. In fact it is perfectly possible for there to be such a simple agreement, under which *X* permits *Y* to use *X*'s patent for a payment of a royalty. But it is far more usual for the agreement to be very much more complicated than that.

- Firstly, a patent is usually not much use without the associated know-how and technical assistance also being revealed to the licensee. Many patents simply cannot be worked without that know-how. There will usually be a package of rights and services dealt with in the same contract, though some people do prefer to deal with matters in a series of parallel single-right licences. Some people find the package licence less likely to lead to inconsistencies, but it is true that there are advantages in having a series of separate agreements where the bundle of rights licensed operate with a variety of field of use and territorial restrictions. Where a bundle of rights are licensed together the contract is going to have to make provision for the relationship between the various rights dealt with. What happens if one right turns out to be invalid – is the licence to continue in respect of the other rights or not?
- Secondly, a mere permission without guarantees of servicing and technical back-up by the licensor may be something of a pig in a poke for the licensee. And a licence which imposes no

active obligations on the licensor may in some cases be of little value to the licensee, who will want more than a permission to use a property right – he or she will often expect obligations of an active nature from the licensor who may well have to earn his or her royalties.

- Thirdly, in more sophisticated licences there may often be a degree of reciprocity, involving mutual obligations to report improvements to the product or process, support in litigation against infringers, or cross-licensing of compatible technology (i.e. the licensing by *A* of technology to *B* in return for licensing to *A* of some of *B*'s technology). These may become extremely complicated and unexpected eventualities provided for – what happens if a rival of the licensor takes over the licensee? Usually one of the provisions in the licence for termination of the agreement will allow the parties to terminate the licence on any change of ownership of the licensed right or of the companies to the agreement.

So, in sum, an intellectual property licence may be a very complicated document even where no long-term collaboration is envisaged, and it will be the product of long negotiations and much professional advice. It is essential to emphasize very strongly the legal implications of a licence agreement, for the perceptions of the businessman or R and D specialist on this matter often differ markedly from those of the lawyer or company secretary, and much mutual misunderstanding results.

Equally, the administrative and personal implications cannot be forgotten. Licences need to be administered as much as any long-term contractual agreement, not just have the dust blown off them at royalty payment dates. And much of the successful working of the contract will depend on personal goodwill and compatibility between those intimately involved, a point we will return to later.

9.3 LEGAL IMPLICATIONS OF A LICENCE AGREEMENT

A licence agreement is a contract. A contract is an agreement which the law regards as imposing binding obligations on the parties to it. Breach of the terms of such an agreement by one of the parties will either entitle the other party to treat himself or herself as discharged from his or her obligations under the agreement and to claim damages for the loss he or she has suffered, plus the costs of going to court to establish his or her rights; or in the less serious cases where the term broken is not an important one or was broken in an unimportant way causing no great loss or inconvenience to the other side, will entitle him or her to compensation for any loss or, where appropriate, loss of profit, he or she may have suffered as a result of the breach plus the costs of recovering that loss by a legal action in the courts.

People often talk about going to court to enforce a contract, but it is very rare for a court to be able or willing to order a person actually to perform the obligations he or she undertook under the agreement (what the lawyers call specific performance) as opposed merely to ordering him or her to pay damages for wrongfully refusing to do so. And if he or she can show that his or her refusal to perform was justified by your failure in some material respect to perform your side of the bargain, he or she may even be excused performance.

Usually, the question of fault does not come into these issues. If you fail to perform your side of the bargain, even if your failure is due to circumstances beyond your control, you will be liable to pay damages to the other side for the loss they have suffered. Only in the highly exceptional case where the fundamental object of the contract has been frustrated by events outside the control and contemplation of the parties such as the nationalization of the assets concerned or supervening illegality of the objects of the contract, will the parties be released from their obligations by the law.

Of course, it is quite possible for the parties to make express provision in the agreement itself for certain circumstances to excuse non-performance. For instance the agreement might provide that whereas a licensee will be obliged to maintain a certain level of production or to guarantee a certain minimum royalty, that he or she will not be so obliged where industrial action interrupts the supply of raw materials and imports rise above the cost of domestic supplies.

9.3.1 Formation and application of licences

A contract becomes binding on the parties to it when one party makes a definite offer of a set of terms by which he or she says he or she is willing to be bound, which the other accepts without qualification. Both sides must expressly or impliedly consent to these terms, but the law looks only at objective appearances. A mere secret intent not to be bound by what you appear expressly or impliedly to agree to will be immaterial – you will be bound.

It is quite possible to make a contract orally and without any formality, though of course in practice the only sensible course of action in this field is to make contracts in writing with all of the agreed terms being expressly set out in detail, so that there can be no argument afterwards as to what the terms were or as to the proof of them. Until the offer has been accepted, the party who made it is quite free to withdraw it if he or she wishes, by giving notice of the withdrawal to the other side. But once the contract is concluded by this offer and acceptance, neither side can withdraw without the consent of the other.

The lawyers distinguish agreement by an offer and an acceptance from what they call invitations to treat. These are proposals or 'offers' in a non-technical sense which are not intended by the parties to be binding, but which are negotiation points. A number of these suggested terms and points will be made in the course of the negotiation process, gradually becoming more and more precisely defined and concrete until they begin to resemble the heads of agreement which will finally be the terms offered and accepted.

The process of moving from invitation to treat to the offer and acceptance is often almost imperceptible. In fact, usually, neither side intends to be bound until the terms are in writing and both have signed the documents so that even very precise terms agreed orally or in the form of a draft document will not bind the parties if they understand that only on signing is the agreement to be concluded. These preliminary documents are often called letters of intent.

In addition to agreement, to be binding a contract must (unless it is one made by deed under seal) bring in consideration from each side: that is each side must 'give' or promise to 'give' something in return for the other's 'giving'. In the usual case the consideration by the licensor will be the promise to grant permission to the licensee to use the right; the consideration from the licensee will be the promise to pay a royalty payment, though it might equally be a cross licence. Without such consideration coming from both sides the agreement does not bind either of them. If one side does not fully comply with his or her promise – for instance, the licensed process does not work as specified or the licensee does not perform an obligation to achieve a certain output – he or she will be in breach of contract, with the consequences already outlined.

It may also be possible to escape a contractual obligation where although the other party has not broken the terms of the contract, you were induced to contract, i.e. to take out a licence or grant a licence by misrepresentations of fact which materially affected your decision to grant or take the licence. For instance you might have been induced to contract by a representation that the licensor had a staff of twenty back-up engineers on call where the process went wrong, or by a representation that the potential licensee had a much larger turnover or production capacity than he or she in fact has, which misleads you over the potential royalties from granting him or her a licence rather than one of his or her competitors.

Even some wholly innocent misrepresentations can enable the side misled by them to escape the contract if they were material, but if the misrepresentations are negligent or fraudulent, the persons misled can claim damages for any financial loss suffered as a result of reliance on them. And again the courts will reserve the discretion to refuse you the right to escape the contract where you have been misled by a misrepresentation unless it is a fraudulent one, and to award you damages alone while holding you to the terms of the agreement. Whether they will do so depends on how serious the misrepresentation is.

It must be emphasized that the misrepresentation must be one of fact, not of law or of opinion or of future intentions unless those intentions are not held at the time so that the person is misrepresenting the state of his or her mind. Thus a mis-statement about the anticipated sales figures of a process or of its intrinsic merit would not give rise to liability, unless it implied that, for instance, market research had been done to support that estimate, whereas a mis-statement about actual existing sales levels would clearly give rise to liability if it was relied on by the other side.

In very exceptional circumstances a fundamental mistake made by him or her for which the other party was not responsible may relieve him or her. If for some reason the contract was literally impossible to perform without fault on either side, perhaps because it was impossible to obtain supplies of a particular substance, for instance, it may sometimes be possible to escape the contract on the basis that it had been entered into under a fundamental mistake of fact (not one of law). But the courts are extremely unwilling to accept this plea if one party seeks to escape the agreement later, and it very rarely succeeds.

A licence will usually not be assignable by the licensee without the permission of the licensor: the rights conferred by a licence are purely personal and cannot benefit or bind third parties. Even if the contract between *A* and *B* provides that *C* shall be entitled to a certain benefit, *C* cannot enforce the agreement to that effect. Care should be taken where dealing with multi-national corporations, where the holding company may wish to insist that it be entitled to exploit the licence through its associates and affiliates.

9.4 OPTIONS TO GRANT OR ACQUIRE A LICENCE

At this point it is probably a good idea to emphasize the difference between a concluded licence agreement and mere agreement to agree or an option. These so-called agreements are often encountered on the way to a licence agreement.

9.4.1 Letters of intent

A so-called agreement to agree usually has no legal significance at all. It may mean that the parties have reached a stage of consensus where they have expressed their willingness to and intention to reach a concluded contract, without actually having done so. That in itself has no legal effect. Even if there was a genuine contract to enter a contract, in the absence of fixed terms in that putative later agreement it would be very difficult to work out what a party had lost by the other party refusing to go on and enter it. Merely to express an intent to accept an offer at some time in the future is not to accept it and is of no legal effect.

That being said, it is not intended to suggest that, for example, so-called letters of intent are of no value. They may be a very good thing. They are often employed as an intermediate stage between a preliminary confidential disclosure of the outlines of a new piece of technology and the conclusion of an option to acquire a licence in it. A general summary of the common ground so far achieved in the early stages of the negotiations and record of what has been conceded by each side undoubtedly provides some tangible signs of progress and must help on the way to settling the smaller or more technical points. The point to make is not to overvalue or misinterpret the significance of these documents.

As a precaution some firms when conducting their negotiations by letter for a potential licence always head letters or draft contracts 'subject to contract' to ensure that they do not appear to constitute an offer or an acceptance of an offer: in other words, the magic formula you often come across when moving house that indicates that whatever the other side may think you are quite clear that you are not contractually bound by the terms contained in the letter even though at some time you fully intend to become bound.

9.4.2 Options

People sometimes say that they have acquired an option to a patent licence. An option of what? Unless the terms of the proposed licence have been fully worked out in advance on paper and are clear to both sides there will be nothing worth having an option in. Even if there is such an agreement the party granting the option will be fully able to withdraw it unless the party granted it has paid something for it so as to provide consideration for the promise to keep it open.

But it is quite possible to grant an option to take out a licence. This will be a full contract under which one party pays the other money for a right at some time in the future to take out a licence on the terms agreed in the option agreement. It is in a sense buying the right to prevent the offeror from withdrawing his or her offer to give (or take) a licence. Such an option will often be part of a preliminary disclosure agreement where a limited inspection of the technology with a view to taking out a licence is agreed.

The significance of this type of agreement should not be underestimated. Often a prospective licensee wishes to examine the technology at length, do his or her own product development, seek regulatory clearance (e.g. for pharmaceuticals and medical products) or develop original complementary technology of his or her own. This requires considerable investment and the prospective licensee needs to be sure that he or she will, if he or she wants, have a licence at the end of the day. Considerable fees may change hands for an option, e.g. upwards of $1 000 000 for a six months option, to which will be attached the terms of the licence which the purchaser of the option has the opportunity to take up.

9.4.3 Licences in applications or future patents

In this context it may be worth mentioning the practice of taking assignments or licences in patent applications or applications for other rights. There is nothing to stop a company from contracting for the right to use a process in respect of which the licensor has not yet obtained a patent but is in the process of applying for one. Nor is it objectionable to assign a patent application. For this reason it is quite possible in the licence of a subsisting patent for the licensor to contract to grant a licence in any future improvement patents granted to the licensor in respect of the same invention, and such a term may be quite common (but see Chapter 14).

This leads us to another important point: the very different nature of the various legal rights with which we are dealing in this field, which may pose some very different problems.

9.5 VARIATIONS IN LICENCE FORMAT

Business people often complain that their lawyer will not give them a simple standard form contract to carry around with them for licensing deals; apart, perhaps, from such consumer agreements as the computer software licences of business software marketed for mass sales.

You simply cannot do so. Each transaction is unique. It is very rare to be able to use precisely the same agreement draft for every contract you make to license a particular product, unless there are very good reasons for doing so and you are in a sufficiently strong position to be able to impose the same terms on everyone – as for instance you might be if you were in charge of a franchise store business, dealing with individual franchises. A patent licence contains terms wholly inappropriate to a copyright licence or even a patent licence for a different type of invention; a licence to a small manufacturer may contain terms inappropriate for the large manufacturer.

Nevertheless, it is true that many licences are made up of a combination of what are relatively standard form clauses and it is possible for a person skilled in the art to knock together such draft licences from an appropriate combination of these basic clauses. We will be looking at some of these in a later chapter. Many lawyers' precedents books do exist with standard clauses. You are supposed to draft your own contract and then compare that with the models to see if any obvious defects exist. Inevitably the uninventive and not so good draftsmen and draftswomen in the profession do tend to work the opposite way around.

9.6 THE ROLE OF THE LICENCE AGREEMENT

It also has to be emphasized that a mere licence agreement is not the be-all and end-all of the final arrangement. It will rarely be satisfactory for both parties to work strictly to the terms of the agreement. The contract provides a safety net for when things go wrong. No one needs a contract while the parties are in amicable relations and things are going well. Rather, a contract provides the minimum standards for when they are not going so well, in effect to minimize the losses which each side is likely to suffer and to provide for the shares of the profits from the transaction and the smooth running of the relationship. Thus the successful implementation of the agreement requires a spirit of co-operation and goodwill which stretches beyond that required by the strict letter of the contract.

From the point of view of the business person, the contract performs three very important functions. It records for the future the basic ground of the relationship between the parties; it provides a yardstick by which disputes may be resolved; and it forms the basis for the day-to-day administration and management of the relationship. Many disputes can be nipped in the bud if one side can point to the contract and say, well, after all, we did agree on this at the beginning.

Equally, what the business people agree and expect will rarely be adequately precise for the lawyers who have to draft this safety net. There is here a fundamental communications blockage.

Few lawyers claim to fully understand technology and business management and few business people understand law (especially those who have had some quasi-legal training!) And in many cases the business people and the lawyers are dealing with technology which neither understand. Thus if what the lawyers demand by way of greater explanation and definition seems at times pernickity or unreasonable in seeking to tie up every eventuality, however unlikely, one must remember that they are there to protect one's interests. The detailed clauses are not put in there for their own amusement or to inflate their fees.

Lawyers are born or trained pessimists. They look at what could go wrong rather than what they hope will go right. And from their viewpoint in law prevention is always cheaper than cure. If the clause of a licence contract covers the event clearly, uncertainty and the risk of litigation is diminished. Uncertainty is the great consumer of funds in the legal world. Any loose definition of obligations in the contract is the cause of uncertainty. With a tightly drawn-up contract both sides know where they stand. Moral: professional advice is cheaper in the long run. Do not stop the lawyers from saving you from yourself. On the other hand, remember that the agreement is only a business tool: the lawyers should be kept in their place and you the business person must take the business decision, albeit on advice from professional advisers.

9.7 TO LICENSE OR NOT TO LICENSE?

Licensing is just one of the ways of exploiting an invention or other right. Do not think of it as the best way simply because in some senses it is the easiest. Assuming that you have the capital and the professional resources behind you other means of exploitation will almost always prove more profitable in the long run.

What licensing does is to provide a means of profiting where some of these desirable resources and capital are lacking. The more people you split the risks of exploitation among, the more people take a share of any profit away from you. No-one is going to take a licence from you which does not give them a fair share of the cake. No licensor is going to give away a licence for little or no return.

But in many cases there will be no option but to license out. If a small R and D based 'hi-tech' company in, say, the UK, were to make a product for worldwide distribution then even on the best possible outlook for a very successful product it would suffer from sheer over-stretching of its financial, marketing and production facilities.

Let us look at some of the considerations which are present when deciding whether to enter a licence arrangement. Of course, different factors affect the potential licensee and the potential licensor, so they should be looked at in turn.

9.7.1 The licensee

A company decides to seek a licence in technology for one of a number of reasons.

(1) There may be a gap in its product range or a loss of competitiveness because of neglect or a development by a competitor creates such a gap and the time factors involved make it cheaper to license in than develop in-house rivals.

(2) An excess of production or sales capacity, a desire to diversify out of a product line which is either producing insufficiently attractive returns on investment or appears threatened by public health legislation, political factors, and so on, may create an environment in which licensing-in becomes attractive.

The underlying question is a simple one: is it economically more efficient for the company to buy in the technical expertise or marketing prestige under consideration than to generate it internally? It should be emphasized that licensing-in covers a wide range of possible activities:

(1) The licensee may want simply to obtain technology or the right to operate a development of his or her own which appears to infringe another's patent or design. The scope of this may vary from an individual component to an entire spread of technical developments involving a large package of related patents. The source of the package may be one market leader or a number of different suppliers. The licensee may take one licence but require a licence from a third party (or for that matter more than one) to manufacture even a specific component. The licence may even be for a whole plant – a so-called turn key licence – plus expert know-how on how to run it efficiently.

(2) It may simply be a desire to gain a share of the market rapidly by using a well-known and well-respected trade name.

Now, of course, licensing-in is not the only way in which to acquire a new product or process. It is quite possible to diversify or expand or develop through internally generated resources, given time and money. A number of factors may be present which cause you to lean in favour of licensing-in technology as opposed to pursuing its internal development.

- Firstly, there is the question of the cost of producing the same technology internally. The cost of R and D work may be very high, and the evidence is that as much as 85% of R and D projects end in failure, either technically, commercially or merely through remaining unused through corporate inertia or financial deficiencies. In a high-cost research field or in a highly competitive industry, licensing in may often represent better value for money. In the pharmaceutical field, which is admittedly an extreme example, it is said to cost anything up to £50 000 000 to develop a new drug and assuming that drug to be patentable and patented as much as the first third of its patent life will be consumed in pre-launch tests to meet the requirements of public safety legislation. Often more full Food and Drugs legislation approval will require a minimum of seven years tests and thereafter on-going monitoring. Similar problems exist, for instance, in the case of biomars for animal foodstuffs, albeit with a shorter lead in time.
- Secondly, licensing shortcuts the time lag into production and eliminates the risk capital which has to be devoted to the project without any guarantee of even long-term returns. Furthermore, it is often extremely difficult to estimate and budget for R and D costs. Such costs tend to snowball, especially where so much has already been spent in making some progress that the company is reluctant to throw it all away lest the breakthrough be only a few months ahead. Licensing substantially eliminates this difficulty. It is comparatively easy, once the process or product is there, to estimate the costs of production, including the cost of taking the licence for the technology.

Thus a licence in provides a degree of certainty lacking in a home-grown product or process. The impact of this factor is of course one which will always vary according to the breadth of the research base in the company in that particular field or in total. Where the company has little or no existing research base, the cost of setting up the project in both staff and equipment will be proportionately higher than for a company which has a large on-going research activity and which is merely adding a further related project to its list of current work. Many companies simply have to license in because they cannot afford to maintain an R and D operation at all.

- Thirdly, there is the question of the time it will take to develop a competitive product internally. Where the industry moves quickly it may be essential to eliminate that development time to stay in the race. If R and D work is carried on within an institution it may be better to concentrate all resources on one area of development and buy in other areas, so that all efforts are directed to the rapid development of marketable technology which can be exploited to pay for the buying in of other outside technology. In that way the firm remains up with the field in all areas but does not suffer too great a deficit in the technology licensing account.

Of course, the taking of a licence is not going to eliminate altogether the requirement for this R and D work even in relation to the licensed product. It is often the case that the licensed product will have to be redesigned to some extent to fit in with the rest of the product range of the licensee. But the basic design work will have been done and the adaptation of the new product will be a task involving the internal technology with which the internal R and D staff will be familiar. Put another way, the licensing-in policy may be adopted to enable a firm to change the emphasis of its R and D work so as to concentrate on product/process development rather than on more fundamental research.

- Fourthly, and related to the two factors we have just mentioned, licensing in gives you a flexibility in meeting changing market patterns. From the point of view of the firm licensing in this ensures a rapid movement to keep up with competitors when public taste changes, and where the practice of freely available licences is present in an industry it ensures that a rapid diffusion of new technology throughout that industry keeps the industry as a whole competitive with the foreign manufacturers. In some rapidly developing industries like microelectronics, by the time you have caught up with a development it is obsolete and a new one rules the day – licensing in becomes essential to survival and to miss a step is to go out of business.

The adoption of a licensing-in policy on this basis is not all pluses, however. Some companies rely very heavily on technology developed by their suppliers for much of their R and D. If a major supplier ceases to trade or moves out of the field the company must then seek alternative sources of technology for future product development, which in a closed market may be difficult.

- Fifthly, licensing in may also enable you to diversify your product or service range and thereby develop a more balanced business operation able to weather the storms of fluctuating markets. If one area of technology is in a trough the other may be on a rise. There may be excess manufacturing capacity in the licensee's factory where his or her share of the market in what he or she already produces is insufficient to keep up full capacity working. Unused capacity is a waste of money, where a new product or process can be marketed by the existing sale force. If it could be applied to the manufacture of some other product for which there is an adequate demand, it may be desirable to buy in that product so as to eliminate development time and lack of expertise in that new field, so as to exploit that capacity and make money out of it. So licensing in can be the tool of an aggressive expansionist company, not merely the retreat of a company in a defensive position, struggling to keep up with the field.
- Sixthly, the licensing-in process reduces risk in the market place for the producer. In some industries only a tried and tested product will sell. In the pharmaceutical industry, for instance, the new product will be subjected to many long testing procedures in each of the major Western countries. It may be cheaper and quicker to fill a gap in the existing product range to license in a competitor's version than to try to develop your own, especially given the difficulties of market penetration once a new drug is established in the minds of the medical profession and its particular characteristics and contra-indications are familiar to them. Especially if resources are slender you cannot afford to make a mistake.

Licensing in of the right sort of product gives you a ready-tested product or process with, hopefully, the teething troubles eliminated, a good picture of the potential market in advance; although in many cases you will still need a separate licence from the local regulatory authorities, much of the authenticated test material will already be available as part of the licence package. With the product will often come the additional market protection of a trade mark or patent, and usually the benefit of continuing development of the product by the licensor will be thrown in too. Since payment will as often as not be made by royalty on sales, the price varies with the success of the venture, and is paid out of income. It is even possible in some cases to test the product out in advance by buying and reselling instead of going straight into production under licence.

In sum, then, the great advantages of licensing in can be the more rapid entry into a market; a possible improvement in the product range of the company, thus filling gaps which have appeared through obsolescence of home-produced items; a more flexible response to the changes in the market; a reduction in the risk capital expenditure and an improvement of cash flow; and a good means of rapid growth for a small or medium company or for the larger company a way of filling spare capacity. A profit–cost profile comparison of the R and D venture developed in house and a licensed-in product might well appear in Fig. 6.

But licensing in, as we have already said, is not all pluses. There are a number of disadvantages to any licensing in and a number of obstacles to its success.

(1) It may turn out to cost a considerable sum to adapt the working of the existing staff and plant to the newly licensed-in technology, costs which are often underestimated.
(2) The product which it is wished to license in, or more often the process, may be incompatible with the efficient production of existing lines and that in turn may lead to a decision between the new and the old.
(3) The time lapse between the initial decision to seek a licence and getting into production may

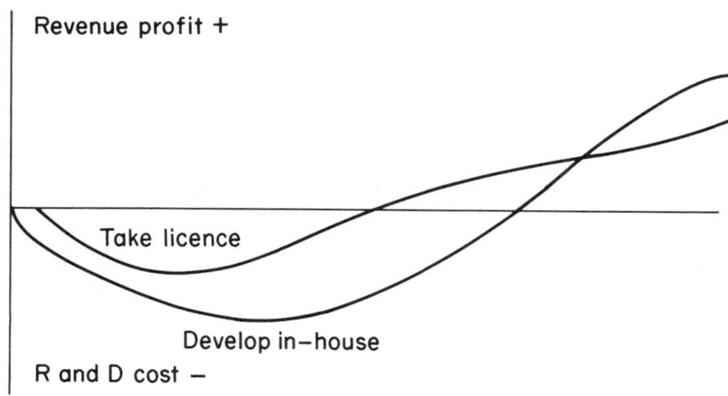

Figure 6 Cash flow profile: in-house against licensed-in

actually be very nearly as long as the process of reinventing in house if the initial negotiations do not proceed smoothly.

(4) To license in may in some cases deprive the licensee of the benefits of the technical expertise which he or she would have acquired through development in house and the fairly superficial level of know-how needed merely to work the process as opposed to developing it may leave him or her in a vulnerable position, over-dependent on the licensor for technical assistance both in the maintenance of the licensed technology and in the development of it further. The new technology base licensed-in may be a rather fragile or artificial one.

(5) It may well be impossible to take out a licence except for a very high royalty if the technology is very new and advantageous, and the level of royalty may even increase if the licensee's dependence becomes too great.

So licensing in should not be seen as an easy way into a market. It has its price just like the internal investment in R and D work. And although there may be a greater degree of certainty in the technology package that you are buying, it is still the case that part of the price is for something much less certain: firstly, the quality of the technical back-up which the licensor is to provide; and secondly, the strength of the patent or other right, if one exists, which protects the invention and which is a commodity dealt in in the licence. Often you are paying for a right to do something which the generality of manufacturers cannot do. The price for that market lead may be heavy, but the strength of the patent, the ease with which competitors may circumvent it, uncertain, particularly where the patent is relatively new.

Licensing in has to be part of an overall policy. The first essential is to decide what the objectives of the company are in development of new products or new processes or expertise. How does the proposed licence fit in to the general policy of the company; how is it to fit in with the general budgeting of the company? We will return to this issue later. There are also a number of obstacles to licensing which are often underestimated.

- The first is the well-known Not Invented Here (NIH) syndrome, the mistrust of externally generated material, of which much is said. How real a factor it is is difficult to say. There may well be difficulties of adjustment to dealing with strangers experienced by the internal administration or R and D departments of a company less familiar with dealing with the outside world than, say, the marketing departments. A company with faith in itself may naturally have greater regard for what it has produced itself than for what its competitors have produced. NIH syndrome frequently has its roots in the boardroom, where the R and D director would feel politically disadvantaged if he or she had to recommend inward licensing too frequently when at the same time constantly seeking increased funding for his or her own department.

- Secondly, the prestige of the R and D department in the licensee company may be felt to suffer from the apparent necessity to get others to do its job for it. This can be a difficult obstacle to overcome. Yet it is not really a rational one. An R and D department cannot do everything, and it is not an admission of failure to take in other people's technology. Many of the most advanced and dynamic companies in the world – for instance in the computer field – are heavily dependent on other people's as well as their own R and D department's output.
- Thirdly, with a very well-known licensor who is also in production in the same line, the reputation of the right owner may be such that the licensee is merely considered an inferior imitator, and much of the benefit of the licence will be lost to him or her with the bulk of sales still coming from the licensor, so that the licensee gains only the crumbs from the rich person's table, the excess over the capacity of the licensor, which may render the investment necessary to start up production something of a waste of money and energy.
- Fourthly, and this particularly affects smaller firms, experience indicates that a major obstacle to licensing in is simply ignorance about a number of important matters. These can be roughly enumerated as follows.

(1) Ignorance of the availability of the technology suitable for exploitation by the firm and available for licensing in, and of the sources of information about it. This particular problem is looked at later on in the book.
(2) Ignorance of licensing *per se* on the part of the firm and its professional advisers. This extends to ignorance of matters such as how to go about pricing a licence, for what period, and so on, questions for which there are no fixed answers. These problems will be looked at later. They can of course be mitigated to some extent through the hiring in of consultants, but that in itself may be an expensive and traumatic step to take and in any case there are not very many such persons around.
(3) Lack of management time to negotiate what may be drawn-out negotiations. The burdens of concluding a successful deal of this nature may be considerable, particularly for the inexperienced person who already has many duties to perform for the company.
(4) Unattractiveness to a large licensor. The smaller company may be unable to guarantee a large enough turnover for the licence to be attractive enough to the large licensor who will have to provide expensive staff in a back-up role, particularly where the small licensee is new to the field of technology concerned.

- All of these problems are exacerbated where the other party to a potential licence agreement is based overseas – even the elementary knowledge picked up in the home market will be of little use in many of these cases, and the many cultural and business practice differences beween the two sides may make a successful negotiation all the more difficult, not least because of a lack of knowledge of local conditions, permissions for licences, taxation regimes and so on.

9.7.2 The licensor

If anything there are even more variables in the licensing-out decision than in licensing in. Why does a licensor want to license out the technology or market reputation which he or she has so carefully built up?

- Licensing out at first sight merely appears to create a competitor when none existed before, by giving another company a lift up to the technology which before he or she lacked. This is of course true. Any licence does immediately create a potential competitor, and attempts to preserve one part of the market to yourself are nowadays often dangerous and ineffective because of competition laws in the EEC and USA. But licensing can actually be a way of controlling the market to some extent. It is a form of co-operation. It can lead to cross licensing and closer relationships, including joint ventures which are considered later.

- Licensing out can be a way of penetrating a new market which for various reasons it is otherwise very difficult to penetrate oneself, perhaps because of the existing competition in that market which would have the resources to squeeze you out.

This may be particularly true where overseas trading activity is considered. There may be financial benefits in licensing a local manufacturer, perhaps through the existence of tariff barriers or import controls, restrictions on foreign ownership of companies, or simply in the low cost of local production raw materials and wages.

On the other hand there may be restrictions on the level or currency of royalty payments or the weakness of the intellectual property regime may make a licence almost useless and valueless. Not only may the country to which you would like to export be in a delicate political position where setting up production yourself locally would be dangerous or difficult, but direct export may be disadvantageous from a tax point of view.

Equally, it may be that local consumers prefer to buy from local manufacturers rather than importers. The licensee may well be more able to respond quickly to changing consumer demands or local circumstances than a distant exporter. Any problems with labour forces or new legislation, internal difficulties with supplies or simply the political situation are somebody else's to deal with and they can in any case probably deal with them better on the spot than you can. Your royalties suffer, perhaps, but they are only one source of income from one place, and no substantial capital resources have been devoted to the enterprise.

So it may be better to take slightly lower profits from royalty payments rather than to undertake the risks inherent in export or self production. The problem may simply be that the nature of the technology or product is one which it is difficult to transport, especially where it is heavy machinery. The market may be a long way away and the maintenance and support facilities necessary for servicing of the product may be such that it is impossible for the licensor to provide them by himself or herself so that it is essential to provide a local licensee who will have a greater incentive to provide adequate service facilities than a mere local agent employed for service contract work or a mere local distributor.

Overseas licensing, particularly in less developed countries, can be a way of extending the income-producing life of technology which is obsolete in Western Europe – see for instance the licence to embryonic car industries in Eastern Europe and the Third World of the parts and know-how to make an old model such as the Volkswagen or Fiat. It may even be that with their lower production costs such products remain internationally competitive with newer but more expensive models and thus extend the life cycle of the technology licensed in Western Europe markets too.

- Licensing out can be a way of satisfying a market demand which self-production facilities cannot meet, and thereby making sales and profits which would not otherwise have been made. A small company with limited production capacity may license other companies to produce so as to meet that demand. With a product which is a world beater or at least very successful internationally the prospect of increasing capacity to meet demand would be expensive even if possible and a daunting prospect too if the market fluctuates markedly or the long-term future of the development is not assured because, as in the computer industry, for instance, the newer developments are relatively short lived in their technical lead.

- Licensing out can reduce the necessity of capital expenditure on plant and therefore improve cash flow, and can reduce the risks of production of a new product or a new market, or concentrate production in other fields, leaving manufacture in this field entirely to others. A licence may be a way of testing a new market, the size of which you are uncertain of, without committing yourself too heavily.

- The licensing process may in some cases of itself increase the tendency away from self production. So a particularly fertile company may gradually run down its own production facilities and switch resources into R and D because the income from licensing turns out to give a higher profit margin than the investment in plant and sales resources. R and D licensing can become a complete substitute for rather than an addition to manufacture on a commercial scale. Licensing can fund on-going research in the same or other areas, if the returns are properly calculated.
- Alternatively, a large company may come up with a marketable piece of technology but a piece for which the market is not big enough to justify a switch in resources or whose scale is just too small for the vast production facilities of that company to produce. It may be just right for a small licensee company. Some of the larger corporations have quite a bit of technology on the shelf for which they have no immediate use themselves or which is just too small beer for them but which they may profit from by licensing it out. To take one example an employee may have come up with an invention which the employer owns but which although marketable is outside the mainstream of the employer's business. A licence out is the ideal way of making money out of it without disrupting the business too much.
- Some companies will licence a product to their smaller subsidiary in order to make that subsidiary a more marketable commodity when they are thinking of selling it off, or to improve its profitability. The licence might instead be a sweetener to help sell other products or services which are necessary in the licensed process and which you supply also, creating a demand for the product by licensing the technology which requires it.
- Licensing may be done involuntarily or at least reluctantly. It may be that you would prefer to produce or keep the technology for yourself, but that it is the only way of persuading another company to license something to you – a cross licence situation. Cross licensing is often a half-way house between straight licensing activity and a collaborative joint venture and in fact that is how many joint ventures start off. Usually the parties are relatively equal in standing and contribute pretty equally. The cross licensing device enables them to spread risk and to concentrate on their own part of the project without exhausting staff resources, making best use of each other's expertise in complementary technological fields. The cross licensing provides a cheap way of developing a broader-based expertise and between them the two firms strengthen their hand against other rivals.

Of course the success of such licensing is heavily dependent on the trust and mutual honesty of the two parties. It is very easy to get one's fingers burnt. Alternatively you may be forced into giving (or for that matter taking) a licence as a means of settling a potentially damaging infringement action. This may be essential where the challenge appears quite strong or you simply do not have the money to fight it with. The licence cuts your losses and those of the licensee. An enemy becomes an ally, in whose interests the validity of the patent may well now be.

- Finally, you may be forced to license out by the law itself or by the threat of the law, whether directly under the compulsory licence provisions which exist in some of the intellectual property regimes or because of the complaints of rivals of abuse of dominant position in the market by the right holder. Or you have to cut the cost of maintaining the right in existence by declaring the right to be available for licences of right, i.e. anyone who applies will be given one on reasonable terms. This reduces the cost of renewal fees for patents and designs.

Thus, in sum, licensing produces income without capital investment and with a reduction of risk. It expands artificially the markets in which one can trade and therefore hopefully increases the turnover over capital asset ratio of the company. But again licensing out is not all pluses. There are a number of countervailing factors to consider before deciding on this particular line of activity.

- The licence may either deprive you of a monopoly in production by setting up a competitor, or indeed even exclude you from the market altogether if it is an exclusive licence.
- Apart from a minimum royalty clause in the licence or an express obligation to achieve a certain level of turnover, there is no guarantee that the licensee will actually do anything profitable with the licence and so make profits for you.
- Servicing the licence may be quite an expensive venture and if you are a small outfit it may deprive you of senior technical staff for considerable periods, especially at the start of a licence venture, staff who you need yourself and who you cannot do without, even if instead of seconding your person to the licensee the licensee seconds one of his staff to you to pick up the requisite expertise. The demands of the licensee may effectively swallow up all the time of these people and disrupt your own activities.
- Furthermore, the licensee is going to want to be sure that you police patent or other rights adequately. The licensee doesn't want to pay for protection which he or she is not getting. Fighting a patent action on his or her behalf as well as your own may be a costly activity.
- There may be promotion costs too. The expense of setting up the licence in the early days can be very expensive and the return long delayed.
- The very relationship with a trading partner makes you more vulnerable to attack under competition laws or anti-trust laws as they are sometimes called. We will be looking at these matters later. The competition laws of a country may restrict your joint scope of freedom of action in relation to pricing policy and the like in a way not possible if the two parties were acting separately, perhaps as supplier and retailer.
- To some extent you lose control over marketing of the product in that country, particularly compared with the alternatives of merger or subsidiary. And however careful you are to draft in quality control checks and the like you do to some extent put your reputation in the hands of the licensee, especially where the rights licensed are or include rights in trading reputations and names etc. This makes an unfavourable comparison with the rights and powers you enjoy in owning a local subsidiary.
- Closely related to this is the divorce from the local market place. Instead of a direct feedback from consumers of the technology you receive information, if at all, through the local licensee, which is definitely second best. Again it is possible for the local licensee to react quickly, but if he or she does not or if the chain of communication between the parties is weak it may be much more difficult to keep ahead of the rest of the field in the development of the product or process.
- You will also lack the benefit of developing a personal reputation in the market unless you take the risk of licensing rights in your trade name and insist on the marketing of the goods or services under that name, which in some countries can expose you to a much enhanced risk of product liability charges and the like if the licensee is not all he or she might be. This may be very important where the licence is regarded by you as an interim measure before penetrating the market yourself, as where the licence is a test run for the product.
- Finally, a small company will face many of the difficulties already referred to in the discussion of obstacles to licensing in, when considering a licence out. Individuals who develop a potentially marketable idea are of course almost totally dependent on the licensing out process to make any money out of the idea, having no production resources of their own. But such people are usually worst placed to provide the technology for a licence out. The reason is that there will have been little product development and little surrounding know-how, which requires the very same facilities which the potential licensor lacks. Without a very strong right like a solid patent the individual is in a very difficult position in negotiating a licence with a large firm.

Similar problems can afflict the university or other academic institution, the contract research establishment, Government research laboratories and public service laboratories. The firm alone has the capacity to undertake and control production of the product and the position of the licensor in these cases is extremely vulnerable to the bad faith of the firm with which he or she has become involved. Policing their activities will be almost impossible. It is often better to grant a number of non-exclusive licences in this case if possible rather than to be tied up with one firm because it is possible then to let each police the other to some extent. But often the initial bargaining power of the first firm you come to is such that a non-exclusive arrangement will not be taken up unless the idea in question is so fundamentally an advance as to be indispensable to any firm in the business.

Thus a number of factors have to be considered before entering into the licensing business, especially when considering overseas business. Most of these factors are going to be relevant whatever the mode of exploitation, and added factors will be taken into account when, for instance, joint ventures or other more complex forms of exploitation are contemplated.

These factors will not come in for consideration just once in the life cycle of any given product or process developed in house. The chance to and therefore decision whether to license may arise right at the very beginning when the idea for the product is first conceived in house. Thus if the product idea seems to be commercially promising in the abstract but the firm has no expertise in that market and has no wish to commit resources to its development, as might often be the case where the idea is an incidental and accidental spin-off from some other project, it might consider licensing. The problem will be that with such an undefined piece of technology the cost and effort of finding a buyer may be such that it is not viable. The employment of a register of technology or a broker might be a low-cost option, but the chances of success may not be too high.

If the idea is developed in house it may become apparent that for many reasons its in-house production is not a viable prospect. At this stage, with some R and D work behind it the product may be much more easily identifiable and matchable with a prospective buyer whose production facilities, cash resources and market experience make it viable for him or her. The profit level on the product at this stage may be much higher – but then so are the initial costs of developing it at this stage and the direct cost of the initial discoveries may have been virtually zero in the case of accidental, spin-off inventions.

After this stage licensing of a newly developed product will be extremely unlikely if the developer has gone into production until near the end of the product life cycle unless one of the reasons mentioned above prompts this such as inadequate production base, cash flow difficulties or impenetrable markets, difficulties of direct export such as a process invention or a new but untransportable product such as a whole plant.

At the peak and beginning of the decline of the product's fortunes the company will be wanting to move on to its next winner and to free production capacity to do so. At this stage licensing of the production may be appropriate, especially as further development of the old product to meet the developments of its market competitors will be ever-increasingly expensive for ever-decreasing incremental gains. Licensing of the old product in so-far untapped markets may produce useful income with which to finance the R and D work at home.

9.8 ORGANIZING FOR LICENSING

We are not here concerned with deciding whether when and who to license, preparation for a licence negotiation, preliminary precautions or terms, all of them matters which will be looked at later in the book but rather with the question of organizing yourself for the policy formulating task. What follows will thus be equally applicable to the question of whether to exploit in some other way, perhaps through a joint venture of the kind described later in Chapter 10.

Two initial points have to be made. Firstly, quite how you organize will depend largely on the precise objectives you have in licensing. Is there to be a long-term licensing campaign, or is this a one-off project? The answer may determine whether you seek to take on the operation internally or choose to use an outside consultant. Secondly, the nature of the organization will be determined ultimately by the size of the firm and the management capacity available. Whereas many large firms have their own licensing department for the smaller firm this function will have to be taken on by management personnel with other responsibilities.

9.8.1 The responsibility for licensing activity

It is certainly a good idea to have one person in charge of licensing activity. In a large company this may be a task which needs subdividing into different executive posts for licensing in different areas of technology or even to different types of firm: a large firm and a small firm contact, perhaps. So, one of the first tasks in formulating that policy is allocating responsibility for developing the firm's licensing-out business or handling the inflow of technology through licensing in. Who is going to be put in charge of the licensing in and out of technology or other rights and what precisely is to be the scope of that person's mandate? This of course is going to be a matter individual to each firm. A few suggestions might be made, quite tentatively.

In terms of choice of the personnel, the licence negotiator or negotiation team is ideally going to have a reasonably high level of technical expertise in the relevant field plus a reasonable basic awareness of the legal background to technology licensing and of course the personal qualities of the sales person and negotiator. So long as he or she is given adequate professional support by the company lawyer plus accountant or finance director or other similar adviser, his or her technical expertise is far more important than his or her legal knowledge. Lawyers are useful animals. But their function is primarily keeping you off the rocks, not finding El Dorado; they have a different role from the commercial negotiator and their training, exceptional cases apart, is not selling or buying, but servicing such transactions of sale and purchase. The lawyer's job is to be the pessimist, to anticipate the difficulties and problems which may lie ahead.

Equally, the licensing co-ordinator or executive should be independent of albeit in close liaison with the patents or R and D departments. The job should extend beyond licensing negotiations to the consideration of joint ventures or subsidiaries and the like, keeping an eye on what competitors in the field are up to, market/expert opportunities and so on. He or she is the trade fairs person. For the smaller firm it may well have to be the Managing Director who takes on these tasks, but it may prove beneficial to employ a specialist consultant in these matters, at least in the beginning of licensing activities, while you learn the ropes, though such persons vary in quality enormously and tend to be something of an unknown quantity in the absence of a personal recommendation from someone in the same technical field as the proposed project.

The marketing and buying techniques required for intellectual property transactions do seem to differ from those employed in end-product sales, and also as between the licensing of the different types of right. This point has already been made in relation to the impossibility of a standard form licence. A product sales person is not necessarily the best sort of licensing executive. But of necessity it will often be staff from his or her part of the firm who are given the job. Assuming he or she has the requisite technical appreciation of the technology this may well be the best bet, provided that a suitable level of co-ordination with patent, R and D and legal departments is maintained.

9.8.2 Licensing technology audit

In assessing licensing capabilities and needs, one of the first tasks for the newly appointed licensing executive or co-ordinator is going to be that of the inventory: examining the company for what licensable technology if any it has, what gaps in the product lines there exist which could be filled quickly and profitably by licensing in rather than by internally generated R and D, what types of technology expertise exist within the company and therefore what products could be easily assimilated into the production capacity and expertise and facilities of the company. The scale of operation, the capability and desire to expand and by how much how quickly, available cash resources for investment, and so on, must be surveyed.

There also has to be asked the question of whether it is desirable to license out technology rather than entering into self production in the light of the company's responsibility to its employees. What are the pros and cons of setting up in manufacture yourself as opposed to creating a competitor whose own employees benefit rather than your own from the new manufacturing venture? What scope is there for the firm extending the expertise and responsibility of staff through secondment to licensees of your technology if you do license out and at the same time provide technical back-up services? What effects will such back-up obligations have on staff?

The other side of the coin is ascertaining what gaps exist which might be filled by licensing in or should be dealt with internally which may complement an existing portfolio. This requires some awareness of the overall business plan of the company and explains why a senior appointment is so necessary: what is the company expected to be doing in the next five to ten years? Will the policy be diversification, expansion, specialization?

In looking at what might have to be done by way of R and D or licensing in, the appointee will be looking at the life cycle of his or her existing products, any spare capacity, changes in the market requirements which the company is going to have to react to, and determining whether it is able to do so in time or whether new products will have to be bought in.

9.8.3 Awareness of the product and market

Next there is the task of familiarizing himself or herself with the available products – in the case of a company seeking to license in – or the available clientele – in the case of a company seeking to license out, the strategy he or she is to adopt in finding them and the means of communicating the willingness of the company to engage in transactions of this nature, and determining the criteria which make potential partners in licensing ventures acceptable or unacceptable, as well as determining which of the characteristics of the company (if any) are attractive or unattractive to outsiders.

In formulating a licensing policy it has to be remembered that the conventional wisdom is most licensing activity is demand led; that it is thought comparatively rare to develop a product or process which in itself creates a previously non-existent demand. Lasers and microchips are examples of the rare exception of a new product which created a demand. One of the problems in licensing much public or university technology apart from its comparatively raw state is that it often represents a solution looking for a problem and you have to persuade the potential licensee that there is a good potential market for the product without the benefit of one already in existence. This has to be borne in mind by the company undertaking R and D work with a view to licensing out.

For a company looking for a new product to license in, the question must be asked, not 'where is a product which we can learn to produce' but 'what market do we want to penetrate where there is a gap to be filled and what skills have we got which can exploit technology on the market to fill that gap?'

This will require an examination of the strengths and weaknesses of the firm and its capacity to adapt cheaply and quickly to any new forms of technology one might be contemplating. Are the staff available to work on the technology or the products concerned? Will the company be able to afford plant in certain fields? What areas of complementary expertise to that already acquired might be exploited? What economies of scale might force one out of the market? At what stage of development – e.g. how far in time away from the market place – will you be willing to take on a product? This process will not enable you to identify a suitable product to license in but it will identify what the firm is capable of considering and go a long way to helping to eliminate unnecessary searches and frustrated negotiations by eliminating a large number of non-starters right at the beginning.

9.8.4 Reviewing policy

Quite apart from these decisions a number of basic policy decisions are going to have to be taken at least at a preliminary level and then probably reviewed when specific project proposals are encountered.

- Suppose a particular gap in the product range is identified. Is the company willing to venture into a field involving the production of a very raw concept or innovative 'hi-tech' product, or only a lower risk level product which is a mature commercial line with an established market? A one-off, or part of a range?

- Is the company prepared to enter a field in which there are already a good many efficient competitors established? Is only an exclusive or sole licence acceptable or is the company prepared to be one of a number of competing licensees? In what geographical or territorial markets does it want to operate in this technology? Does it want to license in from overseas companies?
- How close a relationship does it want with a licensor and is it willing to become involved with a partner very much larger and economically stronger? What profit margins are sought realistically and what levels of risk acceptable?
- One specific problem to be resolved will be not just which products of R and D work are suitable for licensing in or out, but when in the life of those products to license. Obviously the stage in the life of the product will determine in large measure the price of the licence, just as a lease with only a few years left to run will be assigned at a lower value; but equally a very new and untried product will be of less value than one with a proven track record in most cases. It may be a crucial decision whether to opt for the licensing in of a succession of short-term investments or to pin your hopes on a longer life venture.

Conversely, to take a cut in the royalty rate or to expend more money on a development project for a still relatively immature intellectual property right is an option which may face a potential licensor of technology. Deciding between these options may involve very detailed or speculative evaluations of likely development costs, royalty potential at various stages in the life of the right, the time scale on which expenditure and income are to flow, and so on.

These factors will often cut down very narrowly the range of available technology, which may reduce the amount of searching for licensors considerably too. And before leaping into actually seeking out licensees the company or inventor is going to have to be clear as to precisely what it is that it will be able and willing to license. Only then will it be practicable and useful to start looking at the specific products and specific potential licensing partners which a search throws up.

9.9 PACKAGING TECHNOLOGY OPPORTUNITIES

In this context it is worth considering the possibility, subject to the constraints of competition law, of putting together a package of rights which fit together in some way. This may increase income, make the individual elements of the package more attractive in themselves by association with other rights and therefore attract higher royalty rates, quite apart from being more marketable.

It is often the know-how in a licence package which is the core of the commercially valuable technology being licensed but it is the patent which gives the licensor a degree of technical credibility and increases the attraction of the know-how sold; this appears to be especially true of Third World countries, where governments seem to place an undue emphasis on the 'tangible' patent at the expense of the perhaps somewhat amorphous know-how. Furthermore, the packaging of patents and know-how enables you to split royalties into two or more accounts so that if, for instance, the patent is found to be invalid or its scope cut down, there should remain the entitlement to the separate royalty on the know-how transferred, irrespective of the fate of the patent on which it is based.

9.10 ELIMINATING THE OPTIONS

When the decision to look has been made, these efforts can be subject to certain economies. Matters of licensee selection will be dealt with later in the book when we will be looking at sources of information about business opportunities in the licensing field, and also at the question of how to find out about the financial and other information you need to acquire about potential licensees when considering them, much of which is freely available if you know where to look but in sources of information little known by the general business person. But obviously certain types of licensee are likely to be low on the list from the start and not worth the expense of canvassing, exceptional cases apart, for instance:

(1) Direct competitors marketing a technically competitive (as opposed to complementary) product (unless perhaps you want to be bought out by them!)
(2) Leaders in the field you are seeking to license in (with the same caveats)
(3) Companies which have already committed themselves to large capital expenditure budgets on developing technology in the field and who may be unwilling to write that expenditure off
(4) Companies which are already heavily committed in other projects and so which, while perhaps being experienced in this field, could not devote their full or adequate resources to the project
(5) Very large companies within whose walls smaller projects might well be submerged and which might therefore come low down in the list of priorities when allocating funds and marketing efforts

9.11 BUDGETING FOR THE SEARCH FOR A LICENSEE

It is important not to underestimate the cost of seeking and of selecting the right licensee and of negotiating agreements, particularly in management and lawyer time, quite apart from the maintenance costs of a licence such as back-up, technology updating, audit, market research, as well as defence and infringement actions. Short term, licensing is rarely a profitable venture, with a few lucky exceptions. Furthermore, the mere task of seeking out new opportunities to exploit ideas is in itself not without risk – and the same is true of making it known that you are receptive to suggestions for possible developments: we have seen some of the dangers inherent in an open-door policy earlier in the book. But the general sentiment of maintaining one's internal R and D work and licensing in in parallel makes sense, and provides some answer to problems of employee morale in an R and D department when licensing in might otherwise be regarded as some sign of failure on the part of the internal R and D staff.

9.12 THE MECHANICS OF LICENSING AND ASSIGNMENT

In order to either assign or license intellectual property rights effectively usually something more than a mere agreement between the parties to the transaction is necessary. Some of the intellectual property regimes impose certain restrictions on the scope of licensing. Those which are imposed by the rules of competition law will be looked at in Chapter 14. But the law will also impose technical requirements which must be thought of.

Because of the possibility of having to bring legal proceedings for infringement against third parties it is essential that the parties to the assignment or licence have fulfilled these requirements and know both precisely what their respective rights are and what formal requirements are necessary for the transaction to be effective against third parties. The rights of an exclusive licensee, for instance, to sue in respect of an infringement by a third party may differ markedly from those of a sole or non-exclusive licensee. The rights of an assignee may be dependent on the prior registration of the assignment which conferred on him or her the ownership of the right. Moreover, the licensing and assignment of patents, of registered designs and trade marks, and of copyright all have different points to remember. This is lawyer's law: one should always go to a lawyer to have the relevant documents executed.

Trade secrets or know-how licences will often appear separately from licences of other intellectual property rights and properly so, for they have entirely different characteristics from them and they raise a number of difficulties which call for special care and attention.

Whereas most licences let you do something which would otherwise be an infringement of the licensor's or assignor's rights, in the case of a know-how agreement, the agreement discloses to you how to do something which you would otherwise not be able to do simply because you would not know how – if you did you would not be taking out a licence. There are as such no formalities for the 'assignment' or 'licensing' of trade secrets, but really they cannot in the strict sense be assigned or licensed at all.

The obligation of confidence on which the law of trade secrets is founded has not raised the trade secret to the status of a property right like a patent or trade mark: accordingly the only way in which you could in practical terms achieve the equivalent of an assignment would be for the 'assignor' or 'licensor', call him or her what you will, to enter a convenant when imparting the information not to use or disclose that information in business himself or herself in the future or at least for such time as the information remains outside the public domain. The matter remains a purely contractual one as between assignor and assignee and cannot create any rights in the assignee against third parties. Therein lies its very vulnerability, quite apart from the vulnerability of the information itself which is the subject of the agreement.

A second substantial problem lies in the absence of any natural or legal limits on the scope of the agreement in the case of a trade secret, unlike a patent or trade mark which is a national or an EEC right, which has both geographical and temporal limits. This creates a need to ensure that proper steps are taken to define the scope of the assignor's contractual rights very precisely, especially in such matters as payment of royalties.

10

Non-licensing options

10.1 INTRODUCTION TO NON-LICENSING OPTIONS

A number of alternatives exist for the firm with a piece of technology to exploit, or the firm seeking to introduce new technology from outside. One form, that of licensing, is treated as a separate chapter. But a straight licence is not always a viable option for a potential licensor or licensee any more than pure self production. And other means do exist which involve a closer relationship and co-operation than licensing. The remainder are all to be considered together in this chapter. The questions are all the same for each option. What is it? What are its main advantages and disadvantages? And what factors must be taken into account in choosing between them when there is a choice? Factors affecting the choice of business medium will include the secrecy and life expectancy of the project, the element of financial risk involved and tax considerations.

10.2 JOINT VENTURES WITH THE RIGHTS OWNER

The loose term joint venture is commonly employed to indicate a form of close co-operation in a project between two or more firms: sometimes the rather more grandiose term consortium is employed instead. In fact, such joint ventures usually may take one of three forms in law, that of a partnership or that of a wholly owned joint subsidiary company or that of mere contract. And there are fundamental differences between the three as we shall see below.

Why would a firm with an idea and a firm with some money or production facilities contemplate a joint venture instead of a straight licence agreement? The use of a joint venture does after all involve some assumption of co-operation more intimate than other forms of business transaction and a commitment to the project lacking in a more distant relationship.

The answer is that:

- Entering into a joint ventue may be much more attractive because whereas it limits the financial input required of the two parties it creates a degree of stability and a common interest in a highly capital intensive investment which may be felt to be lacking in a straight licence.
- Sometimes, particularly in overseas countries, a joint venture helps to overcome any political and legal difficulties. For example, it may be impossible for a UK company to set up a wholly owned subsidiary in the country concerned because of investment or other trade barriers, and instead a joint venture with a local enterprise will be used as a means of setting up the business locally. Japan provides an example.
- Alternatively in some state-planned economies a joint venture with the government of that country will be employed as a means of penetrating an otherwise impenetrable market.
- A joint venture may have resulted from a conflict between the two parties (for instance over their patent rights in the territory concerned) as a means of their reaching a settlement.
- A joint venture may in some cases present less of a risk of infringing antitrust laws in the country in question than other forms of horizontal co-operation or straight licensing where intellectual property or other technology is involved.
- A joint venture may be established between an individual or small company with an attractive piece of technology and a larger concern with the marketing and other back-up strengths which the inventor needs. This then may enable them to undertake what would for either be an otherwise impossibly capital-intensive project or simply enable a larger company to diversify its risk by investing in a range of projects while enabling a smaller company to enter a venture with the financial and other resources support which it needs to make itself a credible potential supplier of the design or technology involved to others. This perhaps explains why joint ventures are sometimes regarded as second-best options, a course of action taken when none other seems viable.

The dangers of joint ventures of this kind are the loss of sole control over the project and the danger of conflict between the parties, particularly if there are no satisfactory mechanisms for resolving conflict and termination of the venture under the agreement between the parties setting up the business. The initial costs of setting up and administering the joint venture may be higher than a straight licence.

10.2.1 Legal forms available

Let us look at the legal forms of a joint venture available to the parties wanting to undertake such a co-operative project.

(a) Partnership in English, Commonwealth or US law

Subject to certain restriction in some states of the USA on companies entering into partnership, which has never been a problem in England, any joint ventures on a relatively small scale, say between an individual and a small company which are only intended to last for a limited period of time or which are designed purely to cover only one limited transaction or project may often be undertaken by using the partnership model.

Why use a partnership? Partnerships may provide a greater degree of flexibility over legal incorporation in terms of capital management and other financial arrangements. Also, partnership is perhaps more suitable if co-operation is only intended to last for a short time. Although in theory the legal liability of a partnership is unlimited in practice the parties to it can quite simply set up wholly owned limited liability subsidiaries as wholly owned partners in the joint venture. There may be tax advantages in some cases in having this arrangement. A partnership option is often particularly apt for the situation where one half of the partnership is an individual such as the inventor of the idea.

(b) Use of a jointly owned subsidiary company

Most joint ventures however and specifically most joint ventures with an overseas element are undertaken in the form of a jointly owned subsidiary company and registered in the country of operation and production.

This is partly because they are often regarded as more suitable for use as longer term and more open-ended enterprises which therefore require some more definite structuring and partly because at least in most of the common law countries such as the UK, USA and the British Commonwealth, the creation of joint subsidiary companies is so very easy to achieve with few if any of the obstacles some of the civil law countries appear to create.

Furthermore, and particularly where there are some developing countries concerned, use of a jointly owned company will often be desirable to meet requirements of local law which, using domestic nationality of the company as a qualification for concessions, may be designed to encourage local capital contribution and profit sharing.

But very few of these types of joint ventures are ever really truly independent businesses, and most will be heavily dependent on their owners for supplies of raw materials, expertise and manufacturing capacity. They are often merely a more sophisticated way of regulating the transfer or sharing of technology between the partners where a higher degree of mutual independence in matters of technical and financial support and back-up exists than is the case in those situations for which a straight licence may seem to be appropriate. Where the product licence has competition from a number of similar items, the on-going joint venture may encourage licensee loyalty in respect of these other products.

(c) Use of a mere contractual arrangement

Particularly in certain European countries the legal and fiscal environment may well make a purely contactual arrangement with no capital contribution to a joint business or company a better bet though in some countries (and indeed in England) a partnership or its civil law equivalent may be inferred by law out of the facts and thereby subject the parties to the same legal disadvantages as if they had expressly set up a joint venture business of the types already discussed. But such a device is really only practicable for a specific one-off project such as a plant construction venture and not for an on-going development and marketing venture.

Nevertheless, it may be found in the context of a rather looser co-operative venture such as an agreement to exchange R and D results or information or even to establish a patent pool in which all patents of the parties are pooled into common ownership, although this is now fraught with danger from a competition law point of view.

Certainly in English law these looser arrangements are liable to run into the risk of being regarded by the law as a partnership without any express intention to that effect even if only one specific co-operative project is envisaged. In many European countries the additional problem exists that legal liability for a joint venture subsidiary company may be less easily avoidable by the owners than in the UK or the USA.

10.2.2 Partnership options

For small-scale and purely domestic operations a partnership may be practicable. English law imposes no formal requirements for the creation of a partnership, so that an oral agreement or even an agreement inferred from conduct will suffice and in the absence of an express term about particular matters the *Partnership Act 1890* will provide an answer to the parties' obligations and rights with respect to each other and to third parties. But of course it is essential in practice for there to be a detailed written agreement if a serious partnership venture is to go smoothly. The agreement will operate as a focus for planning the full execution of the joint venture in advance. The written partnership agreement will deal with *inter alia* the following matters in a partnership governed by English law:

(a) Duration of the partnership

The partnership may be expressed to endure for a fixed term only, otherwise the partnership may at any time be dissolved at will by either party unilaterally. There can be provision for renewal. A wisely drafted partnership deed will provide for, say, six months' notice of withdrawal on either side, agreement apart.

(b) Capital contributions and profit sharing

The contributions and profit sharing should of course be fixed in advance of entering into the venture partnership deed. In the event of an absence of such a provision the presumption will be fifty-fifty split but in many cases that will be the last thing intended by either party.

(c) Partnership property

If it is intended that any piece of property and in particular of course any existing or resultant intellectual property should belong to the partnership this should be made clear. Otherwise in the absence of any provision the Partnership Act and the common law presumption against common partnership property will apply: this will apply equally of course to any rights in intellectual property acquired during the course of the partnership activities as to any other form of new asset.

(d) Procedures for dispute resolution

The agreement should identify those changes requiring unanimity. Partnership disputes as to those ordinary matters connected with the business are in fact normally decided by a majority vote if there is to be a majority possible, but decisions relating to any change of business, variation of the partnership deed and the admission of new partners can only be settled by a unanimous decision. A wisely drafted partnership deed will also include an arbitration clause preventing disputes from being aired in open court until after an arbitration decision. A partner can only be expelled from the partnership if there is a clause to that effect and his or her conduct falls within that specified in the clause and the procedure specified in the clause is followed. Such clauses tend to be construed strictly by the courts.

(e) Firm name and location of business premises

These should be specified and procedures specified for altering them when desirable. Where a business name is employed there are special rules under the *Business Names Act 1985* as to disclosure of the owners of the business which must be satisfied.

(f) Dissolution arrangements

Provisions should be made for repayment of the capital, ownership of intellectual property and other property rights and sometimes even for post dissolution competition, in the event of a dissolution occurring. Termination of the partnership may of course arise out of a wide variety of causes. Particularly where one partner alone wants to call it a day, provision will be necessary for where the business is to continue in the name of the remaining partners, or whether it is to be sold off as an on-going business with its associated goodwill; what is to happen to trade marks, goodwill, know-how, etc. which has accumulated in the course of the venture and the initial patents and other rights on which the whole venture was first founded.

It should be remembered that all partners are liable for acts of any of them even though that act was not authorized, because every partner has an extensive implied authority to bind the partnership. There is little that can be done about this in the partnership deed. Restrictions in the partnership deed on partners' authority can only be effective against a third party where that third party has actual notice of them. Any partner who withdraws should make this clear by proper public notice to avoid being held liable for acts of the partnership after his or her departure. He or she may be able to claim post dissolution profits in some cases.

- Consideration will also have to be given to the anti-trust and restraint of trade rules on agreements as to competition during and after the partnership's life and especially R and D joint ventures for which there are special provisions.

These then are the specific matters to which much attention should be given when one is structuring a partnership joint venture. More generally, it is vitally important to create a partnership agreement which will ensure that there is a definite direction and objective in mind which will accommodate the often conflicting interests of the partners. The perceptions of the individual – perhaps the R and D – partner who is providing the technical expertise in a partnership as to the desirable role and development of the partnership may well differ markedly from those of the partner providing finance. The terms of the partnership deed should be as unambiguous as possible on these matters so as to limit the scope for internal conflict. But of course no two joint ventures follow exactly the same lines. A precedent deed as such is impossible to set out for this area.

Partnerships are generally unsuitable for large-scale projects and the corporate option is often as easy if not easier in many ways to set up as a partnership. On the assumption that the nature and scale of the joint venture which it is sought to undertake justifies a partnership properly so called rather than corporate subsidiary or use of some other form of corporate co-operation, certain aspects of partnership may make it more or less attractive to the individual dealing with a large and more powerful company.

- Every partner is under a positive duty to provide full accounts and information for his fellow partners and may not make a secret profit out of a venture which should have been collective by the partnership.
- Every partner is entitled to an indemnity in respect of liabilities incurred while conducting the partnership business but not to any remuneration unless the partnership deed specifies this.
- Further, every partner is entitled to have access to the firm's books and to participate in management. These rights are valuable to a small or individual partner who might otherwise feel squeezed out by the larger partner and they can be enforced by injunction or by receivership.
- On the other hand, the courts have also developed the principle that a partner owes a duty of care to his or her fellow partners to ensure that he or she does not cause them loss through his or her negligence and this might be of concern if the partner is dealing with professional matters in which he or she might expose the partnership to liability and find himself or herself having to indemnify the other partner.

These obligations together emphasize that whereas the straight licence is a legal transaction at arm's length the partnership involves a wholly different level of relationship between the two parties, a much more intimate relationship involving mutual obligations of the highest good faith and trust. Mere honesty is not enough. The corporate option by contrast appears to occupy a half-way house between the arm's length licence and the intimate partnership.

10.2.3 The corporate joint venture option

The use of a company as a vehicle for exploiting intellectual property may arise in a variety of ways. Thus the parties may set up a joint venture company from scratch, alternatively two companies may form a joint subsidiary. For the more predatory there is the possibility of a take-over of an existing company which is exploiting a desired piece of design technology, an option considered below. Whichever *modus operandi* is adopted many of the same general principles of company law will govern the arrangement. The reasons for opting for these modes of operation will be looked at later.

If the corporate joint venture subsidiary option is taken then in many countries companies can either be bought off the peg and then modified to the particular needs of the venture or set up from scratch.

The great advantage of the corporate option is that a company is separate from its members in English law, unlike a partnership firm, and so has a separate legal identity. This means that its members are not liable for its debts or its obligations in contract or tort, and a parent company is not liable for its subsidiary's debt. This is true in many other countries too, though both here and there there are circumstances in which this separate legal personality will be ignored by the law or by the regulatory or fiscal authorities if the substance and the form of the arrangement do not fully conform.

The principal reason in the UK for choosing legal incorporation, however, is the ability to use floating charges over the assets for the time being of a company as a means of securing debts, something not available to the partnership.

The rules which govern the internal running of the company will be laid down in what in England are termed the articles of association of the company but which are labelled differently in other countries (so, for instance, in the USA they are called the bye laws). It is these which must be agreed with as much forethought as possible, especially if there is a disparity in the capital input of the two parties. Some of the matters to which thought must be given include the following, dealing both with the firm's relationships with the outside world and with the internal rights and duties and organization of the firm so as to safeguard the position of the members of the joint venture.

(1) What are the powers of the joint venture company?
(2) Duration of the joint venture, objectives
(3) What restrictions on membership are there to be?
(4) What contributions are to be made by the parties?
(5) What is the equity share in the venture to be?
(6) What law is to govern the joint venture?
(7) Who will own the intellectual property rights?
(8) What trade marks etc. is the venture to use?
(9) What are the dividend rights to be?
(10) Salaries for employed staff, pension funds, etc.
(11) Appointment of management, control of policy
(12) Liquidation and termination by one of the parties
(13) Certain other considerations listed below

Let us look at some of these more closely.

(a) What is the equity share in the venture to be?

The parties may contribute cash or technical know-how, etc. in return for their shares in the company. The means of contribution and the method of profit-taking must be carefully considered, especially if there is an individual participant in the joint venture who will be devoting substantially his or her whole time to it and has no other sources of income. If, for instance, an inventor is exploiting his or her invention by a joint company and his or her contribution is basically his or her talent and know-how and other intellectual property but no cash, he or she may well seek to protect his or her position in the venture by being appointed not as an unsalaried director with merely a shareholding out of which profits are to be paid but as executive salaried director with a contract of service with the company, i.e. employed by it, and also choose to participate in the equity of the company, i.e. to hold shares in it which may pay him or her dividends.

If an individual can obtain this sort of deal then even with a minority shareholding he or she should enhance the strength of his or her position in the company. For this may give him or her two lines of approach in defending his or her interests in a joint venture with a much larger corporate shareholder in the jointly owned company, one as a director and one as a shareholder.

(1) As a director, he or she will have access to information and be in a strong position to affect board decisions. He or she may be dismissed by the shareholders or even by his or her fellow

directors, but a large shareholding and certain provisions in the articles such as weighted voting provision can effectively frustrate these measures. He or she is safer than a mere partner from personal liability for the wrongs of the company but not wholly immune, and he or she also owes strict fiduciary duties to his or her company which may well prevent him or her from undertaking entrepreneurial activity on his or her own account in relation to other inventions, the ownership of which he or she may well lose to the very company he or she runs: this was considered in Chapter 7.

(2) As a member and shareholder of the company, his or her position is protected in many countries from unfairly prejudicial conduct by the company or other members even if he or she is a minority shareholder, as he or she might well be if he or she is only able to put in expertise and the other majority shareholder is providing most of the working capital for the company. But it is unsafe to rely on this for the remedies are expensive and time consuming. Far better to bargain for express protection.

Whereas there is absolutely no necessity and often no possibility of a straight half and half split in the equity or even something approximating to it there will be a minimum share of the equity below which a minority shareholder will lack any real control over the affairs of the company and the exploitation of the technology or the marketing rights which it was formed to develop unless he or she has some special weighted voting rights or a veto vote in the company meetings.

A failure to ensure that he or she acquires the basic minimum share or some equivalent entitlement through voting rights etc. can leave the minority shareholder in a very difficult position, effectively locked into the joint company because of the restrictions on any share transfers which are commonly found in the articles of such companies, even assuming that a third party would be willing to buy into such a weak position. In the absence of a weighted votes provision in the articles of the company to defend oneself from this possibility, a minority shareholder's only option would be to seek a court order dissolving the company, which is expensive. Being bought out under s459 *Companies Act 1985* is an increasingly common escape route. Even in the absence of an effective veto some provision should always be made for premature termination of the relationship at the will of the parties, perhaps by any one or more of the following means:

(1) Liquidation of the joint venture on the election of either party
(2) Seeking listing on the public market to enable one party to sell out his or her share
(3) Option to buy with liquidation on failure to agree price on giving notice, or with arbitration
(4) Freedom to sell to a third party on other partner refusing to buy at a set price

(b) What law is going to govern the joint venture?

Apart from the question of the law applicable to any agreement between two parties as to the formation of a joint venture company, it will have to be decided what law is to govern that company. The law governing the conduct of a company will be that of the country in which it is established, which will be final. But in many countries it would be possible for the parties to reach an agreement governed by a different law as to the way in which they will exercise their rights as owners of shareholdings in that company which would be enforceable.

(c) Who is to own the intellectual property rights?

This will be central to the agreement in most such technology ventures. It will have to be decided whether the joint venture company is to be merely licensed or whether it will take an assignment of the existing and future rights and whether the parties themselves will enjoy any preferential position as potential licensees of those rights. The disposal of those rights on the liquidation of the joint venture company for any cause will also have to be considered. Particularly if mere licences only are granted and if there are provisions for grant backs and the like, the agreement will have to be vetted very carefully for compliance with the anti-trust laws of the territory in which the joint venture is to operate.

(d) In marketing the product what trade mark is going to be used?

A trade mark associated with the joint venture company alone or one with some connection with the partners running it? Each course of action has its advantages. The creation of a separate reputation for the joint venture company may make that company a marketable commodity itself. The use of the local partner's reputation in an overseas location gains the benefit of an established name there but like the first option does nothing to advance the reputation of the other party in that market. On the other hand use of the existing parties' names may have adverse effects and create consumer resistance. If the company's marks are to be used these should be licensed not assigned in that territory if possible.

(e) What are the dividend rights to be?

Another aspect of importance to the smaller party will be payment of dividends. This will be crucial to a director who has agreed to forgo a certain proportion of his or her salary in return for equity in the company, which will usually be of little value as a capital asset even if the regulations of the company permit transfer to third parties and third parties are prepared to buy. It may be that some guarantee of dividends to a certain class of shares, held by that director, is made, so that the majority voter cannot lock profits into the company and starve out the minority shareholder who is dependent on these dividends.

(f) Export policy

Should the company be able to purchase from the joint venture parts or products which can be produced cheaper there than at home, even after taking into account costs of transporting the materials back into the UK market or other overseas markets for assembly and sales? What policy will the joint venture have on export to markets traditionally served by each of the partners in it? What anti-trust considerations are there?

(g) Other matters for joint venture consideration

Other matters for consideration and agreement by the potential partners in corporate joint venture, which speak for themselves, will include:

(1) Control of joint venture by owners:
- (i) Transfer of only 100% of an owner's equity
- (ii) Liquidation on take-over of one shareholder
- (iii) Management structure of the joint venture
- (iv) Admission of additional parties to joint venture
- (v) Which issues are to require unanimity or majority
- (vi) Deadlock on board or between owners of venture

(2) Extent of commitment by joint owners:
- (i) Liability of joint venture owners *inter se*
- (ii) Maximum capital calls demandable of the owners
- (iii) Borrowing limit and source for the joint venture
- (iv) Guarantees and security demandable of the owners
- (v) Transfer of business to outsiders or one owner
- (vi) Establishment of subsidiaries of joint venture

(3) Relationship between joint venture and owners:
- (i) Use of R and D activities of the joint venture
- (ii) Scope of territorial activity of joint venture
- (iii) Competition between the joint venture and owners
- (iv) Confidentiality and rights of access to reports
- (v) Terms of trading between joint ventures and owners:
  - Supply of goods and services
  - Distribution network for goods
  - Service back-up facilities
  - Loan of personnel, premises, etc.

10.2.3 Factors in deciding on joint ventures

The factors which will determine the decision to take the joint venture option, whatever its precise legal form and structure, will include the following, many of which will be relevant only to overseas ventures with a local producer. They are divided into the consideration of external and internal factors.

(a) External factors

- What benefits in terms of relaxed governmental regulations, freedom for competition, competition laws, import controls, etc. will a joint venture bring if any? Do there exist any rules which discriminate against partly foreign owned companies in the country in question, e.g. by prohibiting use of such companies in government procurement contracts? Is there any maximum shareholding or law on the management and control of the company? Are there any grants to induce the inward investment of foreign capital in the country in question?
- What will be the effect of being a majority or a minority shareholder in the joint venture in the country in question? What employment and other laws are there? How easy will it be to move home-based staff overseas and back to supervise start-up etc. – e.g. through visa requirements, freedom of movement, income withdrawal, etc.? What corporate reporting requirements are there which might be embarrassing?
- What is the tax position at home and abroad of the company joint venture? How is the

know-how/patent contribution to be valued as opposed to the cash input into the company? Are there provisions for taxation on the basis of group profits rather than only on the basis of the profits of the local subsidiary? Can the owners withdraw royalties and profits and charge any management fees to the subsidiary?

- Is the country in question liberal in permitting import of raw materials, semi-assembled parts for the production, etc.? Will supplies be reliable?
- Given the higher risks for which a discounting allowance will have to be made in making the final calculation, will the profit levels from the overseas venture exceed those obtainable simply by expanding home production and exporting up to the maximum ceiling allowed by the country in question?

(b) Internal factors

- Will the joint venture continue to be properly and realistically funded? On what basis will the parties undertake to input further capital if expansion of the joint venture's scale of activities and growth requires more capital? Can each partner cope with that demand?
- What will be the dividend policy of the joint venture? This will be crucial if the contributor of the technology is being rewarded by a shareholding and is not expected to receive royalties from the company set up. And will the dividends or royalties if there are any be exportable from the company overseas? Often they will not be in Third World countries. How will the parties overcome this?
- What policy decisions are going to be urged in any joint venture with respect to trading policy, to management and marketing plans, etc.? If the parties to the joint venture are unable to agree, what is to happen?
- What will the influence of the joint venture's activities be on the company's own business activities? Who is to supply the senior staff? What confidentiality provisions and degree of co-operation and capital commitment is expected?

10.3 TAKE-OVER OR MERGER OF AN EXISTING CONCERN

On occasion an entire business may be taken over for its intellectual property or other assets. Asset stripping was of course the fashion of the early seventies, but the sort of take-over we are concerned with here usually envisages the company continuing to run in its existing or expanded form rather than being dismantled for its asset value.

10.3.1 Why consider a take-over or merger?

Why would one want to acquire or merge with a company rather than merely acquire its rights and know-how? Obviously, the prospect of acquired control of rights by acquiring their owner may appear to be an unnecessarily expensive way of going about things. But merger or take-over may be an appropriate means of acquiring the technology plus any trade marks and any accumulated goodwill.

Acquisition of a running company, a going concern with an existing complement of trained staff fully conversant with a process and the plant and know-how connected with it, with established sources of supply of raw materials and clientele, may be much more convenient than licensing in the technology and having to start entirely from scratch or at best having to rely on secondment or poaching of staff for technical back-up or advice, especially if all that is needed is a shake-up of sales staff or additions of existing in-house know-how or even complementary lines and the service back-up resources to transform the commercial impact of the technology acquired.

Take-over of a firm lock stock and barrel may be a short cut over these problems, and may eliminate or substantially reduce the competition law and factual competition problems which might be faced if the licensor was also permitted to stay in business as in independent entity, outside the control of the licensee. If the intellectual property rights we are dealing with are marketing rather than technology rights the acquisition of the owner's own business may in some cases be essential, for instance if the trade marks are unregistered and goodwill of the business must therefore be acquired with the business itself to be effectively transferred.

Where the company or business acquired is based overseas these reasons for contemplating take-over may be of equal application and in addition there may be good cause to have a local producer while retaining control at a distance. But in addition, some of the factors already mentioned in setting up a joint local subsidiary with a partner will be applicable. For example, there are benefits in having an established local producer with local knowledge and experience of the trading conditions in the market in question with which to acquire instant access to these advantages.

Mergers and acquisitions are of course somewhat dramatic indeed pretty terminal in effect. This is because if the project goes wrong you cannot terminate it with the lack of formality possible in the case of a straight licence. Disengagement of the parties from the venture can be even more difficult than in the case of a partnership, the legal relations between the parties being so much more involved. Liquidation can only be effected through the formal process.

It is generally accepted that mere take-over of a whole company for its design or technology where that is wholly outside the field of experience of the new owner is a very risky task. Even holding companies tend to specialize in particular areas of expertise at which they become very proficient. The same risks as are inherent in diversification of production exist here in managing the newly acquired company, and if there is to be great reliance on existing management it should be remembered that the newly taken over management staff frequently has the habit of going elsewhere where the grass is greener. Concerns with business in related fields to those of the new parent are better bets.

The assumption is made of course that a proper price is paid for the company and that the technology and other assets have not been over-valued. The great volatility of share prices of high-technology companies on the Unlisted Securities Market (USM) is an indication as to how fragile the value of a company may be in this field, not least because of its propensity to fall prey to deficiencies of budget control and financial management.

10.3.2 Legal problems with take-overs and mergers

The legal aspects of the take-over are really no different in this context from any other and so will not be dwelt on. It is difficult to enforce a take-over of a private company unless the existing owners do want to sell out because the articles normally contain clauses which restrict the transfer of shares to outsiders and give directors an absolute discretion to refuse to register share transfers. Appointment of these people to senior positions in the new company or a generous shareholding may be a necessary sweetener to gain their consent. In the rare case of a public company take-over simply for design and technology – though USM companies are becoming more and more attractive for this purpose – these problems do not exist but there are the rules of the Stock Exchange and Takeover code for listed companies as well as the rules on concert parties and allowing the targeted company to investigate the true ownership of its shareholding.

The sort of company likely to be taken over purely for its technology is unlikely to be big enough to involve the investigation of the Monopolies and Mergers Commission in the UK or the EEC Commission, though in the USA concern over the export of technology might lead to difficulties and the same rules against foreign ownership of companies as exist for starting up a new venture will of course apply in many countries. At the top end of the market mergers do create problems of this kind: in 1979 the UK Monopolies and Mergers Commission prevented a merger of two pharmaceutical companies because of its concern that the research activities of the UK pharmaceutical industry would be harmed.

10.4 USING A SOLELY OWNED MANUFACTURING COMPANY

Although some larger companies may consider the use of a wholly owned subsidiary abroad or at home for their production, particularly where its prospects of success or other factors make it desirable to keep the two businesses quite separate, that situation is not a matter which will be discussed here.

This section is concerned with the case of some relatively small piece of technology or design created by the individual inventor who has decided that he or she now wants to go into production on his or her own and that because of the nature of the product this is feasible, either because he or she chose to exploit this way or because he or she simply cannot find a potential licensee.

The indications are that most companies are simply very sceptical of most individual inventors' ideas and it is much easier to persuade a major manufacturer to take on a tried and tested product already with a foothold in the market, in which he or she can see scope for improved market performance and possible improved design, than to persuade him or her to take on a new product not yet beyond prototype stage.

The classic example of the invention which no-one would take until it was already well established is the Ron Hickman 'Workmate' bench, rejected by all of those he approached, including the Stanley, Burgess, Record, Polycell, Salmens and Marples companies and indeed by the eventual licensee Black and Decker, who eventually took on the bench in basically the same form as that marketed by mail order by Hickman himself, but only after it had shown proved commercial success when he decided to manufacture and market it personally.

For almost all individuals the product marketing organization and production facilities necessary for a very successful product will be far too expensive to acquire while retaining a sensible cash flow. But in the short term of establishing the product in the market place and achieving reasonable sales, while being able to service any loan capital necessary, to establish this early record, the self production option may well be feasible if the product is some computer software or something like a well-engineered technical product such as testing equipment which does not seem to require a very high level of plant investment, but merely relatively cheap assembly by unskilled or semi-skilled staff.

Indeed production may not be by his or her own plant at all: the individual may choose to order the components for assembly from a component producer and assemble them in his or her own new premises or instead to have the whole thing made by another and confine himself or herself merely to sales and service of it. In fact this is the premise of many of the science parks which have been springing up in the UK, especially in the new electronics field where many of the components are now pretty well standardized and cheaply available in large quantity, but are to be put together in different ways.

What should the individual bear in mind when he or she is planning or determining the viability of such ventures? The basic considerations fall into two categories:

(1) The question of the means of entering the market
(2) The question of the legal form of the business

10.4.1 Means of entering the market

Where the product goes through a number of stages e.g. design, component manufacture, assembly, sales and distribution, service and technical back-up, the real expertise of the inventor may be limited to the first and perhaps the last two stages. But the more stages the inventor performs through his or her own business and not through subcontracting, the more profit margins the venture may be able to make at each stage of the process. The level of investment required, the reliability of suppliers, the cash flow implications, the work load, all have to be weighed up against each other to determine the best balance between loss of profit and reduction in capital commitment.

10.4.2 The legal form of the business

The three methods available to such a person for conducting the business are to function as a sole trader, to start a partnership or to incorporate a company and run that as a business making and selling the technology. Although sole traders can start up without any formality at all in practice those who wish to go it alone will be better advised to spend the small amount of money it costs to set up a company as the mechanism through which to do business:

- The company is much easier to sell than an unincorporated business either sole or partnership.
- The tax position of a company is much easier to manipulate than that of a sole trader or of a trading partnership, though if there are to be little or no profits in the earlier stages the higher administrative costs which a company will face such as solicitor and accountant fees may be such that it is advantageous to start off your trading as an unincorporated body and then convert the business into a limited company at a later stage. But the tax benefits are not all one way, and the small venture may at least in the early years be better off unincorporated.
- The borrowing capacity of a company is greater than that of a sole trader because of the possibility of using a floating charge to secure loans which is not available to the sole trader or partnership mode.

- The liability of the trader will be unlimited: in theory that of the owner of a company will be limited to the shareholding value he or she has, though in practice in a one-person company he or she will usually have been required to give personal guarantees and security for loans made to his or her company in the same way as a sole trader or partnership.
- Some people point to the greater publicity for the company's affairs required by the law than for the partnership or sole trader. By this they mean the obligation of the company to submit regular accounts to the Companies Registry in Cardiff, which in theory contains an up-to-date file on the constitution, the officers, capital, business and financial position of the company, which is open to public inspection. In practice this is no great disadvantage: many companies flout this requirement, although this is risky since it can lead to disqualification of the directors and in extreme cases the companies struck off the register, and in any application for funds just this sort of information will have to be given to the lender whether the business concerned is incorporated or not. The Department of Trade and Industry is currently running a campaign to catch the 40% which default. For smaller companies the requirements have in any event been substantially relaxed since 1981.

In deciding to set up a company, much of the basic material for consideration does not differ from that applicable to any new company venture and which can be better read about in many specialist books about starting up in business. Some comments may be of use at this stage however.

If you are setting up a company so as to make the product it may be wise to retain ownership of all the technology rights and license their use to the company for manufacture with a clause in the licence providing for the immediate termination of the licence on receipt of any petition for winding up or receivership. In this way, if the company should fail and creditors appoint a receiver, the receiver or liquidator will not be able to deprive you of the intellectual property rights which he or she might otherwise be able to sell off as one or more of the assets of the business.

In the contract between the director-inventor and his or her company there should be express provision releasing the director from obligations to hand over further improvements and inventions related to the business of the company which might otherwise be lost to the company by the inventor under the rules which we have already discussed in Chapter 7. Once the company is set up it has a life of its own and the director is there to serve it. Since the whole object is to make the company a tax-efficient means of the director-inventor exploiting his or her inventions, he or she should be careful to ensure that his or her position is strong in this regard.

10.5 USING A TECHNOLOGY BROKER

An inventor with a good idea, no cash and no contacts in industry or a lack of response from those he or she has approached may think of using a technology broker. This is a company or other body which acts as an intermediary in finding a potential buyer of technology or even taking in the technology itself and marketing it. In the public sector bodies concerned with acting in this way often take on investment into projects as well but in the private sector generally investment and brokerage are separated.

An initial point to make in respect of them all is that they represent to the inventor what is a low risk but low yield option in the exploitation of his or her idea. The brokers in the public sector in particular run on commercial lines and expect a good return on their risk which is not always compatible with a good return for the inventor. The net revenue is shared by the two parties, only inventions which clearly have a good long-term potential tend to be accepted and usually the broker requires a good intellectual property standing. The minimum threshold for viability may be considerably higher than would be the case in a normal licensing deal with a much smaller company.

In practice inventions which might produce say a turnover of £250 000 per annum and therefore create a royalty account of say £5000 to £10 000 will rarely be considered by the public brokers. The private sector may be more receptive but will take a bigger tranche (i.e. slice) of the royalty.

10.5.1 Public sector – British Technology Group (BTG)

This is the former joint venture of the National Research Development Council and National Enterprise Board, which changed its name to the British Technology Group in 1981. At one time, BTG enjoyed a first right of refusal on publicly funded technology but in 1983 the decision was taken to abandon that principle and now such institutions can choose to go elsewhere with their project. This has caused a change in the role of the BTG. There is much misunderstanding of the role and approach of the BTG in the field of technology transfer and it may therefore be worthwhile outlining what it can offer the inventor. The BTG performs both a broker and an investment role. We are here concerned only with the former.

The principal interests of the BTG do lie in the output of publicly funded institutions such as the UK universities, polytechnics and UK Government research stations, but it can also deal with individuals. Thus in the year 1983–4 about one-quarter of the proposals received by the BTG were from private individuals – defined by the BTG as persons without the backing of a company, academic institution or laboratory. As the BTG has said, for a variety of reasons the take-up rate of BTG seems to be lower in respect of the individuals' proposals than for the more usual applicant for help, the university or other public institution, but several suitable ideas are accepted and exploited through BTG. What can the BTG do? The BTG can:

- Fund feasibility studies (under a so-called 'seedcorn scheme') in which up to £2000 may be spent in looking at the prospects of a proposal being accepted for further funding by the BTG.
- Take responsibility for patenting and other steps necessary to protect technology derived from the public sector sources. This responsibility will extend to defending the rights and pursuing in court suspected infringers of the rights. This will solve the assignment of all the inventor's rights to the BTG as a condition of assistance. He or she cannot merely license the BTG. However, if the BTG rejects an offer, the inventor is free to take the technology elsewhere and if because of lack of any industrial interest or because of some technical problems the BTG feels it best to drop the patents or other rights it took out, it will offer them back to the inventor to do with them what he or she will.
- Fund development work on technology by the public sector institutions to the point where it can be taken up and exploited commercially by industry. This will involve a requirement of assignment of any intellectual property thereby created and the BTG will insist on very carefully drafted research plans and timetables, regular reviews, reports, etc. of the sort which were described as important in any private sector R and D work in Chapter 8.

- Seek licensees and negotiate licence agreements with them. This may involve use of the new BTG computerized confidential database to match up the potential licensees for the technology proposals it receives. Or it may involve the many personal contacts which the BTG staff have developed in the years since the activities of their predecessors developed from 1949 when the NRDC started life.
- Finance the further development and launch of the licensed products by licensees on a commercial basis. Funding is available but not grants: the project must be commercially viable.
- Promote start-up companies to enable the inventor to exploit the technology the BTG is backing, rather than finding a licensee. Apart from of course providing finance the BTG will assist in producing a detailed business plan for the new company, including making the necessary financial forecasts and advising on marketing the products.
- Fund joint venture companies in which the BTG provides up to 50% of the venture capital and then recovers its investment by way of the levy on sales or usage of the technology developed.

The big advantage to the inventor, especially the individual inventor, is that the high initial costs of applying for the patent or other rights will be met entirely by the BTG and its in-house services are free. There is a price to pay for all this of course. As the BTG works on the basis that all its activities must be carried out in accordance with the normal commercial principles, it expects projects to be self financing and profit making within the short- to mid-term and it examines project proposals on that basis. Although the assistance of the BTG as such is without cost to the inventor who approaches them, so that there is no charge for reviewing his or her invention and considering it for support, it does demand a share of any licence royalties as the price of its work though it will often plough these back into the further development of that or other technology on its books. The usual split of revenue from such a profit-earning licence is a fifty-fifty split of net income though in the early stages of the licence the BTG may pay the inventor or the institution which provided the technology a more generous share, the first £5000 gross income without deduction, then 20% cumulative gross income until the BTG has recovered its patent legal and other costs, and only then will a straight fifty-fifty split be made, up to a (rarely achieved) limit of £10 000 000, whereupon special provisions come into force.

Apart from forgoing part of his or her profits as a price of losing some of the risk and gaining the technical, financial and legal support of the BTG, the inventor will also be expected to provide a substantial degree of personal commitment and active involvement in the project. The BTG will not just allow him or her to sit back and enjoy half the profits. Thus it may consult and take his or her views on its proposed uses or on the licensees of the invention, possible applications, or require him or her to join in negotiations with their potential licensees (paying out-of-pocket expenses) and it may on occasion employ the inventor as a consultant, or encourage employment of the inventor as the project consultant to the licensee.

Although the BTG's record has often been heavily criticized it is worth bearing in mind in their defence that they have a very high level of acceptance of the projects proposed to them – of about one in five – and this compares favourably with the one in two hundred acceptance rate of funders in the private sector; this is possible because unlike the private sector they cross subsidize from a very small number of big winners to support a large number of projects and a high level of activity.

10.5.2 Private sector brokerage

A variety of pure brokerage services exist in the private sector for the firm which has been unable to find a licensee (or licensor, for unlike BTG many of these private sector bodies work equally at identifying suitable technology for licensing in to their clients).

(a) Licensing consultants

There are a growing band of people holding themselves out as technology consultants or as licensing consultants, aiming to assist a company in marketing or finding new technology. The roles they play vary widely from person to person.

(b) Database matching services

There is much less intervention here than in the case of the licensing consultant or broker, and to call these brokerage services is perhaps a little misleading though they do purport to match up clients. So far they have proved to be of only limited use.

(c) Trade fairs etc.

These are perhaps the oldest established formal brokerage service in the UK, providing an opportunity for companies to display and seek new technology and design. They are still valuable – or perceived to be so by business – if the level of activity in this field is anything to go by, and publicly sponsored events now operate on a similar basis too. The main cost of these trade fairs is borne by the exhibitor, not by the person attending to peruse the new design or technology exhibited. Again they are really information services rather than active intermediaries, but they provide an opportunity for potential partners to meet at arm's length.

(d) Private brokers

These operations really began in the USA some years ago and have spread to the UK only relatively recently. Of the larger concerns, their role often mirrors the role of the BTG discussed above and indeed in the case of at least one organization, the UK Research Corporation Ltd (RCL) set up jointly by the well-known Investors in Industry (3i) consortium and Research Corporation of the USA, the loss of the BTG limited right of first refusal for public sector technology provided a stimulus for action as a competitor to BTG for acting as a broker for university and other public institution research. The RCL provides a good example of the large-scale private sector broker's role.

RCL is a wholly owned subsidiary of the Research Corporation Trust, which is a charitable body. It now offers agreements to universities, polytechnics and to other independent research institutions under which it provides a free evaluation service for technology developed by the client and then at its own expense files patent etc. applications for that technology where warranted and secures licence agreements for its exploitation, sometimes using its sister organization.

It also provides finance for additional development work on a promising but raw piece of technology and will consider funding a start-up company or a joint venture. It monitors the patents etc. acquired for infringements and the licences granted. Any resultant income is split between the institution and RCL, these profits held by RCL being ploughed back into support and development in other projects, by being convenanted to the Research Corporation Trust whose purposes are the advancement of basic science and technology and which makes research grants to UK science research. From the point of view of potential licensees, RCL offers technology matching services. The terms of a standard RCL non-exclusive agreement for the submission of new ideas and the BTG arrangements make an interesting comparison, each with its own pluses and minuses.

At the other extreme is the technology licensing consultant who acts as a broker and licensing adviser to the smaller firm, making use of both his or her own personal contacts and knowledge of specialized areas of technology and the publicly available sources of information about licensing opportunities.

10.6 ASSIGNMENT OUTRIGHT

The simplest way of exploiting the technology is simply to assign it together with all the intellectual property rights you may have outright to an assignee third party for a fixed price. This has both pluses and minuses. It can free you from the obligation to pursue further product development of a raw invention, though of course the rawness of the technology will cut the price obtainable. It will probably not be a very profitable process unless you take on the tasks of a consultancy for on-going R and D work at a set scale of fees for hours worked etc.: this may be a sound option for an individual inventor who becomes in effect a paid consultant or part-time employee. The payment for this assignment might be by way of an issue of shares in the assignee company or by a share of profits over turnover for a set time. The problem with the latter forms of payment is that they hold the inventor assignor with the knowledge that by continuing to co-operate he or she might be able to enhance profits and therefore his or her payment. With the cash payment he or she is free to go elsewhere knowing that his or her fate is not tied up with that of the assignee.

Legally and administratively the assignment itself can be extremely simple. A deed of assignment is drawn up by a solicitor very cheaply indeed, with the only further obligation on the assignor being that of giving an undertaking to assist in obtaining any rights not yet applied for or under application. The costs of so doing may by agreement be reimbursed, e.g. attendance at the company offices and attendance at any Patent Office hearings etc. However, the other party may well require convenants both as to initial ownership and as to later assistance both in development and in marketing, which will make the agreement much more complicated and which will of course require bargaining for further payment.

10.7 USING CONTRACT RESEARCH AND DEVELOPMENT

Instead of buying in existing technology on a licence a company may well seek to get another company to create *de novo* some technology designed to meet a specific research brief and product specification. This option for a company buying in technology is one which does not much reduce the inherent risk of wasted investment in seeking new technology but does enable it to decide that it is not good in a particular field in which it needs a fairly specific piece of technology perhaps to complement something it has already got, and to subcontract the R and D work to a firm which it thinks is good in that field. This may enhance the odds of it achieving a result at the end of the day which may be quicker than in-house R and D would have achieved.

Much depends on the precise field of research as to whether the firm will use the contract R and D. The level of such subcontracted R and D work is quite high in electronics, computation, and in the development of the specialist instrumentation and related fields. And some firms tend over a period of time to build up a close relationship with a contract R and D firm and send all their work of a particular kind to that firm. It is seen not as a substitute for internal R and D but as a useful complement. It avoids the necessity of the firm developing a costly expertise in fields which are peripheral to the main R and D activities of the company: the overhead costs are less and it is a more flexible research facility because when the money runs out the work can stop without in-house staff being shed or their work being disrupted. It may be necessary to keep the same firm as the subcontractor if the nature of the commissioning firm's work and organization is uncommon or specialized so that to perform work efficiently the subcontractor has to have experience of the way in which the firm works.

It will be rare for the whole R and D of a project to be subcontracted in this way, though it may well be in some cases that a number of different subcontractors each carries out a portion of the R and D project package (servicing of the core work having already been internally undertaken) and acts under the co-ordinating eye of the commissioning firm's R and D department.

10.8 EMPLOYEE TRANSFERS OR POACHING

As an alternative to outright purchase of another business which has desirable technology it may be much cheaper to acquire the relevant staff in that business. Both methods enable a firm to make a rapid advance in its knowledge without the time and the administrative problems of the other options, though the acquisition of personnel without the business can involve problems, both in intellectual property rights and in employee confidentiality obligations and non-competition clauses in their contract of employment. It can be buying a lawsuit. Nevertheless, the practice of employee poaching is very common in certain industries and it should not be ignored. A case example indicates the sort of thing which may be achieved by those bold enough to attempt a wholesale staff transfer, which is considered in more depth in Chapters 2 and 7.

EXAMPLE

G.D. Searle and Co. v. Celltech (1982)

The defendant company had induced a member of the plaintiff company's own research team to terminate his contract in accordance with its provisions, i.e. quite lawfully. But before leaving the old employer he gave the defendants details of the names of some of the rest of the research staff still working with the plaintiffs, which the defendants themselves had to some extent publicized in brochures; and as a result many more – almost the entire research department for the project in which the defendants were interested – lawfully terminated their contracts and went to the defendant company. On the facts there had been no breach of confidence, no breach of contract and no inducing breach of contract. The defendants were not liable to pay any compensation for having effectively taken the entire research team of the plaintiff company without their consent.

10.9 IMITATION AND COPYING

An obvious way of taking in technology which you cannot obtain in any other way, particularly if there is little or no know-how involved but merely an easily transferable design or piece of technology is simply to copy it and infringe any rights which may be protecting it or to seek to design around it.

In the very short term the economics are very good. A lot of R and D work has been saved and the rights protecting the product may in the end turn out to be invalid for some reason. The person whose rights have been infringed may be in a sufficiently weak position for him or her to have to reach some compromise and grant a licence rather than go to court to enforce a right when he or she cannot afford to do so or he or she lacks confidence in the strength of his of her legal protection.

The implications of copying may be very expensive if you lose, however. The morality of this course of conduct is of course another matter. The finances may be difficult to work out when balancing risks. Much will depend on just how badly the design or technology is needed and how strong its owner and his or her rights are as opposed to the strength of the infringer.

10.10 CONCLUSIONS

It is impossible to state any general principles applicable to all firms or all the firms in any given industry. The options of licensing in or out considered in an earlier chapter and the non-licensing options considered in this chapter have to be considered in the light of the particular internal and external factors affecting the firm in question at any given time.

11

Evaluation, negotiation, implementation of technology agreements

11.1 INTRODUCTION

This chapter is concerned with the practical side of reaching agreement on and then managing a licence or a joint venture. It assumes that, after the preliminary survey of technology and parties, one in the course of which all your technology requirements as a potential licensee, or your requirements as a potential licensor, have been defined, a potential match has been found. The search for a licensing or a joint venture match, or rather a group of potential matches, is dealt with in a later chapter when we turn to examining the UK sources of information about technology and design for licensing in and the sources of information about technology, design and commercial opportunities. Once a short list has been accumulated an assessment programme must be conducted in respect of each of the entries.

11.2 EVALUATING A POTENTIAL LICENSEE OR LICENSOR, ETC.

This requires evaluation of potential partners and what they can offer. The evaluation may be reduced to four basic categories of information which may form the subject of an evaluation profile on the company or any other organization under consideration:

(1) Commercial
(2) Technical
(3) Financial
(4) Legal

- Commercial evaluation will require both sides to assess the commercial implications of any such deal for themselves and the prospects of the product itself as a commercially valuable commodity irrespective of who has it. In other words what are its intrinsic commercial merits and can the other side maximize its potential to the profit of the party concerned? What if any external factors will affect his or her potential one way or the other, such as competition, obsolescence etc.?

- Technical evaluation will require assessment both of the technology product or process itself and in the case of a licence involving know-how or manpower the assessment of the personnel of the other side to the transaction and the capability of each side of meeting the demands of the other party. This may require the parties to sign a pre-licensing confidential disclosure agreement to establish any degree of useful evaluation.
- Financial evaluation will require an assessment of the royalties to be demanded or those which the party is prepared to accept or pay as the case may be, the current financial position of the other side, including a review of their accounts, obtaining bankers' references, costing of investment and plant and of the level of the return to be expected on that investment, and any government aid which may be available etc.
- Legal evaluation will involve the review of any intellectual property portfolio involved for a full assessment of its strength, age, any competition law implications, third party rights which may be involved, and conflict and any existing litigation concerning the rights. Any draft contract submitted by the other side should be reviewed and comments on it recorded for the use of the party's negotiating team. Any regulatory authority obstacles or permissions needed such as the controls on export of technology to the East, food and drugs clearance and the like should be investigated, reported on and if it is appropriate initial moves for obtaining the relevant permissions made.

Of course the questions will differ depending on whether you are the potential licensee or licensor, but there is a common core. The following three subsections provide a checklist of, respectively, the questions to be asked by both parties, those to be asked then by the licensee and those to be asked by the licensor, in each case divided into the commercial, technical, financial and legal sections identified above. This is not to say that factors to be considered by the licensor should not also be considered by the licensee, but the lists represent the priority concerns for each side.

11.2.1 Questions common to both sides of the deal

(a) Commercial evaluation

(1) What is the current reputation of the other side in the market place generally and in this field?
(2) What is the size and turnover of the other side?
(3) What negotiating strength does it have? What is its present control of the market in the field in question?
(4) What business and press reports are there on the trading activities of the other side?
(5) What anticipated development plans for the future are there in the commercial market concerned?
(6) What form of business relationship is each side prepared to accept – plain licence, sole licence, or an exclusive licence, joint venture?
(7) What technical and territorial fields does each side want to do business in?

(b) Technical evaluation

(1) What is the technical expertise in the field in question of the other side? In particular what is their management and staff training and their R and D record in this and related fields?
(2) What is their record for quality of production, delivery dates met, etc., design quality, product liability etc?

(3) Who are the key technical staff? What happens to the deal if they leave the company just before or during the implementation of the agreement? Could the company still fulfil an obligation to perform the transfer or as the case may be to take on and exploit fully the incoming technology?

(c) Financial evaluation

(1) What does an examination of recent accounts, dividends and profits reveal about the other side?
(2) What does an examination of its current assets, liquidity, credit rating, bankers' references reveal about the other side?

(d) Legal evaluation

(1) What is the legal form of other side, i.e. company, subsidiary, parent, partnership? What is its legal capacity? What is the identity of the present control and ownership of the business? What are the prospects for take-over or merger and by whom? What would the likely effect of such a take-over be on the transaction or the parties to it?
(2) What government controls are there on the company? What future legal developments, e.g. food and drugs legislation etc. are likely to affect the market in which the product under consideration is to be produced or sold? How will this affect the future profitability of the product?

11.2.2 Questions for licensee on the proposal

(a) Commercial evaluation

(1) Why does the licensor want to license out this product? What is wrong with the product or its business to merit this?
(2) Where does this product fit in with our existing range or our long-term development plan? Will it complement or compromise our present activities?
(3) Do we want the whole package? Can we persuade the licensor to unbundle it? Do we want to tie in any product and will it be legal to do so?
(4) What is the competition for this product and how well established is it? Will we get any marketing rights as well as the technology, e.g. any trade marks or advertising support?
(5) What are its sales figures or its estimated potential sales figures? What is its market penetration and in which markets? Can we improve?
(6) What is the cost of introducing it into our factory in terms of plant requirements etc.?
(7) What technology from our own portfolio could we offer in exchange to cut the cost of the venture?
(8) What is the balance of cost of the in-house development of an equivalent over the cost of our buying in the technology and what fringe benefits result from each course of action?

(b) Technical evaluation

(1) How old is the technology? Is it about to be superseded by some other development? Will the licensor be bringing out a product in competition which will out-perform the old technology?
(2) Is the technology already in a fully transferable state, i.e. are the installation and the running manuals and other documentation ready? What lead time will there be before production can begin?

(3) Are the standards measurements etc. of the product compatible with the market we want to serve or with our own plant?
(4) What performance and test results are publicly available for a preliminary assessment? What results can the licensor give?
(5) What is the present range of technical application of the product of process in question? Will all these technical fields be licensed or will there be restrictions?
(6) What staff training, service back-up and other personnel support is the licensor willing and able to give and how much will it cost?
(7) What improvements are or will be effected to the technology by the licensor and will these too be transferred and if so for how much? What will be the position on any improvements effected by the licensee?
(8) What more do we need to know about the technology to make an objective decision, which is still confidential and not so far disclosed? Will we be able to gain a confidential disclosure agreement to ascertain more information?

(c) Financial

(1) Will the licensor be in a position to assist in the defence of the intellectual property rights licensed as their owners? Will his or her commitment to the product be matched by financial support?
(2) Is there a sufficient profit margin on the product to enable us to manufacture under such a royalty bearing licence and still make a profit on the net returns after deduction of the royalty payable?

(d) Legal

(1) Does the company own the intellectual property rights involved, or are they held by a parent? And what intellectual property rights are involved? What is their strength, width, how many countries are covered? Have there been any successful legal challenges, or any successful infringement actions brought? What is the legal position of a licensee in the countries in which the licence is to be effective? Is there any possibility of extending the scope of legal protection for the technology?
(2) What terms are on offer from the licensor in any draft contract which he or she may have produced? What position would this put us in and what is its standing in competition law terms?
(3) Are all regulatory authority clearances for marketing of the product passed in the countries in question, e.g. food and drugs etc. What other government consents are necessary, e.g. for export control administration and the like?

11.2.3 Questions for licensor about the proposal

(a) Commercial evaluation

(1) What territories are we prepared to license in and how will this affect our existing or future business plans for the product or technology?
(2) What is the marketing record and capacity of the licensee for other products, both those which are in-house developed and those which were licensed in, if any?
(3) What is the competition? What major customers does the company have in what technical and territorial fields?

(4) What will be the future development of the other company, what resources will it devote to the marketing and exploitation of the product? Will it be put in a position to threaten our own sales?

(5) What industrial relations record does the company have? What is its manufacturing and distribution network? Can it make a sufficient turnover in the product to create an adequate royalties income on the licence for us?

(b) Technical evaluation

(1) Does the company have the technical capacity to take on the product and what technical back-up will it need from the licensor?

(2) What technical facilities does the company have and what potential for on-line improvements to the technology which might be licensed back or otherwise beneficial to the licensor?

(3) Does the company have ready access to all of the raw materials necessary and adequate handling capacity?

(4) What quality of work does the company turn out and what manufacturing experience of related products does it have? What staff training in the technical field in question is available?

(c) Financial evaluation

(1) Can all the royalties from the overseas licensees be repatriated? Are there any exchange control rules governing the deal? Are there any extra-territorial taxation implications for the royalties to be paid?

(2) What currency implications are there for an overseas licensee deal in terms of exchange rates, choice of currency, etc? Is the currency of the licence stable, strong, will it be devalued by inflation or other considerations? Will another currency be acceptable? Will payment be in kind, e.g. oil or other commodities from Eastern Bloc?

(3) So far as the party itself is concerned, can its finances and bankers cope with the agreement? Are there any guarantee facilities available such as ECGD or other sources of protection? What private insurance costs will have to be borne?

(4) Given the structure of the market we anticipate, what mixture of royalty and up-front payment do we want of the licensee?

(d) Legal evaluation

(1) What law governs the company? What constraints on the company exist? Can the company be sued easily in the courts of its home jurisdiction? What are the legal implications of any draft contract proposal which it may have submitted? Has it any history of disputes with other partners in licence or other arrangements?

(2) Are there any technology import controls to be dealt with? Are there adequate controls on use of technology imported to the countries in question for the licensor to be clear of trouble with his or her own government under export or technology regulations, e.g. will the importing country pass the technology on to the USSR contrary to law?

(3) What is the legal position of an owner of a right who has granted an exclusive licence in the country in question – what standing does he or she have in the courts to protect his or her rights? What is the legal position of foreign plaintiffs and of local defendants in the local courts? What remedies does the legal system there provide and how efficient is it in terms of procedure and speed of the legal proceedings?

11.3 PREPARING FOR A NEGOTIATION

It should not need saying that there is no real substitute for adequate preparation and homework before going into a negotiation. This falls into a number of heads:

(1) Establishing responsibility and command
(2) Selection of negotiation team
(3) Preparation of briefs to team and team orders
(4) Pre-negotiation homework and analysis of issues
(5) Preparation of agenda and documentation
(6) Agreement on tactics and desired result

11.3.1 Establishing responsibility and command

A team leader for the negotiations who has clear overall command should be chosen. This enables internal disputes to be terminated quickly and gives a reference point to every negotiation.

Who this will be of course depends on who is available within the company and on his or her qualities as an organizer and negotiator. It will be the task of this person to make the final decisions on who else is in the team, on the conduct of the negotiation and its conclusion subject to the board ratification.

Because of this the desirable figure is something of an all-rounder, with a good grasp on the technical financial commercial and legal situation, who is surrounded by people with specialist knowledge on these matters to advise and make specific contributions to the negotiation.

11.3.2 Selection of negotiation team

In so far as there is a choice the selection of the proper negotiating team has already been discussed briefly in Chapter 9 on licensing. Some expansion of this issue may be useful here for having chosen a team each member will be delegated a particular role, and this may affect the initial choice of personnel made. Given that every agreement for a licence or for a joint venture has to cover certain well-defined albeit interacting areas, each member of the team should have been selected with the delegation of responsibility for all such areas in mind. The areas of concern have already been listed in Section 11.2 as the factors which require proper evaluation in any potential agreement, as these will go into the final contract documents. The areas were:

(1) Commercial
(2) Technical
(3) Financial
(4) Legal

If the resources of the company permit it this may lead to the following sort of negotiation team under the team leader being built up for the negotiation of a major piece of technology or design transfer.

(a) Commercial

Often the team leader will be a commercial person and combine the two roles. The commercial representative on the team is concerned with negotiations of price, assessment of risk in the market and in production, capacity of the other side to come up to scratch on the commercial side of the deal.

(b) Technical

This may not necessarily always be the technical expert most familiar with the technology now under consideration, although he or she may on occasion be drawn in to give further explanations. He or she is probably better employed doing what he or she is best at, that is R and D and production line work, not stuck in the rather alien environment of a negotiation suite. Probably a senior R and D management person, well briefed in the technology but with managerial and negotiation experience is the proper person, and maybe even a pure commercial employee with a technical background. He or she may be expected to probe and respond to probing on each side's technical capacity to take on or pass on the technology respectively and to assess the performance standards etc. claimed by the other side.

(c) Financial

The accountant or the finance director will be appointed to consider the financial side of the deal and to check that the figures work out as alleged, that the tax implications are understood etc., and to look at the cash flow and similar factors.

(d) Legal

Probably the company's lawyer will fulfil this role. Having him or her on the spot enables more rapid redrafts of terms and assessment of the competition law and other legal restraints and their implications for any proposed changes to the draft and for assessment of the other side's draft terms and conditions, the legal means by which the transfer is to be achieved, the new joint venture company's constitution and voting etc. if there is to be one, performance bonds and warranties and the like. It may be necessary to have in addition a patent agent fully conversant with the details of the intellectual property if the technical aspects of the patent portfolio are central to the negotiation, for instance, where the licence negotiation stems from an infringement lawsuit compromise.

In addition to this the team must, of course, be compatible. Remember that probably only the commercial representative if anyone will have had any formal type of training in presentation and negotiation techniques. A good lawyer is not necessarily a good negotiator and the same goes for the other team members. The team leader and the team spokesman may not necessarily be the same person. There is a strong argument for the team leader being able to sit back and review what is going on while the team spokesman actually executes the opening statements of principle and discusses options.

If the negotiation does not justify such a large team then of course two members could take on the roles of the missing two, financial and legal and the commercial and technical respectively combining. But a one man or woman negotiating team should be avoided at all cost. No-one can take on all these roles, take notes and listen to and respond to points all at once.

11.3.3 Preparation of brief and team orders

Make sure that each team member has a good grasp not only of his or her own area of responsibility but of the others and of their interrelationship. Thus you must be able to brief the lawyer on the technical content of the material you are seeking to license in or out: this will affect his or her advice on the form of the agreement. Of course the same goes for the other non-technical people who must know what they are negotiating over.

Equally, all the team members must be given a clear idea of the financial position and what is and is not viable. The commercial technical and financial members should be given a run through of the draft agreements which the lawyer has drawn up to tender to the other side to make sure that they understand what the document is trying to do and how it matches up with the basic objectives which the team has set itself: what is important, what may be changed, and so on.

The allocation of tasks, who leads the discussion for the team on which points and so on must be agreed in advance. This creates a degree of discipline and balances the loading of stress on each team member. This may well involve each member of the team preparing an explanatory brief of his or her part of the negotiations to come which can be thrashed out in a meeting of the team. This provides a useful means too of clarifying the team's objectives and approach to the negotiation.

In all cases the team leader is there to take the final decision on their advice and they are there to give that advice. The lawyer can advise on the interpretation of a draft and its implications: the team leader decides whether to accept it or not. The team leader and his or her opposite number decide on what they want: the two lawyers debate whether draft 'A' or 'B' achieves that result. The roles should not be confused. Lawyers never cause the breakdown of negotiations: the failure of the team leader to lead causes the breakdown if legal points become sticking points. The same is true of the other fields of specialist advice on the negotiating team.

11.3.4 Homework and analysis of issues

As part of the briefing each member must be then permitted to do his or her homework on those matters to which he or she has been delegated and report on them before the negotiations begin. These will be on the matters outlined above in the introduction to Section 11.3.

11.3.5 Preparation of agenda and documentation

Out of the analysis and briefing reports conducted by each team member the desired agenda for the meeting and negotiations will be set out, together with any documentation which it is desired to present to the other side at or before the meeting. This will have to be cleared for confidentiality etc. beforehand and each team member should be familiar with the contents of all the documentation. The documentation will then usually include, some of them in a form edited so as to be suitable for consumption by the other side and use in the meeting, others for personal use only and clearly labelled as such:

(1) List of points for discussion and objectives
(2) Agenda for meeting, perhaps sent to other side
(3) Notes of earlier meetings and correspondence
(4) Notes of technical data, report etc. on technology
(5) Notes on existing commercial record, sales, etc.
(6) Draft contract clauses, alternatives available
(7) Notes on previous position held by other party
(8) Notes on any intellectual property right involved
(9) Notes on any existing licences in existence
(10) Notes on costings, other financial data, budgets
(11) Notes on any alternatives available from others

The last item on this list is often forgotten. The team cannot properly evaluate the offers of the other side unless they are also fully briefed on all of the possibilities of alternative sources of technology, of licensees etc., which might by comparison be a more favourable option than the party on the other side of the negotiating table.

11.3.6 Agreement on tactics and desired result

Out of this will come the agreement under the sanction of the team leader on the tactics and desired result of the negotiation; its objectives and how they are to be achieved and at what price. This should be clearly in the mind of each team member. Of course during the course of the negotiation these may change but in that case they will be changed by the team leader after a private discussion during an adjournment or similar natural break.

The list of points to be discussed noted above is of course for personal use by the team members. They should have clearly in mind before the meeting an agreed view on:

(1) What the desired outcome of each point is, what points they are prepared to concede and how far, what points they are not prepared to concede, and conditions of concession.
(2) What the arguments on their side are for taking the position they take on each particular point and in particular what the factual basis for that position is.
(3) What the other side's argument may be anticipated to be on those corresponding points and what the team's agreed response to it is to be.
(4) The limits of authority of the team, what points will need ratification from above.
(5) Who is responsible for minuting the meeting, for handing out the documents prepared for both sides to examine, etc.

11.4 CONDUCT OF NEGOTIATIONS

No one can teach you to negotiate. There are many books giving perfectly good advice and this book will not attempt to do badly in a few pages of one short chapter what others do very well in many chapters of a book devoted to the subject. A few factors may however be kept in mind when deciding on the mode of conduct of negotiations of this kind, particularly with regard to precautionary measures to be taken in advance.

(1) Firstly, the point to be reinforced is that the negotiations are always to be the responsibility of the team leader. This should be clear to the other side too.
(2) Secondly, having got an agenda try to stick to it. It forms the basis for your approach to the whole venture and imposes a structure on the meeting which makes sure that all matters of concern are dealt with properly and in proper order.
(3) Thirdly, make sure the negotiation facilities are properly set up, adequate copies of documents available and unnecessary distractions such as telephone calls, bleep alarms and the like excluded from the room.
(4) Fourthly, make sure that there are adequate lines of communication back to the home base if its sanction is needed for some change of plan and that these can be opened up relatively quickly.
(5) Fifthly, no internal dissent should be permitted in the meeting within the team. If there is a serious prospect of this then the dissentient should find some reason for seeking an adjournment so that the dispute can be discussed privately and the team leader's decision is final.
(6) Sixthly, ensure that the meeting is properly minuted and that any revised drafts of clauses are clearly represented and recorded and if possible copies of both minutes and newly drafted clauses circulated to all before the end of the meeting and even perhaps initialled as agreed variations and minutes.

11.4.1 Disclosure of confidential information

In many technical licensing negotiations it will be necessary to disclose confidential information to satisfy both sides of the technical and the financial suitability of the other for the deal in question. This means that proper measures will have to be taken before beginning the negotiation process for guarding the integrity of the information which will have to be disclosed, either on a factory visit or in the opening bilateral discussions.

The legal side of this problem has been considered in Chapter 2 earlier. This section is concerned with the practical measures involved at the pre-licensing negotiation stage. Such precautions may have already been taken by a licensor at the earlier stage where, in response to an enquiry, he or she permits a potential licensee to have access to the information as part of his or her evaluation of it. What form should a pre-licensing disclosure agreement take? The main requirements are stated below, and a sample form is reproduced in the Appendices to this book.

11.4.2 The implications of agreement

This raises the question of whether the whole bargain is to be signed, sealed and delivered at the negotiation. Generally this is a bad idea:

(1) Both sides will want to be able to review in more leisurely surroundings the precise import of the draft agreement, to check that the figures add up correctly and that the sums work out, that

the agreement involves no adverse legal consequences under competition or other laws, and that it fully reflects what is agreed.
(2) Comparisons will have to be made with the other possible partners in a transaction of which each side is aware. Is the package on offer as good as could be obtained elsewhere? Was any point missed out which should have been covered?
(3) The requisite consents for the agreement will usually have to be obtained from the boards of both sides.

But many negotiators feel that there is the need for some tangible sign of success and this could be fulfilled by the signature of a letter of intent or a heads of agreement document which goes a little beyond a mere minute or memorandum of what has taken place. The legal implications of such a letter of intent have been discussed elsewhere. If it is decided to grant an option in the technology there should be a clear time limit on this, otherwise the freedom of a licensor to license it out to others may be considerably impeded, at no further cost to the option taker.

11.5 DRAFTING TERMS OF LICENCES OR JOINT VENTURES

This is ultimately the lawyer's responsibility, to put into a suitable form the intentions of the parties, in accordance with their instructions. This section is not concerned with the individual terms of a licence or joint venture which are considered in some depth in the next chapter but with a more general question of the approach to be taken to such contract drafting in this context. This raises three issues.

(1) Getting clear orders
(2) Putting it all down
(3) Checking the result

11.5.1 Getting clear orders

One of the best ways for lawyers for either side to ensure that they are given clear orders is for them to tell the team leader what those orders are to be! It is not such a silly suggestion as might at first seem. It is possible for the lawyer to provide a checklist of all of the essential things on which there must be some agreement between the parties for the type of licence or other transaction in question, with if appropriate a draft clause which is available for use in the licence negotiation to achieve a particular result.

A suitably experienced team learns to use both its checklists and its drafts very efficiently. In this way the minimum field of the licence or other agreement is set out from the start and any additional obligations agreed can be fitted in. The lawyer and the rest of the team should have a pretty good idea of the scope of the eventual agreement to be negotiated before they enter the negotiation room at all.

11.5.2 Putting it all down

It is of course essential that the lawyer has a complete record of just what he or she is expected to finalize in the form of a legally binding document. This is the reason for the use of full minutes of the meeting to be agreed on at the time and not some time after the event. It may be a wise move to delegate the task of minute taking to the lawyer attending the negotiation – it is or should have been part of his or her training as a lawyer to take accurate concise notes of such proceedings and to draft appropriate forms to meet the needs of the parties.

11.5.3 Checking the result

As already stated no full binding agreement should be made on the spot. Both sides should then have time to inspect a full clean copy of the document as finally drafted up and inspected by their lawyers. The legality of the agreement under competition law, all the precise details of the patent and other registered rights, the schedules with copies of copyright drawings referred to in the agreement, lists of registrations, trade secret manuals and so on must all be checked for completeness and accuracy. Only then should a full binding agreement be signed and authenticated by both sides.

11.6 IMPLEMENTATION AND MANAGEMENT OF AGREEMENTS

The management of one's newly negotiated licence agreement will face different problems in the shorter term implementation stage of the licence and the longer term on-going maintenance of the licence. These should be looked at in turn.

11.6.1 Implementation of licence or joint venture

Obviously both parties hope to start the licence or similar venture in a spirit of harmony and without major distractions. But often things appear to go very wrong at this stage either through defects in the negotiation process which rapidly become apparent or through some mere unforeseen difficulty in adapting the technology transferred to the transferee. There will also be some tidying up to do on the legal side such as procuring the registration of an exclusive licensee as such or as the registered user of a trade mark, or in confirming clearance for the deal from the regulatory bodies before executing it etc. This may require the following measures.

(1) Establish a timetable for implementation and a checklist of all the procedures which must be followed to complete them. This plans out the transfer and may well have been the subject of detailed agreement at an earlier stage, during the negotiation itself.
(2) Budget for time and money making an allowance for things to go wrong. The licence agreement may well have made provision for extra payments to the licensor in respect of delay in installation needing any additional help from him or her which can be attributed to defects on the part of the licensee.
(3) Make sure that the necessary documentation is really clear and usable by the licensee. This is a major problem for the person responsible for all the technical data being transferred. Very often the data in the possession of the licensor, built up as he or she goes along, are in a wholly unsuitable state for transfer and it may involve considerable administrative duties to get them all in order. This is equally true of the lists of copyright drawings etc. which have to be incorporated into the schedules of the agreement and handed over by the licensor. Of course if the proper R and D management techniques such as those described in Chapter 8 have been followed, the problems should be much diminished.
(4) Ensure that the licence documentation package for the licence or joint venture is fully made up and that identical copies are kept by both sides. The content of this package is discussed in Section 11.7.

11.6.2 Maintenance and review of the agreement

Do not merely let the licence agreement once it is executed lie gathering dust in the filing cabinet. The licence itself or some external agreement might well be drafted to provide for the regular consultation and ongoing review of the agreement in a manner beneficial to both sides of the bargain, not with a view to revision of its provisions but rather so as to enable both sides to learn from it and develop the business relationship more fully.

This could be made the responsibility of the same person to whom the leadership of the negotiations was delegated in the first place. He or she will be best placed so far as awareness of the background to the licence itself and any unwritten understandings of the parties is concerned and also will hopefully have already established some rapport with the other side. Review of the agreement at regular intervals will have a number of benefits.

(1) As a result of a regular follow-up and assessment programme, any defects in the product or process itself which should be improved upon and improvements effected by the other side may come to light which otherwise may be missed. On the other side of the coin, any ongoing developments can then be reported and discussed more freely.
(2) Any market developments external to the product itself may be noted which could be taken advantage of by either side or which require responses of them. The reaction may be quicker if the parties have an ongoing habit of communicating as a matter of course.
(3) A better business relationship may be developed between the two sides and possible conflicts resolved at an earlier stage. This is cheaper and quicker than the alternatives.
(4) Supervision of any quality-control requirements imposed by the licensor become more palatable to the other side as part of a wider venture and take on less of the aura of an annual inspection: they are also more efficient from the point of view of the licensor than relying on self testing and reporting of results by the licensee himself.

11.7 THE LICENCE DOCUMENTATION PACKAGE

Both sides should on the implementation of the agreement have an identical full package of documents, which will be made up of three major items, discussed below.

(1) The licence or other contract agreement
(2) The technical specification
(3) The reporting and assessment documentation

There should always be a clear obligation on both sides to take the greatest possible care of the package as a piece of confidential information. As soon as the package is completed and received by the licensee it should be subjected to a proper security regime to avoid the dangers of any industrial espionage by third parties or accidental unauthorized disclosure.

It is in the interests of both sides, both for the continued business relationship between them and for their wider reputation, to ensure that these security precautions are fully complied with. It may even be thought to be desirable to set these obligations out in writing when agreeing the licence. Since the ownership of the documentation will remain vested in the licensor it is quite proper for him or her to require the licensee to take such measures, consistent with his or her being able to use the documentation as intended by the agreement to implement the technology transfer he or she has paid for.

11.7.1 The licence or other contract agreement

This will be the licence agreement itself in full as signed by the parties and complete with the schedule lists etc.

11.7.2 The technical specification and schedules

This will contain all the information to be transferred in the form of know-how and trade secrets, the formulae etc., results data and the like and will remain the property of the licensor. This is the sole detailed account of the technology and its associated information.

The precise content will of course vary with the technology or designs being transferred but it will always conform to certain characteristics, whether it be a turnkey project or a simple product or process licence. It must contain an index of the contents and be properly laid out so as to be of use in a practical way, for it will form the backbone of the licensor's obligations under the agreement. It will of course be provided that it is to be returned to the licensor on the termination of the agreement and that unauthorized copies may not be made by the licensee.

The specification will usually contain *inter alia*:

(1) Patent specifications etc. as published
(2) Background information to the know-how
(3) Methods and processes employed
(4) Design calculations, measurements, formulae
(5) Safety controls and testing procedures
(6) Security and confidentiality measures
(7) Performance criteria and maximum outputs
(8) Raw material, power etc. supply requirements
(9) Staff training manuals and procedures
(10) Maintenance procedures, inspections
(11) Legal and other regulatory information
(12) All drawings blueprints and other diagrams

Any firm developing a project of this kind will involve different people who will have different ideas about arrangements etc. It is a good idea to establish a set system and format for these documents so that when a licensing package is put together they can be assimilated into the package easily. This will involve such mundane standardization as all using the same size paper, and all having consistent drawing numbering and filing references but also making sure that terminology and all technical descriptions are used consistently in all the contract documentation with a standard glossary being used within the firm.

11.7.3 The reporting and assessment documentation

This is a desirable optional extra. It enables the licensee to record in the way which experience of the licensor has shown most effective, his or her implementation of the licensed technology and its performance. In this way any defects or failure to meet any of the warranted performance standards can be investigated together to determine the cause of the failure and to establish how it may be rectified. This should diminish the chances of any failure of the promised and actual performance to correspond leading to a full-blown dispute between the parties. It also serves the purpose of providing a ready-made discussion document and agenda for the later regular review meetings recommended above in Section 11.6.2.

11.8 RESOLVING LICENCE OR JOINT VENTURE DISPUTES

Should any dispute arise out of the transfer of the technology or the interpretation of the document, it will often be essential to sort things out outside the courtroom, and of course always desirable to do so. The disputes between licensor and licensee or joint venture partners classically arise out of the following causes:

(1) Failure to meet promised performance figures
(2) Inability to implement the technology licensed
(3) Defects in the negotiated agreement itself
(4) Incompatible technologies or partners
(5) Changes in the external commercial environment

The means of resolving the issue are three:

(1) Negotiation
(2) Arbitration
(3) Disengagement

11.8.1 Negotiation

This speaks for itself. The renegotiation will require reassessment of the calculations made in the pre-licensing stage, discussed above, and in the light of the experiences which have shown those to be either over-optimistic or defective for some other reason. But whatever the legal position on breach of contract etc., this is a small or even counter-productive bargaining tool if there is a genuine desire on the part of the wronged party to continue in the relationship which may be profitable for him or her even on less favourable terms and which may simply not be viable for the other party on the terms as agreed, for whatever cause.

11.8.2 Arbitration

Arbitration is discussed in Chapter 12 below. The decision to go for an arbitration may follow on after unsuccessful or inconclusive attempts at renegotiation of the agreement to rebalance the transaction. Equally, however, it may be used as a genuine means of escaping or simply of delaying litigation by the party who is legally in the weaker position. But the possibility of abuse is not easy to eliminate if an arbitration clause requiring action on the part of one side only to invoke it has been incorporated into the licence agreement.

11.8.3 Disengagement

Ultimately if negotiation and arbitration fail, it is possible that the parties may simply agree to bring the agreement to a complete close without litigation by a disengagement. A contract may well have provided for this eventuality if the sides were well advised by the lawyers: it is one of those matters which negotiators bent on a successful collaboration may prefer not to discuss but which their own lawyers should make them discuss. Just like a marriage it is often better if the parties can effect a complete clean break from each other.

Disengagement provisions in a licence agreement will make provision for the return of all confidential documentation, ownership of which should have been retained by the licensor; treatment of any seconded employees; payment of all outstanding royalties and other payments owed; disposal of licensed products made or semi-assembled before the date of disengagement; and consequential destruction of trade marked and other advertising material created for the purposes of the licence agreement and its execution. The question of a disengagement from a joint venture or distributorship agreement will be all the more complex, especially if a venture is not to be sold off as a going concern but simply dismantled.

12 Common licence agreement terms

12.1 INTRODUCTION AND THE GENERAL APPROACH

This chapter goes through the most common licence agreement terms to explain in lay people's terms the role they play in a licence, pointing out and commenting on some of their principal characteristics and the things to look out for or pay particular attention to. This chapter thus serves as an introduction to explain what goes into any licence and things to think about when negotiating.

12.1.1 Preliminary steps in drafting a licence

The first few points are very general. This is certainly time to go to a lawyer or patent agent. The licensors and the licensees will know the essence of what they want. They will have discussed this and the licensing team will have prepared some draft proposals.

The lawyer or patent agent will be in a position to put one or two things in that they either did not think about or they did not think about enough. The lawyer or patent agent is also another pair of eyes on the transaction. He or she will put the agreement into the terms precise enough and clear enough to eliminate so far as possible the uncertainties from which disputes arise – we hope.

To repeat a point made already, if a dispute does arise the court will look first not at what you thought you had agreed, but at what the common intention of the parties appears to have been from the text of the written agreement. Only if the words used in that written agreement are uncertain in meaning will the court hear any evidence of what the two parties thought they were agreeing at the time.

12.1.2 The appearance of the licence agreement

Often a licence agreement does not contain much punctuation to speak of. Often you will find that it says that headings and commas used in the agreement are for the purpose of ease of reference or reading only and shall not affect its interpretation. The origin of this practice lies in the courts ignoring punctuation when construing a document. You must remember that the rules and significance of punctuation were for a very long time simply not fixed at all and they gave very little meaning which could be agreed upon. And the common lawyers were then not very distinguished for their literary or linguistic prowess. This practice led the lawyers to draft their documents, including all the Acts of Parliament, so that they would all be clear and unambiguous without punctuation. It is true that more recently the courts have begun to pay some attention to commas and the like and punctuation can make complex clauses easier to read – but why make them complex?

The best compromise is to cut the agreement down into easily manageable clauses each divided into a series of numbered paragraphs as uncluttered as possible. A collection of whereas's hereins and the like may sound legal but they only serve to confuse. What do they mean? Good draftsmen and draftswomen do not use them more than is necessary. Also beware of set legal phrases. Set legal phrases have precise meanings. Their use by people unfamiliar with those meanings is a dangerous thing. It will be difficult to persuade the court of a meaning different from that with which it is familiar. The lawyers do – or should – know what those meanings and uses are.

12.1.3 Precision in grammar and language

Finally of course this is an area where sloppy grammar and language is fatal. To take an example, compare:

(1) 'Within 30 days of the termination of the licence the licensor shall have returned to him confidential plans and drawings of the licensor's plant'
(2) 'Not later than 30 days after the termination of the licence the licensor shall be entitled to demand the return to him of the confidential plans and drawings of the licensor's plant'

Let us assume that the parties have taken the trouble to define what precisely is meant by the licensor's plant for the purposes of the agreement, which is by no means always the case even in quite sophisticated agreements, if possible.

In the first clause the parties have agreed that in the period beginning with the date on which the permission under the licence ceases to operate the licensor shall be entitled automatically and without demand to the return to him of certain documents – it is not clear whether the drawings as well as the plans are confidential. Nor is it clear whether all the copies or only, for example, one copy of each relevant document has to be returned to the licensor.

In the second clause the parties have agreed that in the period following the commencement of the licence and up to 30 days after the termination of the permission granted under the licence the licensor shall be entitled to claim the return of the confidential plans (all of them) and either the confidential drawings or all the drawings of his plant – it is not certain which – but he is not entitled to them automatically and if he does not put in a demand within that period he cannot claim to be entitled to them at all thereafter. On the other hand he can claim them at any time during the continuance of the licence, even before its termination, when the licensee will presumably still need them.

12.1.4 The structure of the licence

The overall structure of the licence should flow in a logical order: from explaining the background to the transaction in the so-called recitals, to stating precisely what is meant by certain phrases which are defined early on; then specifying the duration of the agreement; dealing separately with each of the rights licensed and the scope of permission; providing for fees and royalties in return for those permissions and accounts; dealing with challenges to the validity of the rights licensed and how the parties will respond; the way in which improvements will be dealt with; how the licensee is to behave when in possession of these rights and how he or she is to do his or her best to exploit them so as to maximize the royalties; making provision for termination of the licence and its consequences; and finally dealing with administrative things.

In looking at the standard sort of clause you can expect to see in any licences this basic order of treatment will be used in this chapter.

12.1.5 Using precedent books for drafting

The use of the many available precedent books for drafting an agreement is useful – if they are used properly. The point of a precedent book is to enable you to compare your effort with an expert's effort in the same field in dealing with particular points. It is not intended to be a book from which you just cobble together a collection of directly copied phrases and to do such a thing is extremely dangerous. The approach to take is to set out a list of the things you want in the licence, to set out as precisely as you can the desired result and then to get a professional draftsman or draftswoman lawyer to formulate these provisions and compare his or her efforts with your original wishes.

12.2 RECITALS CLAUSES

Many licences you will come across begin with what are called recitals. Recitals generally either tell the story of how the parties have come to this point or set out the basic purpose and objectives of the agreement. Some draftsmen and draftswomen clearly regard this as unnecessary. And in a sense they are right of course. But recitals are not meaningless. They can get you into trouble.

Firstly recitals state the facts. And so where the recitals have unambiguously represented that one of the parties is or has something that party, who has signed the agreement, is estopped from denying that he or she made that representation even if it is not true. If it is not true that can mean a breach of contract warranty giving rise to an action for damages and even allowing the other party to escape the contract altogether.

Further, where any other parts of the agreement appear ambiguous the courts will often reserve a right to construe the licence agreement in the light of the recitals in order to be able to come to determine the true meaning of the agreement. And so carelessness in the recitals can be a very expensive mistake. Since they are not strictly necessary, why have them at all? The short answer is convention.

12.3 PARTIES TO THE AGREEMENT

These must be fully defined in a schedule or they might equally well be placed at the beginning of the licence agreement if desired. Two points should be made.

Firstly, it is important of course to make sure that the full correct name and address of the parties are set out and, especially where the company or other party is an overseas one, also to establish clearly what sort of trading unit it is and the law under which it has been established or incorporated. You are going to have to be sure that they have the legal capacity to enter the agreement and that they can be sued if things go wrong.

Secondly, it is important that you make sure that you know who is the licensee in the sense that it is just one entity and not also its associated group of subsidiaries etc., whom you will generally not want to be able either to exploit your technology or take assignments of your technology.

Apart from all this, knowing the address of the other side and therefore where to serve writs etc., is equally important! All of this information should have been sorted out at evaluation stage, but checked again here.

12.4 DEFINITIONS OF CLAUSES

It makes life much easier and much reduces the risk of internal conflicts if you define the key terms right at the beginning and stick to them throughout the agreement. Some draftsmen and draftswomen scatter the definitions about in the agreement. That is bad practice. All the rights relating to the licensed products or processes must be clearly identified, preferably in a schedule, but also – even though this is extremely difficult – the extent of the know-how to be disclosed should be defined as clearly and precisely as possible, preferably by some reference to the drawings etc. which contain it which is supported by a retention of copyright in the contents of these drawings etc. and a retention of ownership of the papers on which these drawings are made and recorded. Where the know-how included manpower the best that can be done is to specify the nature of the duties of the staff seconded or loaned under the licence.

12.4.1 Territory

The definitions section will also specify that territory for which the licence is to operate. Again precision in drafting is essential. To say the North of England will not do. Support the definition by a map where possible so that conflicts between adjacent territories can be avoided.

Do not forget that when dividing up the licensed markets territorially you can get into trouble with the competition law provisions, especially in the EEC and USA, considered in Chapter 14. A patent is only valid for the territory in which it is granted, hence the necessity for applying overseas for patents where you want an overseas market. Similarly you will have to license those patent rights.

The USA view is that once you have consented to the grant of a licence to use, manufacture and sell in one country, the licensee then should be free to export to any other country where there is no patent protection and any restriction on so doing in the licence will be void as an abuse of the patent. We will be dealing with the limitations imposed by competition law on certain types of patent and other licence clauses in Chapter 14.

12.4.2 Packaging rights

Often a licence gives a package of rights. You may find separate licences for each right. The advantage of this is sometimes said to be that the package contains less risk of inconsistencies between the terms which govern those rights licensed than a series of licences, though a good draftsman or woman should not be troubled by this. The danger of a package, dealt with later on, is what is to happen if one element of it fails or proves to be invalid.

There are arguments for and against packages of this kind. There is no difference in the cost of having a series of agreements drawn up by the lawyer for each right or having them bundled up in the same agreement. Much depends on the personal preference of the lawyer advising the parties.

12.4.3 Field of use restrictions

If appropriate there should in the definitions section be reference to any field of use restrictions which are being imposed. This may involve competition law problems, discussed in a later chapter.

12.4.4 General points about definitions

When filling in the schedules to identify the intellectual property rights licensed, there is relatively little difficulty in the case of the registered rights. You simply specify the country in which they are registered, the number, date of grant and title.

In the case of trade marks, especially where common law trade marks are involved, but also where there are registered marks, the mark itself should be incorporated into the schedule of the agreement with a permitted form of the mark and a statement defining in what circumstances the mark can be applied by the licensee.

In the case of unregistered copyright material copies attached to the licence or held somewhere safe and identified in the licence agreement, i.e. on deposit, are the best ways of defining what is meant to be covered by the agreement, whenever possible.

12.5 SPECIFYING THE DURATION OF THE AGREEMENT

This is a very important part of the agreement, dealing as the title says with its duration. The full duration of a licence will vary with the nature of the licence and its commercial purpose. Thus generally the duration of the licence agreed on by the parties is going to depend not only on the nature of the market and product but also on the capital investment input on the part of the licensee, so that he or she has an opportunity to make some profit out of the transaction.

When know-how is involved, one possible approach is for it to provide for either side to pull out at any time after the expiry of the first period of the agreement by giving written notice after the expiry of an initial fixed period so that the parties are both locked in for a certain time.

Clearly it is in the interests of the licensor to try to avoid a smash and grab operation over his or her know-how. But it is not essential. This is a fundamental commercial decision to be made by the two parties – the licence simply reflects what the parties decide. Many agreements simply provide for the early termination of the licence by either side at will so long as adequate notice is given and arrangements are made in the licence for what is to happen to any of the existing stocks of licensed products not yet sold at the time of the termination etc. Another way of looking at such a 'two time zone' agreement is to regard the first time zone as probationary or a trial period, depending on whose point of view you are looking from, followed by an automatic option to carry on subject only to one side opting to back out.

The second thing to note is that where you have a package of rights which are interdependent of each other the licence will usually be expressed so as to terminate simultaneously in respect of them all. But not always. Usually we would have to deal with this matter in some detail.

As a matter of the EEC and US competition law where rights of limited duration are the subject matter of a licence, i.e. a patent which endures only twenty years, the licensor cannot demand royalties after the expiry of the right licensed. So if you license a patent plus know-how, the royalties will have to be adjusted to take account of the expiry of the patent even if the licence continues in respect of the know-how.

Know-how has no fixed life, but although the UK law in itself does not prevent a contract to pay royalties for know-how which has become public this will again be subject to more stringent provisions in EEC law. Trade mark licences and therefore royalties can in theory go on for ever.

In fact it is fairly common to provide that even after the expiry of the know-how licence, the know-how licensee will be able to continue to use the know-how, albeit without the right to claim any technical back-up or support, simply because it will be virtually impossible to prove his or her continued unauthorized use and in any case after a short time the know-how is likely to have become almost inextricably mixed up with his or her own products and be quite impossible to disentangle. This is why the question needs so much care in a licence.

Whatever the provision for termination it is generally a good idea to have a fixed term. Not only does it force the parties to a long-term relationship to review it periodically, but in some countries fixed-term contracts are apparently required by law anyway.

12.6 COMMENCEMENT DATE

Note the specification of a commencement date. This need not be the date on which the document is signed or comes into force. It might be some time ahead. Given the obligation often set out in licences, to communicate know-how immediately on the licence agreement coming into force, the commencement date will in fact probably be set a short time after the date of signing. Very occasionally you may come across licences which have a retrospective commencement date, where the licence has been agreed in settlement or compromise of an infringement dispute out of court. It is very bad practice to allow someone to go into production before the agreement has been signed or fully concluded, not least because it takes the pressure off the licensee to conclude the bargain.

12.7 THE SCOPE OF THE LICENCE

Four points may be made here.

The scope of the permission to exploit might vary enormously, depending on the nature of the licence we are dealing with. It might be a licence to use for one's own purposes, thus incorporating it in one's own products, but not to sell the invention separately as such. It might be a licence which restricts the range of commercial uses to which the licensed product was put, reserving other uses to the licensor or perhaps to other licensees, though such so-called field of use restrictions are very difficult to enforce and may have undesirable EEC competition law consequences. They are useful when it is possible to gain higher royalties in some fields than in others because of different market conditions.

It is here the choice between exclusive and non-exclusive licence is made. Generally the choice of form will be determined by the market forces and the nature of licensed product. We have already discussed in Chapter 9 the implications of the different types of licence commonly employed which for convenience's sake we labelled exclusive, sole and non-exclusive. It is quite possible to mix up the types of licence in respect of the same right and territory over different time scales. For instance, you might grant *X* an exclusive licence under your know-how for the first five years of its life in the UK but a non-exclusive licence in respect of other countries of the world, with a non-exclusive licence everywhere thereafter. There might be exclusive licences for some of the rights in the package but non-exclusive rights in respect of others.

The licences granted may be packaged up but may be said to be separate and to be independent. Again this can be an important safeguard for the licensor. There may of course be cases where the licensor would not want any of the provisions of the licence to continue in force if one of the rights licensed proved to be invalid – say the patent. But equally it might well be the case that he or she did want the agreement to continue in force. For instance where the know-how communicated was an important part of the licence he or she might want to continue to extract royalties in respect of the licencee's continued exploitation of that know-how in connection with the now invalidated patent invention and would not thereby lose all control over the licensee's activities.

The licence will often provide for execution of formal licences or agreements for the purposes of any registration which may be so required in the licensed territories. These formal requirements will differ from country to country and it may be quite essential to have such formal licences executed in order to make the licence agreement effective in that country. A number of formal provisions of this kind exist for the UK which it is important to stick to. We looked at these so-called mechanics of licensing in Chapter 9. Then provision should be made for the execution and the 'annexation' of formal licences of the patent etc. licence in accordance with the statutory provisions. You might expect to find express provision for allocating the expense of these matters as well as preparing the licence agreement itself. It is not very expensive and usually the party wanting to register pays. But patent licence registration can be very important in some countries, where exclusive licences cannot be enforced without registration.

12.8 KNOW-HOW DISCLOSURE TERMS

The licence may require disclosure of know-how. It might be found that this obligation is fulfilled by the provision of a document which would occupy two or three pages of a book or technical manual annexed to the licence. Alternatively the two parties may shift all of this into a separate know-how consultancy agreement providing for payment on a person day basis but allowing a certain number of 'free' days at commencement of the licence.

This may be an extremely expensive term of the licence for the licensor to comply with, especially where it involves the training of the licensee's staff, especially in technical back-up – hence the requirement often found that the licensor should consider it to be necessary before he or she is committed to help.

In practice it will often be necessary to specify fairly precisely what the obligations of the licensor are in terms of provison of staff, training time, qualifications of staff, conduct of staff, allocating responsibility for insurance in the event of the accidents, provision of meals etc., and confidentiality. Although this is very difficult it may in some cases be possible to specify a certain number of hours of service free, with a charge for the excess. But particularly where the expertise of and therefore the precise scope of assistance required by the licensee is uncertain, it is virtually impossible to draw up fully watertight terms on technical assistance and back-up and it is at this point where mere trust and the licensor's desire for royalties arising out of a successful licence are the only real guarantees. The licensee will generally be highly dependent on the good faith of the licensor, especially in the early days of setting up production.

12.9 DEALING WITH INFRINGERS

Again this is a business decision. Though this might be appropriate for a sole or exclusive licence, especially given the procedural advantages which the exclusive licensee enjoys, where there are a number of licences it makes sense and convenience for the obligation to be on the licensor, and for provision to be made for a reduction in royalty if he or she fails to take steps to pursue infringers of whose existence he or she is notified by the licensee or of whom he or she becomes aware through other means. But there are a large number of variations on the rights and obligations of licensees and licensors in this area, which can lead to some very complex licence provisions concerning indemnities, court costs, damages, etc.

On the usually unfounded assumption that the third party does not make the licensor a co-defendant, in any proceedings in which he or she challenges the validity of the right licensed, there should be some provision under which the licensor is required to give the licensee assistance.

Where there is any challenge to the basic validity of the rights a very strong licensee might be able to extract a term in the licence providing for suspension of royalties or at least payment into a suspense account to be held pending resolution of the dispute or even to be used exclusively in the defence of the right challenged.

Even though we have advocated trying to place some responsibility on the licensor to take action, he or she may find continual action simply uneconomic and to insist on action in every local case might be unreasonable, so the licensee ought to retain some level of competence to bring actions on his or her own initiative with assistance from the licensor; and if the licensor refuses to take the initiative after a demand by the licensee, the licence might give the licensee the right to do so together with an indemnity to be paid in respect of the licensee's costs in acting in the licensor's name.

12.10 TRADE MARK CONTROL CLAUSES

We saw in Chapter 5 how vulnerable trade marks can be under UK law to misuse and this is often even more true of overseas countries. Hence a well-drafted agreement provides for particularly strict control of the licensee's use of the trade marks licensed. Indeed in many licences the licensee is not granted any rights at all in respect of the licensor's trade marks and has to market the goods produced under his or her own name and promote entirely at his or her own expense.

However, in most licences the parties each have sufficient to gain commercially from the use of a well-known licensor's mark to be included with appropriate safeguards and some licensors may require use of the mark on all trade literature and invoices etc. connected with the licensed products.

Although it is not needed, you might often find an article in this sort of an agreement expressly prohibiting use of the marks in respect of other than the licensed products or components or in respect of products failing to conform with the standards of quality set by the agreement. Again, often this sort of article will be closely linked in with obligations to do with marketing.

12.11 CONFIDENTIALITY

In most licences there will be an imposition of an obligation of confidence in respect of the know-how and copyright material which is being licensed under the agreement, restricting both use and disclosure to third parties. This will provide expressly for the return of all tangible materials on which this information is to be found at the end of the agreement. These clauses can and often are much more complicated, making detailed provision for procedures to be followed ensuring that the integrity of the know-how is preserved.

12.12 QUALITY CONTROL, PROMOTION, ETC.

A licence may impose three very heavy burdens on the licensee which we must examine in turn: maximizing demand; quality control; and product liability.

12.12.1 Maximizing demand

What may broadly be described as assurance of performance clauses undertaken by the licensee cover a vast range of clauses from the vaguest 'best efforts' clause to very precise obligations of minimum turnover or royalties clauses with obligations of promotional budget expenditure and the like.

What exactly is undertaken will depend on a number of commercial factors. Best efforts is of course all that the parties may expect commercially but it is rife with dangers from a legal point of view because of the difficulty of proof of best efforts or the lack of them should any dispute arise. The clause becomes legally almost meaningless.

There is quite a lot of complicated case law in the USA on what best efforts are, but above all it is clearly a variable quality and in some ways it is a cop out – the parties have used it as an excuse for not investigating clearly enough what they may anticipate the realistic market is, or alternatively it has been used as an escape route when negotiations for minimum standards of performance or target figures appear to have broken down, and ultimately the parties are burdening the courts with the job of deciding what those proper figures should in fact be.

In England the early cases imposed very high standards on licensees undertaking these obligations but with the passing of time they have been watered down to what a reasonable licensee would have been doing in the circumstances, and furthermore all such clauses do tend to be interpreted in favour of the licensee in any case of doubt.

For example in one American case the parties agreed that the licensee would employ his best endeavours to satisfy the demand in the licence area for storm windows which he was to market under licence. The court held this to mean existing demand only and that it imported no obligation to seek to expand that market or to develop interest in the product beyond that which arose of itself.

If such a clause is to be used it is a good idea to supplement it with an obligation to make regular reports on what the licensee has been doing to further the success of the products licensed, since this will provide useful evidence in the event of a dispute arising at a later date and also give an early warning of any problems to the licensor. It will be much more difficult to extract this information in retrospect once you get to the stage of litigating the point.

The draftsman or woman should be careful not to draft his or her clause in such a way as to impose an obligation not to deal in the products of a competitor, for fear of invoking competition law problems: a mere best endeavours clause has been held not to imply anything more than an obligation to give the licensor's products no better than equal treatment and promotion to that accorded by the licensee to the licensor's competitors' products.

In any case, reasonable efforts will usually be implied by a court into a licence where the royalties are not fixed, at least in case of an exclusive licence. Perhaps in some ways it would be a good idea to eliminate the clause altogether from a commerical point of view if it makes you do your homework a bit better.

From a legal point of view there is no doubt whatsoever that a best efforts clause is best replaced with a clause laying down either minimum turnover obligations or a minimum royalty backed up by precise obligations to advertise and promote. If the market is particularly difficult to predict these could be supplemented by provision for a review of the figures at regular intervals – in itself a good thing anyway, to ensure that the agreement is kept under review in general – with provision for arbitration in the event of disagreement, but this is subject to competition law issues.

From the courts' point of view all that has to be done to establish compliance with or non-compliance with the obligation is to produce the books. A realistic target imposes of itself the same levels of activity as a good faith and diligent performance or a best efforts clause. The licensor may wish to go further and direct the nature of the advertising material and its mode of presentation, invoking the right to veto material or require advance approval of it and will frequently do so if the trade mark is being licensed along with the product so as to protect his or her own goodwill – the Trade Marks Act itself requires quality control of the licensee's product and it is only sensible to extend the quality control to the advertising and not to remain content with merely financial obligations. Again these are an improvement over best efforts clauses in terms of certainty. Both parties know where they stand.

There is of course the risk that the licensee may feel hamstrung by such an agreement and if the licensor exercises a great deal of control over these matters it may be less easy to justify imposing high turnover of sales figures when part of the means of getting these has to some extent been taken out of the hands of the licensee who might have done things differently.

These sorts of clause will always be found in franchising licences where it is really goodwill and image which is being licensed and where such controls are essential to the value of the licence for other licensees: one of the rare cases where the more licensees there are the more valuable the licence will be.

In sum, the best efforts clause although it can be useful as a supplement to minimum standards clauses can do only a little more than the court would in any case imply into the contract and on its own is extremely unwise. It should not be used as a means of avoiding the difficult negotiations which minimum standards clauses inevitably involve and which give both the parties and the courts something objective to look at and judge by.

Firstly, the licence may impose a rather nebulous obligation to maximize sales and contacts. But many licences go much further in imposing minimum royalty clauses under which an assumed minimum turnover is guaranteed. The main advantage of a fixed level is that there is no problem about whether the licensee is in breach or not. The intermediate stage is to specify the minimum size of marketing budgets and so on. The licensee will quite rightly be unwilling to accept either a minimum production clause or a minimum guaranteed royalty, without qualification at the very least. He or she might also be unwilling to disclose the names and addresses of successful contacts though these are sometimes required by strong licensors.

Of course, where there is a joint venture much more extensive provision might be made especially with regard to the development of the business goodwill.

If there is a minimum royalty provision it might well be backed up by some form of guarantee of performance together with a bond with a bank or similar institution.

12.12.2 Quality control requirements

Secondly, a licence may impose certain quality requirements. Once again this is a very simply drafted clause indeed. One might in practice expect to see a more detailed clause, detailing what these standards etc. are. There are a number of methods whereby a clause of this nature can be given more teeth, but some of them involve unpleasant consequences to do with the competition law authorities.

One way is to confer rights of inspection of the product and even a veto on material which falls below an agreed standard or in extreme cases of imbalance of bargaining power or incompetent negotiators, a standard which is imposed unilaterally by the licensor. Clearly this will not be very attractive to the licensee, but if the goods are going out under the licensor's trade mark he may have a legitimate interest in this type of clause.

Another way is to require the licensee only to use approved suppliers of components or raw materials. If these suppliers turn out to be the licensor or companies in some way associated with him or her he or she can be in big trouble with the competition law authorities.

A third way, often combined with the first method mentioned is to agree in a schedule a very specific set of standards and specifications for the product and its components and raw materials, combined with guarantees of certain quality control procedures to be followed by the licensee. There may be provision for any changes in standards to be notified. The licensee may be more willing to swallow clauses of this kind where there are reasonably comprehensive provisions for the technical back-up and assistance from the licensor in ironing out production problems in the licensor's plant.

12.12.3 Product liability

Thirdly, a licence may allocate all liability for defective products. This may be an important provision in some states where liability for defective products can bypass the retailer and the manufacturer and go on to the head of the licensor as in USA and Australia. The cost of these, including for instance recall of defective products and payment of damages to third parties might be vast, quite apart from the loss of business and set-up costs, etc. A new product liability directive is about to be implemented in the EEC too.

A possible distribution might well be to allocate loss arising out of the inherent nature of the technology licensed to the licensor and liability for loss arising out of the manner of its use to the licensee although there may then be great difficulty in resolving what the cause is in many cases.

One point to bear in mind is that the restrictions in the UK on excessive exemption clauses, which the UK's *Unfair Contract Terms Act 1977* applies in certain contracts, can never apply to contracts to the extent to which they only relate to the creation or transfer of any of the intellectual property rights: this for our purposes includes licences of intellectual property rights. If liability is distributed in this way rather than being thrown onto the shoulder of one or other of the parties, it may be important to include in the agreement a provision making it clear just who has to show who was responsible.

The distribution of liability in this way may make it easier for each side to insure at a reasonable premium and that may lead to express requirements to firstly notify each other of any claims in respect of product or process liability as soon as possible, and also to take out insurance to a certain level, if it is available, perhaps even with provisions that the insurance company notify the one party when the other party has taken out the appropriate cover. Certainly the licensor could legitimately insert a clause prohibiting continued production by the licensee if the products are actually defective or dangerous and under his or her trade mark as opposed to below his or her specified quality minimum. There may have to be provision for some independent arbitrator before such a clause can be invoked.

12.13 TERMINATION

There are four initial points to be made.

Firstly, a licence may provide that even quite trivial breaches by the licensee may occasion a right in the licensor to terminate the agreement immediately. Usually no such provision exists in the agreement in favour of the licensee, as perhaps one might expect in a licence agreement drafted in a licensor's favour. For any such termination for failure to pay it is common to insert a provision requiring thirty days notice to rectify.

Secondly, the termination of the licence is often expressed to have no effect on the licensee's continued obligation of confidentiality or of liability to account for any existing royalties owed and to permit access to the accounts. These are agreed to survive. You might expect in a licence to see express provisions for the obligations of indemnity in respect of claims arising after termination to survive in a similar fashion.

Thirdly, the sanction of termination for even trivial breaches used in the draft is something of a sledgehammer to crack a nut. It is probably better to provide for a very high rate of interest on payment defaults and for similar liquidated (i.e. agreed) damages clauses for specific failures; though in drafting these one has to beware that they are not drafted in such a form that they are taken by the courts to be so-called penalties and consequently invalid and unenforceable, and for that reason prudent licensors do not usually try this route.

Fourthly, in practice one would expect to find fairly considerable provisions of an administrative character dealing with the consequences of termination, especially a premature unilateral termination, thus enabling each party to disengage relatively cleanly and painlessly.

12.14 THE INTERPRETATION CLAUSE

Often you will find that the draftsman or woman attempts to make the contract as written down pre-eminent and to exclude from a court's consideration any other oral or written statements which may have been made earlier and which may have induced or accompanied it. Such a clause can (in English law at any rate) only be partially effective in excluding external matters from impinging on the parties should a dispute arise, both in regard to representations made before and terms agreed at the time of the contract. It is usual and quite sensible to provide for any amendment to be in writing and to be signed only by authorized parties. Again this cannot under present English law be completely watertight but it is good practice.

If there is to be a copy of the agreement in more than one tongue it is sensible to make one only the authentic version of the contract and to expressly deny the other translation any significance at all in the interpretation of the agreement so as to eliminate the possibility of ambiguity arising out of conflicting documents.

12.15 ADMINISTRATION

Any licence must make provision for some everyday administrative matters, of which little need be said. One might at this point make provision allocating the expenses involved in preparation and engrossment of the agreement's formal documents. Practice seems to vary as to who bears this cost of preparation which is usually a relatively small sum compared with the sums changing hands for the licence grant.

No mention has been made of payment of renewal fees for the patents and so on. If the licence is an exclusive or sole licence you might well expect to see the licensor insert provisions requiring the licensee to take the responsibility for all of these. In practice some do in fact omit the question altogether: for completeness's sake you might ask the licensor to take responsibility and then make provision for the licensor to invoice the licensee for the renewal costs paid, perhaps including some administrative overheads. The point is that if you insist that the licensee pays direct and he or she forgets you are in trouble, so the licensor may prefer to be sure that the job is done even if he or she wants to claim a full indemnity or reimbursement later.

Usually the licensee is placed under an obligation to put patent numbers and other suitable marks on the licensed products, which is of course sensible. In some countries it may even be required as a precondition to obtaining damages for the infringement.

12.16 WARRANTIES AS TO VALIDITY OF RIGHTS LICENSED

Often too the licensor expressly disclaims any representation warranty that the patent or Registered Design Registrations (not for some reason the trade mark) is valid, so that he or she will not be in breach of contract and liable to pay damages to the licensee if one or more of them turns out to be invalid. At most the licensee will be entitled to stop paying royalties in respect of the invalid right or altogether, depending on whether the whole agreement hangs on that right. This should be compared with the recitals where often a licensor does claim to be the owner of patents etc., whatever their validity might be.

Clearly it would be very difficult to extract an open-ended warranty from the licensor but a more limited one might be available which indicated that the invention is to the best of the licensor's knowledge his or her original invention and properly patentable, properly drafted so as to be valid and not subject to any third party rights (such as those of an assignee, employee inventor or secret user). They do not constitute an absolute warranty, but one only of facts known to the licensor who should therefore make known to the licensee any claims or challenges or oppositions of which he or she is aware. Provision should also be made for such communication of claims after the commencement of the licence.

12.17 SUBLICENCES

If sublicences are to be permitted we would have to consider much more complicated provisions concerning almost all of the licence, including payment, accounts, policing the agreement, and so on. This is too advanced a topic for discussion in an elementary book such as this.

12.18 CHOICE OF LAW

This can be an extremely important provision, one dealing with the choice of law governing the contract. Many licensing transactions are international, involve parties from more than one country and trading across boundaries. Some laws are more favourable to licensors than others. It is usual to make applicable the law which the stronger side in the licence negotiation prefers! This will usually be that of his or her own country.

Failing that the place where the contract is mostly to be performed and where therefore the greatest likelihood of actions involving third parties arising might be chosen. Some countries actually require this. If the law of the agreement is other than English law then of course it is essential to take expert advice from a local practitioner on the meaning of the agreement and its implications in the local law before you sign. Make sure you know what the local law is.

There might also be an express contractual provision consenting to the courts of the country whose law is stated to be applicable having jurisdiction in such cases. It is generally not a good idea to name a law of a country which has no genuine connection with the contract parties, or which varies according to who is defendant and who plaintiff in the action. This may sound a neutral stance but adds unnecessary risk and uncertainty.

12.19 ARBITRATION

Obviously, anyone with any sense is going to attempt to negotiate away the dispute first, but failing that litigation is not necessarily the next step. Where the kernel of the dispute is some fact, rather than the interpretation of the agreement or a point of law, arbitration can be quicker and cheaper than litigation and it is usually a sound idea to make provision for arbitration.

Although business people often claim to prefer such an arbitration, if only because they think that it may go some way to excluding the lawyers, it is not necessarily cheaper or quicker or less formal, although it usually is more private. The disadvantage of arbitration is that arbitrators usually will simply not have the powers of enforcement that the courts enjoy and they can be used effectively by someone who thinks himself or herself in the wrong to delay the inevitable. From the point of view of the innocent party it is usually best to win in court if there is any risk of non-compliance. The fact that arbitration awards will have to be enforced against an unwilling party leads to clauses in most such licences providing for automatic enforcement in court.

Nevertheless, the general consensus is that such arbitration clauses are a 'good thing', though some people refuse to have them in at all. So far as those who support arbitration are concerned, the main reasons for their support are said to be the speed, the privacy and the cheapness of arbitration compared with court proceedings, though this will not always be such an unequal equation, and depends on which country we are talking about.

As far as England is concerned court fees in themselves are nominal and have to be set off against the usually high cost of paying the arbitrator, as well as hiring the hearing accommodation, shorthand facilities and so on. There is also the point that experts in the field may then be appointed as the arbitrators, i.e. people who know the business well, whereas the courts may be mere amateurs in the field.

There is no doubt as to the advantages of such an arbitration when the dispute is one of pure fact. But where the dispute includes questions of law or mixed law and fact, there may be good reason to go straight to the courts and to waive the right to arbitration because any decision on the law by an arbitrator will usually be open to appeal to the courts anyway, so you might as well go there in the first place.

It is often said that arbitration over matters on which both sides can afford to lose is fine but that as soon as the stakes are really high, the abitration merely becomes a preliminary to litigation, a rehearsal at best to help each side work out the odds. Moreover there are some people who believe that the practice of arbitrators tends to compromise and reconciliation rather than finding the rights and wrongs of the dispute. Many people find the arbitration both less hostile than a full-blown court case and the procedure more flexible.

On the other hand this may lead to less certainty of result and the arbitrator's interim powers are not as great as those of a court in England, although his or her final decision may in many cases be more final than that of a court, since there is a greater opportunity to exclude appeals from arbitration than from a court of first instance.

Accordingly it is important to have an arbitration clause which is well and precisely drafted doing what you want it to do. It is far more difficult to induce someone to submit to arbitration after a dispute has arisen than to put in a clause right at the beginning of the agreement's life, when drafting the rest of the licence. This is not conciliation, an attempt to get the parties to reconcile their differences by negotiation: it is a finding for or against one of the parties, a finding that one is in breach of contract.

Arbitration clauses come in a wide variety of formats and they are going to have to provide solutions to a number of problems. How is the appointment of the arbitrator to be effected? Is arbitration to be made compulsory, initiated unilaterally, or only by mutual consent? Are there to be appeals? What about the legal enforcement of any awards? On what matters is such an arbitration to be available? And so on. This is a matter both of business judgement and professional legal advice.

You may for a number of reasons find yourself dealing with an arbitration regime which is not to be English: usually because the other party will not accept English arbitration procedures or English law and is in a stronger bargaining position. For those agreements involving parties from more than one state there are a bewildering number of regimes to choose from, of variable quality and cost. Still, at least there is a choice. By and large the best advice is not to try your own drafting but to use one of the standard sets of regimes of the Arbitration Act or one of the larger and reputable arbitration organizations. Do think of the cost of overseas arbitration proceedings if they require oral evidence and attendance abroad, requiring local advisers and experts, hotel and travel costs, interpreters, etc. Some of the best known of the organizations and regimes are the London Court of Arbitration, the International Chamber of Commerce, the American Arbitrators Association, and the UNCITRAL (United Nations Conference) rules; in the Soviet Union there is the Soviet Foreign Trade Commission. Some of these may be more acceptable to the other overseas party than an English clause and arbitration, but some of the problems in taking on a neutral territory should be borne in mind when looking at these clauses. Whatever you do decide you should look at any clause you do select to examine what it says about the following things:

(1) The law and place of the arbitration
(2) Nature of disputes in arbitrator's jurisdiction
(3) If arbitration is supposed to exclude the courts
(4) The number and method of selection of arbitrators
(5) The method of voting – unanimous or majority rule
(6) The language of the arbitration proceedings
(7) The enforcement of the award of the arbitrator
(8) Costs and expenses of the arbitration
(9) Witnesses and legal representation

The advantage of one of these institutional types of arbitration regime will often be cheaper facilities for the conduct of the arbitration: the penalty is very often a reduction in the flexibility of the procedure.

12.20 ILLEGALITY OR INVALIDITY

It is common to find a licence clause which deals with the possibility of one or more of the licence's provisions being found invalid or ineffective through illegality – usually competition law problems. It is a safety valve clause designed to mitigate the damage done by the finding. In the event of one clause being found invalid the remainder is to continue in force and the parties are to attempt to negotiate a provision to comply with the requirements of the law governing the contract. The contract is to terminate only if they cannot reach agreement but as always with the licence provisions on secrecy and confidentiality surviving even termination. Some licences make provision for one party to hold the destroyed clause essential and to terminate at once.

12.21 *FORCE MAJEURE* CLAUSES

The *force majeure* clause serves to excuse both parties from performance for as long as the factors enumerated subsist. These might be drawn in a very vague broad way or be enumerated very precisely and narrowly. Obligations may be suspended but do not suspend the obligations of confidence! There should always be provision for notice to be given to the other side before the clause can be relied on and for a time limit to operate at the end of which the other side may elect to terminate the agreement if the conditions which permitted the invocation of the clause still persist.

12.22 WAIVERS ETC.

The waiver clause clarifies the effect of a waiver on the part of one party of another party's inexcusable failure to perform. An absolutely staggering number of cases come before the English courts each year in which the dispute surrounds the effect of a waiver of rights by one party or the scope and interpretation of waiver provision, in which the party alleged to be in default argues that the other party has waived his or her rights for good and cannot go back on that waiver – in other words is estopped from relying on his or her rights – and in which the other party either denies the waiver altogether or maintains that it had some more limited effect only. Many disputes can be eliminated by including a sound procedure to deal with such events.

12.23 GOVERNMENT APPROVALS

Often licences are made operative conditionally on the transaction gaining some necessary governmental approval. Government approval clauses may be relevant where there is an international agreement. Regulatory authorities in the case of pharmaceuticals needing Ministry of Health or perhaps some Ministry of Defence consents in the case of exports of armaments of significant technology may by the law of either state have to consent to the transfer of the technology. Also exchange control barriers may have to be overcome. Thus provision should always be made for who is to procure that consent, what time he or she has to obtain it in and the consequences of failure to do so, or granted subject only to unacceptable conditions, and so on.

12.24 INDEMNITY CLAUSES AND PRODUCT LIABILITY

The licensee may require an indemnity in respect of two causes of action: indemnity for damages paid out to third parties whose rights have been infringed by use of the product or process; and damages paid out in respect of product or process liability, especially where the liability of the licensee arises out of the inherent defects of the product or process or instruction manual itself rather than from the licensee's manner of use of it. Here the licensor may not wish to give anything, except inasmuch as it reassures the licensee, but there are limits as to how far he or she can exclude liability of this kind.

So far as indemnities in respect of third party infringement claims are concerned, if a clause is to be included in the agreement to this effect, it should be very carefully drafted to determine precisely its scope. The licence might limit the amount of liability to indemnify to the amount of royalties received under the licence or to a fixed amount, so as to enable adequate insurance to be taken out.

The clause should in preference exclude liability arising out of activity not directly related to the licensee's use of the rights directly under the licence and perhaps in a more narrowly defined set of circumstances. The clause should make provision for the immediate mutual notification of any infringement claim received by either party, and perhaps for the licensor to be able to insist on termination of the allegedly infringing activity save at the sole risk of the licensee pending resolution of the claim, and precise agreement as to who is in control of the conduct of the defence to any challenge to the rights upon which the licence is based, and with provision for the costs of the defence. Failure of the licensee to comply with these requirements of notification, cesser of operations etc. should be made to trigger annulment of the obligation of indemnity.

In respect of product or process liability, one need only mention Bhopal or Seveso to suggest the potential importance of such a provision in a licence.

So far as direct actions brought by members of the public against the licensor who did not actually make the product or perform the process but licensed it to a licensee who did, thereby causing the public harm, are concerned, this is so far more of a problem for the licensor in jurisdictions overseas than in the UK because of legislation in some overseas states holding licensors liable for licensee product defects in respect of products which have been permitted to be associated with the licensor, for instance in the USA and in Australia.

In those countries consumers will now have a much easier time establishing direct liability against the licensor himself or herself than consumers in the UK who will still have to prove both that the licensor owed them a duty of care and that that duty of care was broken, thereby causing them loss with no intervening events breaking the chain of causation.

Indeed, in those countries where such strict liability regimes exist it may be that the licensor will require an indemnity clause from the licensee for improper administration of a turnkey plant or process which gives rise to liability in the licensor to third parties. There is no strict liability for defective products as between persons in no contractual relationship in the UK as yet. Thus, in the UK the consumer will still tend to proceed against the licensee, who is an easier target than the licensor, unless the licensee is unlikely to be able to pay up. However, with the advent of the new EEC Product Liability Directive, due to be implemented in 1987, the position in the EEC is likely to come much closer to that in the USA.

What if the licensee does pay up and wants to pass on some of the loss to the licensor? Where the licensee has taken and used inherently defective products from the manufacturing licensor there is no problem in claiming an indemnity if those goods do not meet the requirements of merchantable quality implied by the *Sale of Goods Act 1979* sl4, thereby passing on the loss.

When the licence is one of know-how or of a patented technique then the licensee would probably have to establish a duty of care owed to him or her by the licensor – which would probably not be difficult – and a failure to exercise that degree of care by the licensor, either in the tort (i.e. private or civil wrong) of negligence or by establishing that there is an implied duty to take reasonable care and skill in licensing the process.

Thus (in the absence of any express warranty as to the safety of the process or an indemnity clause compensating the licensee for damages he or she has had to pay out to consumers) the licensee would have to show that the licensor had not exercised reasonable care and skill, which in the context of a state of the art process or product, one at the frontiers of technological advance, might be a severe task to accomplish.

If lack of adequate skill or care – in for instance producing the instruction manuals or making provision for failsafe systems – can be shown, to what extent can the licensor exclude liability to indemnify the licensee for such losses? This will depend on what is held to be reasonable under the *Unfair Contract Terms Act 1977* so far as English law is concerned. Much care has to be taken drafting such indemnity and exclusion of liability clauses.

It should, however, be stressed that a licence may with impunity carry an exclusion clause or indemnity in respect of negligence or breach of contracts connected with infringement or loss of patents etc. rights, for the Unfair Contract Terms Act has no application whatsoever to this type of clause. The Schedule I of the *Unfair Contract Terms Act 1977* provides that ss2–4 of the Act have no application to any contract so far as it relates to the creation or transfer of a right in any patent or other intellectual property, or relates to the termination of any such right or interest.

13
Pricing technology transfer

13.1 INTRODUCTION

This chapter is concerned with two things:

(1) The pricing of designs and technology
(2) The drafting of payment and royalty clauses

There are many very learned books about the microeconomics of technology transfer and its evaluation but this chapter takes a rather more limited approach to what is a most difficult task for a company seeking to exploit through licensing design or technology.

13.2 EVALUATING A TECHNOLOGY TRANSFER PRICE

This requires a certain amount of homework before going into the negotiation to determine not only the break-even point but the estimated implications of the various combinations of payment available for licence negotiation. Of course if the licence is just one more of many in respect of the same product there will be a fairly fixed idea about what is right. But this is a luxury which comes only later in the development of the product licensing campaign.

The object of the homework exercise is to establish a negotiable 'band' outside the limits of which the negotiators will not attempt to go and will not be forced out of. But the width of this band will of course depend on the degree of certainty which surrounds the profitability of the product. Where the negotiation bands of the two parties overlap is the negotiation territory, which may be broad or narrow depending on their respective perceptions of the value of the technology.

Two factors make this evaluation process different from that of an ordinary contract of supply.

Firstly, the transaction is likely to be drawn up to bind the parties over a much longer period of time – say the life of the patent or other protection subjected to the licence – than in contracts for component supply and the like. This increases the risk compared with other commercial undertakings where even in the high risk commodities markets, mechanisms such as the futures markets can be employed to smooth out the level of risk undertaken by the parties.

Secondly, the marketability of the material licensed is more difficult to predict than is the case with other tangible assets: the legal protection it will afford will be largely untested as will its likely potential market penetration.

So full production costings and as far as possible market projections should be made before entering into the negotiation or forming a preliminary view as to what the party can afford to buy or sell for.

The first stage is working out the cost of the venture before deciding on the acceptable profit level compared with alternative business.

The licensor will have worked out the full cost of R and D and the time scale over which he or she wished to recoup the expenditure already incurred, plus the costs to him or her of transferring and then of supporting the transfer of the technology under consideration.

The licensee will examine the costs of setting up and then running the venture and of adjusting to the new product as well as the longer term costs.

In this context, a further cause of uncertainty and one which often leads to breakdown of agreements both before and after the negotiation is complete is the simple factor of the licensee's failure to understand properly the nature of the technology at hand and what it is to do.

Obviously, the homework involves a detailed study of just what the party wants out of the technology and whether and to what extent he or she can get that. A degree of mitigation of this uncertainty may be present if there has already been a preliminary disclosure on a confidential basis of this information, and equally if the product has already been marketed successfully by the licensor in this or in another country. But a full evaluation is simply not as such available in some know-how transfers. All that can be done in those cases is to rely on the mere test performance figures etc. assured by the licensor and to have them written into the agreement.

Ultimately the problem is that of striking a fair balance which enables both sides to make a reasonable profit. There is absolutely no point in squeezing an unduly high licence royalty: there will be no incentive on the part of the licensee to exploit if he or she is getting little or nothing out of it. Equally too low a rate will not encourage the expense of back-up support and technical advice and assistance from the licensor. Both sides will have to compromise on their ideal positions.

13.3 FACTORS AFFECTING THE VALUATION OF A LICENCE

The following factors should be examined when you assess the valuation of a licence agreement price:

(1) Is the licence the first for you or this product?
(2) What precedents are there if it is not?
(3) What is the prevailing industry rate?
(4) What makes this licence different from the rest?

(5) What is the territory covered by the licence?
(6) Is the licence exclusive sole or plain?
(7) Are there restrictions on the scope of licence?

(8) How valuable is the technology and market lead?
(9) What back-up services will the licensor provide?
(10) What is the start-up cost and capital investment?
(11) What cost was incurred by the licensor in R and D?

(12) Are substitutes available for product licensed?
(13) What is the proven sales record or performance?
(14) What is the prospective income on the technology?
(15) What have trade rivals offered for the transfer?

(16) How strong and long lived are the rights granted?
(17) The different life cycles of the rights licensed
(18) What is the expense/income structure of the deal?

(19) What is the commercial strength of the parties?
(20) Can both sides make a fair profit at that level?
(21) What is the stability of the market environment?

In particular on the part of the licensor a full comparison should be made between the profit to be made on the licence at any particular rate of royalty or other payment combination with the profit which would be made if the licensor himself or herself were to manufacture and sell the product licensed. In making this comparison it is important not to forget certain direct and indirect sources of income which may be obtained out of the licence apart from the royalty and which may make the licence appear a more profitable prospect than it would otherwise seem at first sight. These are:

(1) Any materials or parts the licensee may buy
(2) Any production expenses saved by buying back
(3) Fees payable in respect of improvement patents
(4) Fees payable in respect of technical assistance
(5) Fees payable for testing and assessing licensees
(6) Fees payable for personnel training

Set against this the costs of:

(1) Secondment of staff and loss of their services
(2) The necessity of splitting some of that profit
(3) Loss of opportunity to acquire goodwill in market
(4) Loss of exclusivity in the product licensed

From the point of view of the licensee he or she should also be looking at the cost of getting into the market himself or herself without infringing the licensor's rights by the use of designing around etc. or otherwise duplicating the technology and design and catching up on the time lead of the licensor; his or her dependence on the rights for basic market penetration and his or her capacity to exploit it better than the licensor, especially if the licence is to be competitive not collaborative, i.e. if the licence will not prevent the licensor from marketing the design or technology himself or herself or through others, rather than by allowing the licensee to make all the running and then just sharing his or her marketing profit through the mechanism of a royalty, without any attempt at direct sale market penetration himself or herself.

And apart from these factors both sides should of course take into account in all of these assessments the anticipated changes in the market structure or the product profile of the market licensed. This should not be too difficult at this stage, for the same questions should have been asked when doing one's homework before reaching the decision whether or not to license or take a licence in principle, assuming that such a choice did exist as a practical question.

13.4 DRAFTING PAYMENT CLAUSES IN LICENCES

These are very difficult clauses to draft. And, of course, at all points we are going to have to bear in mind the taxation implications of the various modes and timings of payments available, and these may well have a very considerable influence on the structure of the final draft of the payment clause.

There are a number of alternative patterns for the remuneration which may be employed in this field and of course it is open to the parties to choose combinations of them operating at different stages of the life of the licence as opposed to one form. We are not here concerned with the earlier question of assessing the royalty payments, that is of the pricing of the technology or design transfer, but rather with all the possible structures available. The basic division is threefold. There may be:

Downpayments in advance
Royalties or levies
Special fees

Special fees may be payable for services such as technical assistance beyond that provided free with the contract, maintenance fees, staff secondment and the like, and options on further technology.

13.5 DOWNPAYMENTS AND DISCLOSURE FEES

Downpayments may be made both for a preliminary disclosure agreement under which a potential licensee is enabled to inspect and assess the technology and other secret know-how to decide whether he or she wants to take a licence out in respect of it, in any option in technology which may be granted by the owner and in the full licence itself if a deal is subsequently struck between the parties. Obviously this will be a very attractive form of payment from the licensor's point of view and the very opposite for the licensee.

Once gone, know-how has gone for ever. So the most appropriate mode of payment for disclosure of know-how is a one-off payment preferably in advance, whether or not supplemented by royalties in respect of sales of equipment made by the processes etc. disclosed.

But even apart from this there are a number of good reasons for adopting an immediate downpayment or an immediate lump sum royalty format of a relatively high sum:

(1) It obliges the licensee to do something with the licence and acts as a deterrent to those who want a licence simply to exclude other potential licensees from the market – a minimum royalty clause will do the same.
(2) It provides some money up front from a licensee who might otherwise be entering the agreement with a view to depending on the success of his or her sales of the licensed product to finance the payments and therefore ensures that your licensee is financially a reasonable bet.
(3) It also gets the licensor some return on his or her investment in the development of the licensed product fairly early on in its exploitation life.

The taxation implications of a lump sum as opposed to spread royalties are not that great. It is possible to draft the downpayments clauses so as to impose the obligation on the signing of the agreement or on the delivery of the know-how. They may be independent of the royalties or an advance on royalties and set off against future obligations to pay, a form of guarantee of payment, with cash flow implications for both sides. But quite apart from that in the rare case where an up-front single sum payment, whether or not divided into instalments, is possible, it has the advantage of much diminishing uncertainty and disputes as to what sum is due and both reduces the risk borne by the licensor and frees the licensee from many of the obligations of disclosure of accounts and the like, making reports etc. etc. Of course, from the licensee's point of view he or she is making payment which under a pure royalty per unit or sale regime might never become payable although this disadvantage is usually balanced out by minimum royalty clauses.

13.6 ROYALTIES AND RELATED PAYMENTS

At its simplest the royalty employed may be very straightforward, a simple lump-sum payment followed by a royalty. This may be calculated on a percentage commission on the sales of units or as a flat sum per unit sold irrespective of price (a levy). But there might be many elaborations even on this, for instance a low royalty on the first 10 000 products made with an upward curve in the rate as the level of production and therefore the level of profitability increases. There might well have been an advance on royalties instead of or as well as this, which operates both as an insurance policy for the licensor and also as an incentive to the licensee to recoup his or her expenditure which the licence agreement has put the licensee to.

13.6.1 Defining the royalty payment

So far as royalties are concerned the usual method is a straight royalty per unit of production, or a flat percentage on retail prices, but there are a number of difficulties surrounding this which call for care. They are basically simple questions of defining what you mean.

It is essential to make it clear whether unit produced or unit sold is the basis of calculation, and then if unit sold what constitutes a sale for these purposes, when the obligation to pay arises – on contract date, delivery date or invoice date of the units sold – whether the price upon which the unit is calculated if it is a percentage is on the net or gross price, and if net whether that includes deduction of discounts for quantity sales, inclusive or exclusive of VAT etc., rejects, and whether any category of sales is to be excluded. What apportionment of the sale price is to be made if the licensed component forms part of a greater whole sold as such?

A further point to bear in mind is the case of a sale at a price which is not a genuine arm's length and commercial transaction – a problem faced equally by tax authorities and licensors. Suppose that the licensee sells the product licensed to associated companies, which trade in those goods, the sale price being below what a bona fide transaction to a third party would have involved. This possibility may require some short provision in the licence agreement, providing for arm's length transactions of fair market prices calculated along the same lines as those used by national revenue authorities for imputing the tax collectable in respect of inter-group transactions.

13.6.2 Minimum royalty payments

Minimum royalty payments are very common now and have the advantages both of deterring the non-genuine licensee, especially when he or she is an exclusive licensee for a territory, and of ensuring a certain basic return and establishing a certain presence of the product in the market place which is potentially beneficial to one's goodwill. To take account of fluctuating markets some clauses make provision for carrying over excess sales to the following financial year as credited to that year's sales.

There are also sometimes found maximum royalty clauses, whereby the licensor is never entitled to royalties above a certain figure no matter how many units are sold and this again operates as a financial encouragement to the licensee to maximize production if the level set is a realistic one.

Minimum sales figures seem to be at about anything from one-third to three-quarters of the licensees' own estimates of potential sales, with the bulk of clauses hitting the one-third to one-half marker, perhaps rising after the first full year. The minimum payment is usually calculated by the reference to an annual figure, since shorter periods are too open to market fluctuations.

Such obligations are not found in the early stages of the licence where the licensee has high setting-up costs to absorb or where a downpayment on the licence has been made. The payment is often required to be made in advance and then credited against the end-of-year totals.

Often there will be some provision exempting the licensee from such an obligation in exceptional circumstances in which it is unreasonable to expect the performance specified – the sort of circumstances in which a strong licensee might have won a suspension of all royalties, such as a challenge to the legal right licensed or the subjection of the product in question to health checks by regulatory authorities and its suspension from sale.

13.6.3 Variations in royalty rates

Royalty rates may of course vary according to the level of production, diminishing or escalating as the production levels go up. In particular, if the licence makes some provision for the licensee to take the benefit of any improvements effected to the invention by the licensor, the clause is going to have to make provision for an increased royalty payable in that event, perhaps at a fixed additional 'supplement' or by mutual agreement with arbitration in the event of disagreement.

Although supplement clauses are easy and certain, they cannot hope to take account of all the possible types of improvement which might be effected and by and large are not particularly advantageous to the licensor. From his or her point of view it is better to draft a clause which makes the right of the licensee to take the benefit of a licence in any improvement to the product conditional on the prior agreement of the parties or acceptance of an arbitrator's decision as to the new level of royalty to be payable in respect of the improved product or process.

It should also be remembered that an in-house improvement made by the licensee may enable him or her to cut production (and therefore resale) costs and thus reduce the royalty payable if a percentage basis is chosen: the increased sales may not compensate for the reduced price. This is the black side of the apparent flexibility and responsiveness of such percentage royalties to the market environment. It may be necessary to make provision for this in the licence agreement.

13.6.4 Division and apportionment of royalties

Also there may be a provision for a variation of royalty payments in the event of one or more of these intellectual property rights licensed proving to be invalid. We have already mentioned this in passing. Obviously, if the whole licence hangs on the validity of the patent, then the invalidation of the patent will be terminal. But if the agreement is expressed not to be dependent on the continued validity of every right licensed, there would have to be consideration given to declaring what proportion of the royalty payment etc. is allocated to each of the rights licensed, so that on the expiry or invalidation of one of the rights (assuming it is still possible to manufacture and the patent invalidated is not in fact infringing someone else's patent rights) the licensee continued to pay royalties at a lower rate, taking that into account. These comments apply equally to the case where the patent application is licensed and in the end that application fails.

EXAMPLE

A good example of the use of these techniques is the licence royalty structure adopted for Ron Hickman's 'Workmate'® bench licensed to Black and Decker, where the UK royalties were split into two tranches, of one and half per cent each, one tranche for the patents and one for the UK's design copyright and know-how: the second tranche was of course immune to the fate of the patent rights covering the workbench. This placed Hickman in a very much stronger position in the UK than was the case abroad, where to maintain his full one and a half per cent royalty he had to attain certain agreed levels of patent protection in Europe, and in the USA the three per cent sum was all entirely linked to patent protection and thus was entirely dependent on continued US patent protection.

13.7 SERVICE AND OTHER SPECIAL FEES

Service fees are another common formula for such payments, especially in countries where for political reasons high royalty fees are not acceptable. These may pass under a number of names and may be structured in a number of ways, according to what is cosmetically and politically convenient or according to the nature of the help and degree of commitment involved on the part of the licensor. Relatively high fees can be justified on this basis which would not be acceptable as advance or start-up fees or as lump-sum royalties.

An important point is that these payments are not royalties paid under a licence but payments for work done, which in some jurisdictions may be important from a taxation point of view. Strategically it may be in the interests of the licensor to make such service fees reasonably low for the servicing of a licensee's plant keeps him or her well in touch with development problems and potential improvements as well as helping to ensure that quality control is kept up. For this reason some licensors keep the payments down to cost price or cost plus small profit.

Of course, the level of the service charge will be somewhat different where you are using that method of payment in a country where high royalty figures are impossible and where the basic technical expertise of the licensee – perhaps the project was a turnkey licence – is so low he or she is or then becomes totally dependent on your continued technical support and this demands a heavy commitment in terms of your personnel.

13.8 CURRENCY PLACE AND METHOD OF PAYMENT

Provision would normally have to be made for the currency in which the payment is to be made and for the payment procedure. As rapid movements of sterling and the dollar have shown this can be of much importance.

You might stipulate for payment to be calculated by reference to the exchange rate on the thirty days expiring at the end of each quarter so as to eliminate attempts to play the exchange rate at your expense. In many high inflation countries, particularly for example in South America, there is going to have to be some sort of consideration of a price escalation indexing clause. Again provisions about the place and method of payment will have to be agreed.

13.9 PROTECTION OF ROYALTY RATES AGAINST INFLATION

The licensor is of course going to be concerned to protect the value of his or her royalties, especially in a long-term involvement and where an international transaction is concerned and the exchange rates may fluctuate. Local inflation *per se* can to a degree be overcome by using a royalty expressed as a percentage of the retail price of the product licensed, where that is possible for then the royalty income rises with the retail price and thus with inflation. But the relative exchange rates cannot be provided for in that way. Such depreciation in royalty values is a serious problem.

In the old days one method sometimes used to overcome this was the gold clause, but now in most of the important countries in the world (including the UK since as long ago as 1931) such clauses are no longer enforceable and in any case since the value of gold is no longer as stable as it was the use of the gold clause may be a self-defeating exercise. Two ways of getting round the problem now are:

(1) The escalation or indexation clause and
(2) The mutual consent to variation of rate clause

13.9.1 Escalation or indexation clause

The escalation clauses tie the amount of payments due to some established index such as a retail prices index or similar standard. Provision should be made for what is to happen if the method of calculating the index is changed as governments are wont to do where the index appears to be rising too quickly or too high for political comfort: this is said to be a problem in a number of countries. Obviously escalation clauses are essential in many high inflation zones.

Drafting a clause is extremely difficult and it may be difficult to find appropriate indices especially in a specialized industry where the rate of increase and inflation is different from that of the norm on which a broad-based index is founded. For that reason, for instance, in the British publishing industry, long-standing subscription services have escalation clauses calculated by reference only to that portion of the UK retail prices index made up by the economic class in the index of books and periodicals, since inflation in that sector has tended to be higher than in most because of problems with both raw materials and labour.

The use of a royalty based on the price of the article itself may be sufficient protection against inflation to eliminate this but it does not help with fluctuating exchange rates. Tying the value to a strong currency like the dollar may be of use. Where it is necessary to calculate by reference to exchange rates a fixed reference point such as the rate employed by a certain well-known and solid bank at a certain time of day and place of payment should be used.

13.9.2 Mutual consent to variation of rate clause

The mutual consent clauses acknowledge that values may change and provide for review of the rates at regular intervals so as to ensure that by their mutual agreement the true value of the agreement to the parties at the beginning of the transaction may be preserved, with provision for arbitration in the event of a failure to agree. From a lawyer's point of view these are rather unsatisfactory in being a little open ended, and looking a recipe for trouble ahead.

13.10 NON-MONETARY PAYMENTS

Of course, direct monetary payments are not the only form which a remuneration might take.

13.10.1 Offsets and production orders

There might, for instance, be an offset, whereby the licensee agrees to place work to a certain value with the production resources of the licensor or of one of its subsidiary companies. This will be relatively infrequent except in deals involving the Eastern Bloc.

13.10.2 Payment in kind

There may be payment in kind – selling back the products made under licence at cost price, enabling the licensor to concentrate its production effort in other areas, where bulk production of a few lines is more efficient that a diverse range of products. Payments may also be made by way of other commodities such as oil or ore. This is common with Eastern Bloc deals, where such countertrading is a major phenomenon, but one beyond the scope of this book. There are now a good number of useful guides on countertrading.

13.10.3 Issues of equity in licensee

There might instead be payment in the form of an issue of equity – an issue of shares in the licensee company. There has been a pattern in the USA of this sort of licensed technology arrangement which often eventually leads to the licensee losing its separate identity and becoming a production subsidiary of the licensor. This is an alternative form to the up-front payment where perhaps ready cash is not available but both sides are confident of the future prospects for the licensee in developing the market for the licensed technology or design: the licensee might well be a new joint venture subsidiary.

In this way the licensee is often able to acquire the technology for a lower net outlay than would have been required if cash were to be payable and the licensor benefits from gaining a return on the technology which is not purely income but includes on the one hand a share in the income of the company licensed through its dividends if any and an increase in the value of the equity paid out for the technology through the growth of the company. This may in certain cases be more attractive than a royalty income on the licence. The payment might even be by an option to buy an equity stake at a discount later, when the fortunes of the company are better known.

13.11 VERIFICATION AND ACCOUNT KEEPING

Although it is common to co-ordinate the access to licensee accounts with the right to verify them, the disclosure of accounts and reports need not be framed in this way unless you want them to be so. But even so, generally it is more convenient to do so. The licensor might want to specify a certain minimum level of detail in the accounts, perhaps specifying procedures in the schedule to the licence. Often an auditor's certificate would be required. The standards for each such exercise should be the same.

Again, quite how much the auditors will be able to extract of such extremely valuable commercial information will depend, as in all of the provisions of the agreement, on the bargaining power of the parties. There might well be an additional report at less frequent intervals to be made to the licensor without the necessity of a prior demand providing a report perhaps annually on the total sales, the price structure throughout the period, existing stocks and production levels and an assessment of the market for the next period.

The power of verification of the figures is an extremely valuable one. It reduces the temptation to cheat, even though in practice it is not particularly regularly exercised and may well cause some resentment. This power should as here be restricted to that which is strictly relevant to the licence. And provision for properly qualified third parties must be included.

Attention should also be paid to obligations on the licensee relating to keeping the accounts over a period of time. Many such licences require the licensee to keep the accounts and records for a period after the year in question, and it will be wise to make it clear that they should be kept for a period of say seven years and that the right of inspection continues after the termination of the licence agreement permission to produce, so that should there be a breach leading to an abrupt termination the licensor will still have a right of access to the past records for the purposes of investigating what is owed to him or her.

13.12 SECURING PAYMENT OF THE ROYALTIES AND OTHER FEES

The licence should make some provision for the licensor to be sure that he or she receives his royalties in full and promptly. Payment is not conditional on the licensee himself or herself being paid for the goods he or she has sold. The licensee's bad debtors are his or her problem, not the licensor's, who has no direct contact with the ultimate customers and who does not act as an insurer for the licensee.

Provision is usually made for the licensor to have reasonable access to the licensee's accounts: this has just been discussed. It enables him or her to see that he or she is getting what is due to him or her and the licensor should not be reluctant to invoke his or her inspection rights on a cordial basis. Apart from this the means of securing payment has nothing special about it which differs from that of enforcing normal contractual debts.

13.13 CONTROLLING PRICES OF THE ITEMS LICENSED

One thing the licensor should never seek to impose on the licensee in the payments clause without detailed legal advice in the country concerned will be any form of control over the minimum or the maximum price of the design or technology licensed. The price at which the licensee sells the licensed products will always be left entirely up to him or her.

Even assuming that you could ever persuade any licensee to accept a requirement of obtaining a prior approval of the prices he or she proposes to charge, you would encounter severe difficulties in the EEC and the USA competition law. This is dealt with in more depth in a later chapter.

13.14 NEGOTIATION FEES AND OPTION FEES

One distinction which should finally be made is to distinguish option fees for technology from negotiation fees which are sometimes encountered in the USA and in Japan but which do not appear to be regarded with much favour in Europe or the Commonwealth. These are a kind of earnest payment demanded by some licensors before entering into negotiation whereby a fee is charged for the privilege and which is non-returnable if then the negotiations break down. It may be credited towards any later royalty or lump-sum payments if the negotiations do succeed, but not always. These are generally used only by strong licensors for obvious reasons and they may merely cover the costs of organizing the venue or may be much more, amounting to nearly the cost of granting an option in the design or the technology under discussion. The underlying idea is to deter those enquirers who are not already serious and partially committed to the idea of taking out the licence. The payment of the fees as such gives the party making the payment no rights or options at all in the technology under discussion.

14

Restrictions on exploitation

14.1 INTRODUCTION

This chapter is concerned with legal limitations on one's freedom of contract in licensing and otherwise exploiting intellectual property rights. It looks at the structure of control over the use and licensing of intellectual property rights in UK, EEC and USA law. A list of dangerous practices and clauses is set out to provide a checklist of matters to be cautious about. This is definitely lawyer's law and no attempt will be made in a mere thirty pages to give the reader a comprehensive guide to the possibilities in an area which occupies thousands of pages in both lawyers' textbooks and economists' monographs.

14.2 UK RESTRICTIONS ON EXERCISE OF PATENT RIGHTS

Only UK patents are extensively covered by UK law in this field. Trade marks, copyright and designs are much less carefully dealt with, though in many cases the EEC law will now achieve substantial restrictions on the use and abuse of such rights even wholly within the UK itself.

14.2.1 Prohibited contract and licence terms

The *Patents Act 1977* imposes certain restrictions on the scope and duration of a contract relating to the supply of patented goods or a patent licence designed to achieve a narrow objective of eliminating the most gross abuse of patent rights, that is of tying the licensee to other non-patented products supplied by the patentee or for a longer time than the duration of the patent itself. s44 renders void any condition or term in so far as it imposes restraints on the licensee or person supplied from obtaining anything other than the patented product or product of the patented process or prohibits him or her from using or restricts his or her use of any articles (whether patented products or not) which are not supplied by or any patented process not belonging to the supplier or licensor and his or her nominee. Moreover, the very existence of such a condition in a contract or licence renders a defence to anyone (not merely to the person supplied or licensed) accused of infringement of the relevant patent.

However, s44 provides that the clause will not be void if the licensor or supplier can prove that at the time of making the contract or granting the licence the licensor was willing to supply or to license the other party to the contract or licence on reasonable terms under a licence or contract without such an offending condition; and further, that if the other party did contract into one of those offending terms, he or she was nevertheless entitled under his or her contract or licence to remove the offending condition on giving three months' written notice and subject to such compensation as may be determined by an arbitrator appointed by the Secretary of State. What will constitute offering an alternative licence on reasonable terms will, of course, be a pure question of fact and it should be remembered that in any proceedings the burden of proof will lie on the supplier or licensor to establish that a reasonable alternative was available.

One restrictive clause which may be encountered is expressly exempted from these sanctions. s44 provides that a condition or term of a contract or licence shall not be void by virtue of this section by reason only that it prohibits any person from selling goods other than those supplied by a specific person, or reserves to the hirer in a contract of hire on to his or her nominee the right to supply such new parts of a patented product as may be required to put or keep it in repair. The limited scope of this exemption should be noted. It only protects replacement spares necessary for the patented product itself. The possibility of implied licences to repair and restore original parts otherwise protected by patent or copyright should be recalled.

14.2.2 Rights of termination of licences

s45 of the Act provides that three months' notice in writing may be given by either party to terminate a patent licence or contract to supply patented products on the cessation from any cause of any or all of the patents (or patent applications) by which the product or invention was protected at the time of the making of the contract or granting of the licence and notwithstanding anything to the contrary in the contract or licence or any other contract; but only to the extent that the contract or licence relates to the product(s) or invention(s) protected by the patent(s) which cease. Furthermore, either party can apply to the court to have varied any terms which, as a consequence of the cessation of a patent, it would be unjust to require the applicant to continue to comply with, and the court may if it is satisfied that it is so make an order varying those terms and conditions if it thinks it just as between the parties, having regard to all the facts and circumstances of the case.

14.2.3 Patent licences of right

Under s46 it is possible to halve all the costs of renewing your patents by declaring them to be available to allcomers for licensing as of right. Where such a licence has been granted the licensee of right may call on the patentee to bring proceedings against any infringer and if he or she fails to do so the licensee himself or herself may sue in his or her own name, joining the patentee as defendant, though the unwilling patentee can avoid any liability for costs unless he or she then enters an appearance and takes an active part in the litigation. The grant of a licence under s46 takes effect as soon as it is applied for notwithstanding that the full terms under which the licence is to be held have yet to be settled between the licensor and the licensee.

If a patent subsequently becomes more successful than was anticipated the patentee can then cancel his or her declaration but to do so must pay all arrears of fees as if the declaration had never been made – probably a small price to pay if the newly found success of the patent justifies such a move – and provided that there is either no licence in existence or all the existing licensees agree to the cancellation. The cancellation must be advertised and may be opposed by any person interested.

14.2.4 Compulsory patent licences

Under s48 of the Act, at any time after the expiration of three years from the date of grant of a UK or European (UK) patent anyone may apply to the UK Comptroller of patents for a compulsory licence in the patent to be granted to him or her; or for an entry in the register of a notice that licences under the patent are available as of right; or, if the applicant is a government department, for the grant of a licence in favour of any person specified in the application. Anyone, including an existing licence holder under the patent, may apply. There are five grounds, set out in s48, upon which one may base an application for grant:

(1) The patent is not being worked in the UK or not being so worked to the fullest extent that is reasonably practicable.

(2) Where the patented invention is a product that a demand for the product in the UK (i) is not being met on reasonable terms, or (ii) is being met to a substantial extent by importation.

(3) Where the patented invention is capable of being commercially worked in the UK, that it is being prevented or hindered from being so worked: (i) where the invention is a product, by the importation of the product, (ii) where the invention is a process by the importation of a product obtained directly by means of the process or to which the process has been applied.

(4) By reason of the refusal of the proprietor of the patent to grant a licence or licences on reasonable terms: (i) a market for the export of any patented product made in the UK is not being supplied, or (ii) the working or efficient working in the UK of any other patented invention which makes a substantial contribution to the art is prevented or hindered, or (iii) the establishment or development of the commercial or industrial activities in the UK is unfairly prejudiced.

(5) By reason of conditions imposed by the owner of the patent on the grant of licences under the patent, or on the disposal or use of the patented product or on the use of the patented process, the manufacture, use or disposal of materials not protected by the patent, or the establishment or development of commercial or industrial activities in the UK, is unfairly prejudiced.

Where the application is made upon the ground that the patented invention is not being commercially worked in the UK or not being so worked to the fullest extent reasonably practicable, and it appears to the Comptroller that there has not been sufficient time for any reason for the invention to be worked, he or she may ajourn the application for such time as will in his or her opinion be sufficient for the invention to be worked. The Act provides a set of principles in accordance with which the Comptroller is to exercise his or her power to grant a compulsory licence when settling the terms of the licence to be granted. He or she is to secure a number of general purposes:

(1) That the invention should be worked on a commercial scale in the UK without undue delay and to the fullest extent that is reasonably practicable
(2) That the patentee should receive reasonable remuneration having regard to the nature of the invention; and
(3) That the interests of anyone working or developing an invention in the UK under the protection of a patent should not be unfairly prejudiced.

In reaching a decision whether or not to make an order, the Comptroller must take into account (1) the nature of the invention, the time which has elapsed since the patent grant protecting it and the measures taken by the patentee to make full use of it; (2) the ability of the person seeking a compulsory licence to work the invention to the public advantage and (3) the risks taken by him in providing capital and working the invention if an order is granted. In considering an application the Comptroller is not obliged to take into account events which have occurred after the date of the application, though he or she may do so in his or her discretion.

14.2.5 Crown user of patents

The Patents Act makes provision for the Crown to use patents granted in the UK, as well as patent applications and European (UK) patents and such applications in a wide variety of ways. Any government department or any person who is authorized in writing by a government department may use a patent without the consent of the patentee for 'the services of the Crown' in the UK and such use will not be an infringement of that patent.

The owner's entitlement to compensation depends on a distinction as to when the invention was recorded by or tried by or on behalf of a government department.

If these events occurred before the priority date of the patent concerned, otherwise than as a result of a confidential communication made in confidence from a patentee or his or her predecessor in title, directly or indirectly, no compensation is payable.

If such trial or recording was made only after the publication of the application or as a result of a confidential disclosure then use of the invention must only be on terms agreed between the patentee and the government and in default of agreement to be decided by the High Court which may refer the whole matter or any issue of fact to an official referee or arbitrator.

Compensation will then become payable from the date of publication of the patent application or the date of communication of the invention if that was earlier and the basis on which compensation is to be assessed is as remuneration for use made by the

Crown, and not on any other basis. Any benefit already received or which may be received by the claimant from the government department for use of the invention is to be taken into account and the court must have regard to whether the claimant or his or her predecessor in title has earlier and without reasonable cause failed to comply with a request of the department to use the invention for the services of the Crown on reasonable terms in assessing the compensation payable.

14.3 UK RESTRICTIONS ON USE OF REGISTERED DESIGNS

14.3.1 Compulsory licences

At any time after grant of the registered design (not merely after a three year wait as in the case of the *Patents Act 1977*) any person interested may apply for a compulsory licence on the grounds that the design is not applied in the UK by any industrial process or means to the article in respect of which it is registered to such an extent as is reasonable in the circumstances of the case. The registrar may make such an order as he or she thinks fit. Generally, the principles applied in the case law surrounding the designs legislation are basically the same as those set out in the much more detailed *Patents Act 1977* procedures.

14.3.2 Crown use

Once again, the provisions on Crown use are basically identical to those of the Patents Act.

14.4 UK COMPETITION LAW AFFECTING LICENSING

The general competition law of the UK as such has little to say specifically about licensing of intellectual property rights, though generally these are covered in the same way as other contracts except in so far as they directly concern the licensing of the right itself, which enjoys some degree of protection in UK law from competition legislation.

14.4.1 The Resale Prices Act 1976

The *Resale Prices Act 1976* expressly extends its provisions prohibiting the use of resale price maintenance to patented articles, including those articles made by a patented process, as it applies to other goods (though note that books and pharmaceutical products are expressly now exempt from the prohibition) and any notice of any terms or condition which is void by virtue of that Act or which would be void if included in a contract of sale or agreement relating to the sale of any such article, shall be of no effect for the purpose of limiting the right of a dealer to dispose of that article without infringement of the patent.

But the Act provides it does not affect the validity as between the parties and their successors in title of any term or condition of a licence granted by the proprietor of a patent or registered design or by a licensee under any such licence, or of any assignment of a patent or design, so far as it regulates the price at which articles produced or processed by the assignee or licensee may be sold by him or her. After all, that is not a resale but an original sale. And a contract infringing the Act is only void to the extent of that condition or term, not as regards the remainder of the agreement.

So it is quite possible for a UK patent or designs licence or assignment to include a term or a condition regulating the price at which the assignee or licensee sells the articles made by him or her under licence.

But where the articles are supplied to the dealer for a resale, no term fixing the minimum resale price may be included and any such purported restriction will be void. The Act has no effect on purely export agreements although these must still be notified to the Director of Fair Trading if they include such terms and they will very likely infringe the terms of the EEC's competition law in Europe and the USA antitrust law in the USA.

14.4.2 The Restrictive Trade Practices Act 1976

The *Restrictive Trade Practices Act 1976* is mostly inapplicable to the licensing of intellectual property rights but does have a role to play. The Act provides that, subject to certain exceptions, any agreement or arrangement, under which restrictions in respect of any of a number of matters within the Act are accepted by at least two persons carrying on business within the UK in the production or supply of goods or in the application to goods of any process of manufacture, must be notified to the Director of the Office of Fair Trading who must consider whether they are contrary to the public interest, with provision for the matter to be heard by the Restrictive Practices Court in case of doubt.

If the practices are contrary to the public interest, the agreement automatically becomes void in respect of those restrictions and the court may order them not to be given effect to by the parties.

There are criteria setting out guidelines as to what is justifiable and therefore in the public interest just as the types of restriction covered by the Act are set out. To have an agreement approved the parties must show that one of these justifications applies and that it outweighs any detriment to the public caused by the agreement.

The Act is highly technical and formalistic in its approach – it looks to the form of a transaction to decide on registrability, not the economic effect of the agreement. If your agreement does not fit into the form prescribed to be registrable you do not have to register.

Patent licences, trade mark, know-how, design and copyright licences etc. are not much affected by the Act, which provides that restrictions only affecting the goods protected by the intellectual property right are discounted in looking at the registrability of the agreement. But patent and design pools are not exempt, however, and nor are those licences, or assignments granted directly or indirectly under such pools. An agreement or arrangement between a group of three or more persons each of whom has an interest in one or more patents or designs to grant licences to one or more of them or to assign the right to grant licences to a trade association is registrable and any licence so granted is itself also registrable under the Act where such restrictions exist.

14.4.3 The Fair Trading Act 1973

Under the *Fair Trading Act 1973* where a single firm or a number of interconnected firms in a group account for more than one-quarter of the supply of goods or services in one of the markets with which the Act is concerned, or if one or more agreements exist which in any way prevent, restrict or distort competition in relation to one-quarter of one of those markets, there exists a monopoly situation, which may be found to be or to be operated in, a manner contrary to the public interest.

Intellectual property rights may clearly be exercised in a manner contrary to the public interest within the meaning of this Act. There is no exemption for use of patents and other rights to establish a dominant position and abuse it. Reports on the exploitation of patents in tranquillizers and in Xerox photocopying machines have been subjected to adverse reports under this Act.

14.4.4 The Competition Act 1980

The *Competition Act 1980* now enables the Director General of Fair Trading to investigate or refer to the Monopolies and Mergers Commission any anticompetitive practice. It is clear that the exercise of intellectual property may constitute an anticompetitive practice. In the Ford Body Parts Report by the Office of Fair Trading in 1984, for example, the Director General thought Ford's conduct in restricting competition in the manufacture and sale of body parts protected by copyright worthy of a reference to the Monopolies and Mergers Commission.

14.5 EEC COMPETITION LAW AFFECTING LICENSING

Much more important for our purposes is the EEC's competition law. There are now four major provisions which affect the exploitation of intellectual property rights: Articles 30, 36, 85 and 86 of the Treaty of Rome of the EEC. The EEC competition law affects the way in which all lawyers now think in drafting licences or any other technology transfer agreements and it cannot be ignored by any business. Even activity purely within the UK may be caught by these provisions.

14.5.1 Article 85 and restrictive agreements

Article 85 provides that all agreements between undertakings, decisions by associations of undertakings and concerted practices which may affect trade between Member States and which have as their object or effect prevention, restriction or distortion of competition within the common market, are forbidden and in particular those which:

(1) Directly or indirectly fix purchase or selling prices or any other trading conditions
(2) Limit or control production, markets, technical development, or investment
(3) Share markets or sources of supply
(4) Apply dissimilar conditions to equivalent transactions with other trading parties, thereby placing them at a competitive disadvantage
(5) Make the conclusion of contracts subject to acceptance by other parties of supplementary

obligations which, by their nature or according to commercial usage, have no connection with the subject of such contracts

Any such agreements or decisions prohibited are automatically void by virtue of Article 85(2), although under Article 85(3) it is possible to gain an exemption from the prohibition if you can prove that an agreement or decision is one which contributes to improving the production or distribution of goods or to promoting technical or economic progress, while allowing the EEC consumer a fair share of the resulting benefit, and which does not impose on the undertakings concerned any restrictions not indispensable to attainment of these objectives or afford such undertakings the possibility of eliminating competition in respect of a substantial part of the products in question.

Now it is not only possible to apply for an individual exemption but in the case of certain classes of agreement there are block exemptions in which the Commission has published a list of clauses which it will automatically give exemption to so long as no other offending clauses appear in the agreement. One of those block exemptions is for patent licences; and another for joint R and D ventures. These provide a good indication of what is and is not permissible in other types of intellectual property licence too. These are looked at below.

In enforcing all these provisions the European Commission has extensive powers both in relation to investigating and vetting all restrictive agreements and practices, ordering the termination of infringing agreements and practices, hearing parties and imposing fines and periodic penalty payments, publishing its decisions and maintaining secrecy in respect of any confidential information discovered in the course of inquiries etc.

Both Treaty provisions on competition, Articles 85 and 86, are directly effective in UK law. This means that, for example, an English court would under EEC law be bound to uphold a defence based on proof that a plaintiff's action is based on an agreement which infringes Article 85(1) of the Treaty and which has not been exempted by the Commission proceeding under Article 85(3); and to deny a plaintiff who was met with a convincing defence based on abuse of dominant position under Article 86 of the Treaty. Indeed, Article 85(2) expressly requires the courts of the Member States not to enforce any agreement which is contrary to Article 85(1) and because only the EEC Commission can grant exemption under Article 85(3) to such an agreement it will be unenforceable in national courts until such an exemption has been obtained, even if one is likely.

The scope of the restrictions which EEC law places on the use of your intellectual property rights is quite extensive. Thus EEC law merits careful consideration in any transaction involving such rights or any attempt to enforce them.

14.5.2 Article 86 and abuse of dominant position

Article 86 provides that any abuse by one or more undertakings of a dominant position within the common market or a substantial part of it shall be prohibited as incompatible with the common market in so far as it may affect trade between Member States. Such abuse may, in particular, consist in:

(1) Directly or indirectly imposing unfair purchase or selling prices or other unfair trading conditions
(2) Limiting production, markets or technical development to the prejudice of consumers
(3) Applying dissimilar conditions to equivalent transactions with other trading parties, thereby placing them at a competitive disadvantage; or
(4) Making the conclusion of contracts subject to acceptance by other parties of supplementary obligations which, by their nature or according to commercial usage, have no connection with the subject of such contracts

The mere possession and use of a monopoly right such as a patent or a lesser right such as a copyright or trade mark will not constitute an abuse of dominant position but its use to achieve certain objectives or in the ways outlined above may do so. Thus charging loyalty rebates or imposing conditions such as patent packaging may well amount to an abuse. Some of the practices which have been held to be abusive are listed below:

(1) Refusal to supply retailers without justification
(2) Tying in patent packages or non-patented products
(3) Charging excessive royalties on licensed products

14.5.3 Articles 30 and 36 and exhaustion of rights

Article 30 of the Treaty prohibits quantitative restrictions on trade between Member States. Article 36 tolerates the exercise of intellectual property rights to prevent the import or export of goods so long as they are not a means of arbitrary discrimination or a disguised restriction on trade between Member States. This has in practice meant that whereas one may rely on intellectual property rights in the EEC to prevent the importation of your protected inventions without your consent from somewhere where they were not protected, once some goods have been put on the market anywhere in the EEC with your consent, you cannot make use of the intellectual property rights you hold in other Member States of the EEC to prevent those same goods being resold in other parts of the EEC, or made the subject of parallel imports to compete with you or with another licensee of the invention in his or her territory. And any attempts to prevent this by any contractual arrangement with a licensee will be unenforceable against him or her.

This is known as the doctrine of exhaustion of rights. Once there has been a first sale of a particular item with the consent of the right owner in the EEC, his or her rights over that patented or trade marked item through the trade mark or patent etc. are exhausted and he or she cannot control its subsequent progress in the common market. It is not therefore possible to use such intellectual property licensing to divide up the common market into individual protected zones, subject to some limited exemptions provided by the patent licensing exemption regulations discussed below.

14.5.4 Intellectual property licences and Article 85

The licence clauses commonly found in intellectual property licences have now to be examined in the light of Article 85. They have been subjected to examination by the Court of the EEC and by the Commission, which has now produced a block exemption for certain types of patent licence term, but not for the other rights which must be considered purely by reference to the existing body of decisions of the Commission and case law of the Court. There is also a block exemption for joint R and D agreements, one for exclusive sales agreements and one for specialization agreements, and a block exemption for trade secret licences is proposed. These will be not looked at in detail.

The starting point must now be the patent regulation since that will be the keynote of the Commission's thinking on intellectual property licences for some time to come. The Regulation now providing for block exemption of patent licences in the EEC came into force on 1 January 1985 and endures until 31 December 1994 but applies retrospectively to agreements made or amended so as to conform with its terms.

The basic scheme of the patent block exemption is that it provides lists of three types of clause:

(1) Clauses which may be included in bilateral patent licences either because they would benefit from the exemptions in Article 85(3) of the Treaty or because they would not infringe Article 85(1) in the first place. Agreements containing only clauses of the first type need not be notified.

(2) Clauses which may not be included in any licence benefiting from a block exemption because they violate Article 85(1) and also could not be presumed to benefit from an exemption in Article 85(3).

(3) Clauses which are in neither of the above two categories, which are to be subjected to an opposition procedure under which they are to be notified to the Commission and will be regarded as falling within the scope of the block exemption unless and until the Commission decides within six months to oppose them. Where agreements are to be notified provision is made for the confidentiality of the information thereby supplied to the Commission.

The Commission must oppose an exemption if within three months of the notification being transmitted to the Member States, one of them requests opposition on the basis of the competition rules of the Treaty.

Although the Commission may itself withdraw opposition at any time it may only do so after consultation with the Advisory Committee on Restrictive Practices and Dominant Position where the opposition was made at the request of a Member State which still maintains its request.

If the Commission's opposition is withdrawn because of the parties' proof that Article 85(3) is met the exemption applies from the date of its notification but if only by their amendment of the licence then from the date the amendments take effect.

The benefit of exemption may be withdrawn in respect of particular agreements which the Commission decides have effects incompatible with the criteria for exemption in Article 85(3), for example, where

- Such effects arise from an arbitration award.
- The licensed products are not exposed to effective competition in the licensed territories from identical products or services or from products or services considered as equivalent by customers in view of their characteristics, price and intended use.
- The licensor is not entitled to terminate the licensee's exclusivity at the latest five years from the date of the agreement or annually thereafter, if the licensee unjustifiably fails to exploit the patent adequately.
- One or both parties refuses without objectively justified reasons to meet demands from customers where the users or resellers would market the products in other territories within the common market (except where the licence includes a clause prohibiting such passive sales

during the first five years of marketing of the product).

- One or both parties make it difficult for users or resellers to obtain the products from other resellers within the common market and in particular to exercise intellectual property rights or take measures so as to prevent users or resellers from obtaining outside or from putting on the market in the licensed territory products which have been lawfully put on the market within the common market by the patentee or with his or her consent.

The Regulation applies not just to patents but to patent application licences, to similar rights such as certificats d'utilité and certificats d'addition, and to sublicensing agreements, and also to assignments where the consideration is calculated on a royalty basis and agreements in which the licensor's or licensee's rights are assumed by connected undertakings. It also applies to those licences granted by a licensor who is not the patentee but who is a licensee of the patent authorized to grant a licence or sublicence.

The patent block exemption does not apply to those agreements between members of patent pools relating to the pooled patents; nor to those licences arising out of joint R and D ventures between the partners or one of them and the joint venture (for which there is a quite separate block exemption); nor to reciprocal licences in trade marks or patents or reciprocal sales rights in unprotected products or exchanges of know-how between competitors in relation to those products (unless they contain no territorial restrictions); nor to licences of plant breeders' rights. Nor does it apply to mere exclusive sales distribution agreements for which there is a separate block exemption. But it does apply to the licences dealing in proprietary know-how forming part of the patent licence.

The regulation expressly states that it applies to agreements involving obligations which extend beyond as well as within the EEC and also to those agreements with provisions dealing only with obligations outside the EEC if they have an effect in the EEC.

14.5.5 The permitted clauses in patent licences

Article 1 provides a list of exempt clauses. These are exempted because the Commission feels that they encourage investment and contribute to improvements in goods and to the promotion of technical progress facilitating the licensing of new inventions and processes, thus giving consumers the benefits of improved quality and supply of goods on the market. The article permits the following:

(1) The licensor may undertake
   (i) Not to license other undertakings in the licensed territory and
   (ii) Not to exploit the invention in the licensed territory himself or herself for so long as one of the licensed patents remains in force.

(2) The licensee may undertake
   (i) Not at any time so long and so far as parallel patents remain in force to exploit (and exploitation includes both manufacturing and marketing) the licensed invention in those territories reserved to the licensor.
   (ii) Not to manufacture or use the invention in those territories licensed to other licensees so far and so long at protected by parallel patents.
   (iii) Not to pursue an active sales policy in those territories of other licensees at any time e.g. by advertising, by putting it on the market or setting up a distribution depot or

branch there in the territories reserved to other licensees after the first marketing of the product by the licensor or a licensee, so long as parallel patents remain in force.

(iv) Not to pursue a passive sales policy in those territories for a period of five years from the date of first marketing by the licensor or one of his licensees within the common market, so long as patents remain in force.

(v) To use only the licensor's trade mark or get up, so long as the licensee may identify himself or herself as the manufacturer of the procuct. The object of this last qualification is to enable the licensee to acquire a goodwill in the product so that he or she does not have to negotiate a fresh trade mark licence on the expiry of the patents, which would otherwise be a means for the extension of the effective commercial life of the old patent.

(3) These block exemptions in favour of the licensee relating to sales in his or her licensed territory only apply where he or a subcontractor or connected undertaking actually manufactures the licensed product.

Article 2 of the Regulation deals with clauses which would generally not infringe Article 85(1) of the Treaty at all. It is not an exhaustive list but is put in by the Commission to ensure legal certainty prevails because they might exceptionally fall technically within Article 85(1) even though they would almost certainly be exempted under Article 85(3). They include the following:

(1) Tie in clauses necessary for the technically satisfactory exploitation of the licensed invention
(2) Minimum royalty and minimum turnover clauses
(3) Technical field-of-use clauses
(4) Obligations not to exploit the patent after termination of the agreement while the patent remains in force
(5) Obligations not to grant sublicences or assign
(6) Obligations to mark products with the patentee's name, licensed patent or the licensing agreement
(7) Obligations not to divulge secret know-how ever after expiry of the agreement
(8) Obligations to inform the licensor and take any action in respect of intellectual property rights infringers
(9) Minimum quality clauses necessary for the technically satisfactory exploitation of the invention licensed and allow checks to be carried out by the licensor
(10) Reciprocal obligations to exchange information about their experience in exploiting the licensed invention and to grant non-exclusive licences in improvements patents and know-how. Exclusive licences are not within the exemption.
(11) Most favoured licensee clauses

Article 2 clauses are unlikely to fall within Article 85(1) of the Treaty at all, but if in certain circumstances they did do so they are none the less given exemption under this Regulation. The exception to that general principle is the most favoured licensee clause which does appear to be within Article 85(1) but would usually easily qualify for Article 85(3) exemption.

But it must be remembered that the Commission retains the right to withdraw the benefit of the exemption from any particular agreement which has certain effects which are incompatible with the conditions laid down in Article 85(3). This may seem to concentrate on parallel imports but its potential scope is much wider. For example it might apply where the minimum royalty clause was designed to put the licensee at a competitive disadvantage compared with the licensor.

14.5.6 The forbidden clauses in patent licences

Article 3 sets out clauses which are forbidden to the licensor. The inclusion in an agreement of any of the clauses set out in Article 3 precludes the gaining of an exemption under the block exemption. It includes the following clauses:

- No-challenge clauses (but without prejudice to the right of the licensor to terminate the licence if the challenge occurs).
- Clauses automatically prolonging the agreement by inclusion of new patents, unless each party has an opportunity annually to terminate the licence after the expiry of the original patent rights licensed under the agreement (but without prejudice to the rights of the licensor to royalties in respect of secret know-how which remains secret after the expiry of the patent on which it was based). The parties are of course free to enter into separate agreements later as the new rights arise.
- Non-competition clauses going beyond the clauses in Article 1 of the Regulation and without prejudice to the obligation on the licensee to use his or her best endeavours to exploit the licensed invention.
- Clauses requiring payment of royalties on non-patented items or for the use of know-how which has entered the public domain otherwise than by the fault of the licensee or of an undertaking associated with him or her, without prejudice to the arrangements whereby in order to facilitate payment by the licensee the royalty payments for the use of a licensed invention are spread over a period extending beyond the life of the licensed patents or the entry of the know-how into the public domain.
- Maximum quantity clauses.
- Price setting clauses, restricting one party in the determination of prices, components of prices or discounts for the licensed products.
- Restrictions on the classes of customer which the licensee or licensor may serve, the use of certain forms of distribution or, with the aim of sharing their customers, using certain types of packaging for their products save as provided for in permissible clauses on trade marks and quality control.
- Non-reciprocal grant-back clauses, whereby the licensee must assign all rights in or patents for new improvements to the licensed patents. These in effect give the licensor an unjustified and unfair competitive advantage.
- Tie-in clauses involving patents products or services the licensee does not want, unless they are necessary for the technically satisfactory exploitation of the licensed invention. These again may impose an unfair advantage on the licensor. These have always been regarded as excessive use of the rights licensed. But where the tying clause can be justified by reference to the quality control argument, whether the right be a patent, trade mark or other right, then the Commission has consistently accepted it as justified on the facts and exempted in under Article 85(3) of the Treaty.
- Clauses having the effect of impeding parallel imports, e.g. export bans. The regulation prohibits clauses whereby one or both parties is required to refuse without an objectively good reason to meet demand from users or resellers in their respective territories who would market products in other territories within the common market, or to make it difficult for users or resellers to obtain the products from other resellers within the common market and in particular to exercise intellectual property rights or take measures so as to prevent users or resellers from obtaining outside or from putting on the market in the licensed territory products which have been lawfully put on the market within the common market by the patentee or with his or her consent. Equally such concerted practices having this effect or intent are said to be prohibited.

14.5.7 The potentially permitted patent clauses

Agreements which fall outside the scope of the regulation because they include a clause under Article 3 might still benefit from individual exemption under Article 85(3) if they contribute to the technical or economic progress or improvements in the production or distribution of goods. To be eligible for an individual exemption the agreement must be notified to the EEC Commission to be vetted. Notification even if it does not lead to an exemption gives a certain degree of immunity to fines for acting on an unlawful agreement.

14.5.8 The R and D regulation and block exemption

There is now also a block exemption for certain joint R and D ventures. While joint R and D agreements do not generally violate Article 85 of the Treaty, they may do so if they deprive the parties of opportunities of gaining competitive advantages over each other by restricting the right to engage in R and D of the same field, or if they restrict the freedom of the parties to exploit the fruits of research. On the other hand such agreements generally do promote technical and economic progress and consequently benefit consumers.

Accordingly, the EEC's Commission seeks in this Regulation to encourage the use of joint R and D and the joint exploitation of its results which is a natural development of it in so far as it does not impede their effective competition.

Agreements between undertakings having a large combined market share could seriously restrict competition and are therefore excluded from this regulation but they, like agreements outside the scope of the patent block exemption regulation, may still gain an individual exemption under Article 85(3).

Again the regulation operates retrospectively but it entered into force on 1 March 1985 and expires on 31 December 1997. This longer period is justified by the Commission on the basis of the longer term nature of such joint R and D ventures particularly where they extend to the exploitation of results.

The notification and opposition procedures are identical to those of the patent block exemption. Again agreements automatically exempted by the Regulation need not be notified but in particular cases parties may request a decision from the Commission giving them an individual exemption.

The basic scheme of the Regulation is as follows. The Regulation exempts from the scope of Article 85(1) agreements for joint R and D products and processes and joint exploitation of that R and D as defined in the Regulation's Article 1, namely:

(1) Joint R and D products or processes and joint exploitation of the results of that R and D
(2) Joint exploitation of the results of R and D of products and processes jointly carried out pursuant to a prior agreement between the same undertakings; and
(3) Joint R and D of products or processes excluding joint exploitation of results in so far as such agreements fall within Article 85(1)

The scope of R and D agreements covered by the block exemption is however very closely confined. It applies only to the R and D agreements which fulfil the following characteristics:

(1) They must comprise a programme within a defined field and objectives
(2) Allow all the parties access to the results
(3) Leave the parties free to exploit the results of R and D and any pre-existing technical knowledge necessary to it if the agreement relates only to R and D
(4) Regulate exploitation only of results protected by intellectual property rights or know-how which makes a substantial contribution to technical or economic progress in each case only where the results are decisive for the manufacture of the resulting processes; and
(5) Require any joint undertakings of third parties charged with making the resulting products to supply them only to the parties, while requiring undertakings charged with manufacturing by way of specialization to supply all the parties.

The Commission takes the view that such a joint exploitation is not justified where it relates to those improvements which were not made within the framework of a joint R and D programme but under an agreement which has some other principal objective such as licensing of intellectual property rights, or joint manufacture or specialization, merely containing ancillary provisions on joint R and D.

The R and D falling into this category may either be carried out by the parties themselves through joint teams of researchers, organizations or undertakings, through allocation or specialization in the research, development or production, or through tasks jointly entrusted to a third party.

The exemption applies only to agreements between parties who are not competing manufacturers of products capable of being improved or replaced by the results of the R and D programme, or who even though competing have a combined market share in the EEC or a substantial part thereof of under 20%.

The exemption continues for five years after the first marketing of the products and thereafter until the parties' combined market share exceeds 20% subject to a six month period of grace after the limit is exceeded and to a provision catering for fluctuating market shares by allowing exemption to continue for two consecutive years despite a combined market share of up to 22%.

The same division into three types of agreement clause applies here as in the patent regulation. All these agreements are again subject to withdrawal of the block exemption in any individual case and there will be a withdrawal of the exemption if the Commission feels that the agreement is outside Article 85(3) and in particular where it substantially restricts R and D by non-parties because of scarce R and D facilities or substantially restricts non-parties' access to the market for the products or where the products are not exposed to any effective competition. Again notifiable agreements are treated with complete confidentiality by the Commission.

14.5.9 The permitted clauses in R and D ventures

Article 4 exempts the following clauses:

- Bans on the carrying out of R and D in the same or closely related fields either independently or in collaboration with third parties during the execution of the agreed R and D programme
- Obligations to buy the developed products resulting from the joint R and D programme only from the parties or subcontractors of the parties
- Territorial manufacturing exclusivity

- Field-use-of restrictions, except where the parties are competitors in products or processes replaceable by the R and D at the time of the agreement
- Prohibitions on active sales within the reserved territories of the other parties for a period of five years from first marketing of the products so long as parallel importing by other resellers and users is not impeded. The same meaning as that in the patent block exemption applies to passive and active sales.
- Reciprocal obligations to grant non-exclusive improvement patent licences and exchange information about new applications etc.

Article 5 gives a non-exhaustive list of clauses which probably do not even need exemption because they probably do not infringe Article 85(1) anyway.

- Obligations to communicate any patented or non-patented technical information necessary to carry out the R and D and subsequent exploitation of its results
- Prohibitions on using know-how received from the other parties outside the scope and purposes of their agreement
- Obligations to maintain intellectual property rights for the resulting products or processes
- Obligations to keep know-how received or jointly developed secret even after expiry of the agreement
- Obligations to inform other parties of an act against infringers of their intellectual property rights, including sharing the costs of the action
- Obligations to make payments to compensate for unequal R and D contributions or unequal exploitation of its results
- Royalty sharing clauses for royalties received from third parties
- Minimum quantity or quality clauses concerning the supply of developed products to the other parties

14.5.10 The forbidden clauses in R and D ventures

Article 6 provides the forbidden clauses, which take the agreement outside the scope of the block exemption. Agreements containing a forbidden clause must be notified to the Commission for the possibility of an individual exemption at best.

- Restrictions on the parties' freedom to conduct other R and D unconnected with the agreement or after the expiry of the agreement whether or not connected with the field of the agreement
- No-challenge clauses after the completion of the programme in respect of rights which the parties hold in the common market and which relate to the programme or after the expiry of the agreement which relate to the results of the R and D
- Quantity restrictions on exploitation of the results of the R and D
- Price-fixing clauses on the resultant products or their components
- Restrictions on the types of customer to be supplied, subject to those permitted obligations in Article 4 discussed above
- Prohibition on direct and active sales into other parties' territories after a five year period permitted in Article 4 of the Regulation discussed above
- Prohibitions on allowing third parties to manufacture the resultant products or to apply the resultant processes in the absence of joint manufacture
- Clauses having the effect of impeding parallel imports, e.g. export bans. The regulation prohibits clauses whereby one or both parties is required to refuse without an objectively good reason to meet demand from users or resellers in their respective territories who would market

products in other territories within the common market, or to make it difficult for users or resellers to obtain the products from other resellers within the common market and in particular to exercise intellectual property rights or take measures so as to prevent users or resellers from obtaining outside or from putting on the market in the licensed territory products which have been lawfully put on the market within the common market by another party or with its consent.

14.6 USA ANTITRUST LAW AND INTELLECTUAL PROPERTY

The present framework of American antitrust law seeks to enforce competition by three means: the criminal law invoked by the law enforcement agencies; civil actions brought by individuals injured by anticompetitive activities but also in some cases by the public authorities for injunctions; and by regulation in the form of the Federal Trade Commission. There is, however, a considerable overlap between these sources of enforcement.

These three sources of antitrust law can all limit the scope for using intellectual property rights to advantage, which have traditionally all been regarded with hostility by the American judiciary even though they are enshrined as lawful under the American Constitution.

14.6.1 The Sherman Act sl

The Sherman Act sl provides 'Every contract, combination in the form of trust or otherwise, or conspiracy, in restraint of trade or commerce among the several States, or with foreign nations, is declared to be illegal'.

The Sherman Act is designed to cover collusion between parties designed to restrict competition unreasonably. It cannot affect unilateral market activity. The Act applies to cases where three things can be proved.

Firstly there must be a contract combination or conspiracy. Secondly, this must be in restraint of interstate (i.e. within the USA) or foreign trade or commerce. Thirdly, it must create an undue restraint on trade.

14.6.2 The Sherman Act s2

The Sherman Act s2 provides 'Every person who shall monopolize or attempt to monopolize, or combine or conspire with other persons to monopolize any part of the trade or commerce among the several States, or with foreign nations, shall be deemed guilty of a felony'. s2 of the Sherman Act is designed to deal with unilateral action, as well as multiple parties, and covers any attempt to monopolize, whether successful or not, including attempts to use monopoly power in one market to influence events in another. Three points may be made.

Size or dominance of itself it not sufficient, there must be an intent to monopolize, although in the case law size and *de facto* dominance is sometimes taken as circumstantial evidence of the requisite intent.

Attempts by individual concerns to monopolize and attempts by conspiracies and combinations are just as unlawful as actual monopolization but there must be proof of a specific wrongful intent in such cases.

The market to be monopolized can in practical terms be defined extremely narrowly by the antitrust authorities so as to bring a particular action within the net of s2.

Apart from criminal proceedings, brought by the Antitrust Division of the Department of Justice, the Sherman Act s4 empowers both the Department of Justice and the Federal Trade Commission to institute proceedings with a view to preventing future violations of the Act, and the courts faced with such actions have power to grant detailed injunctions and even dissolve businesses or split up mergers and parent-subsidiary relationships, and these powers have in practice been of more significance than the criminal penalties prescribed under the Act. The Federal Trade Commission has no criminal powers. Civil and criminal proceedings may run concurrently or consecutively. Private individuals may also institute proceedings under the Sherman Act.

It will be noted that the Sherman Act itself applies only to arrangements which might affect interstate trade. This is of very little significance in the light of judicial interpretation of the Act which found that even purely intrastate activity could in certain circumstances affect interstate trade. It is also true to say that a number of the States have their own State legislation matching the Federal provisions for purely internal activity – something encouraged by the provision of funds for drafting and enacting and enforcing those State laws as a result of the enactment in 1976 of the Federal Antitrust Improvement Act.

14.6.3 The Clayton Act

The Clayton Act s2 outlaws predatory pricing and the abuse of purchasing power to squeeze out less powerful competitors. The Clayton Act s3 deals with tying arrangements, exclusive dealing arrangements and agreements requiring a certain proportion of output to set aside for the other party, but only where the restrictions are placed on the purchaser of goods; it does not extend to services. Restrictive arrangements outside this scope have to be dealt with under other provisions, often the Sherman Act or the Federal Trade Commission Act.

The section deals with three types of agreement: that which incorporates a tying clause, an exclusive dealing agreement or a total requirements clause. The section does not condemn all exclusive dealing and total requirements contracts, but only those which can be seen to have an appreciable effect on the channels of distribution. Justifications such as the need to provide a dealer with a push-up into a new market, to introduce a new project, to protect goodwill etc. are permissible defences where of reasonable duration and not beyond what is needed to achieve the desired effect.

The Clayton Act s7 appears at first sight to be concerned only with monopoly and merger cases, but in fact the Act has been held to extend to intellectual property rights and therefore the Act does have some direct relevance to the assignment and licensing of intellectual property. The section prohibits both the acquisition and holding by any person engaged in commerce unless solely for the purpose of investment of any intellectual property right where in any line of commerce in any section of the country the effect of such acquisition may be substantially to lessen competition, or to tend to create a monopoly.

There are no criminal offences created by the Clayton Act, which may be enforced by the Department of Justice or by the Federal Trade Commission in civil proceedings. But there is also the possibility of private actions. Private individuals may also institute proceedings under the Clayton Act. In civil actions, on proof of infringement of the Sherman or Clayton Act, the Clayton Act s4 now provides that the person bringing the action may be awarded treble damages in respect of his or her loss.

14.6.4 The Federal Trade Commission Act

The Federal Trade Commission Act potentially covers an enormous field far beyond that of the other provisions so far discussed. The Act s5 provides that unfair methods of competition affecting commerce, and unfair or deceptive acts or practices in or affecting commerce, are declared unlawful. The Federal Trade Commission itself, subject only to supervision by the courts, has the power to decide what it regards as unfair methods of competition.

In practice, s5 has been interpreted to cover actual infringements of the other legislation, and acts which infringe the spirit of that legislation but which for some technical reason do not infringe the letter, or which are potential antitrust infringements if allowed to continue, and finally a number of practices which are otherwise seen to be anticompetitive or unfair and should be stopped.

In fact, because of the breadth of the other Acts there are few types of practice not already covered and most of the practices which the Federal Trade Commission has held to be unfair under s5 of the Act which are not also infringements of the Clayton or Sherman Acts have been practices directly prejudicial to the USA consumer interests, such as misleading advertising practices.

The mechanism of enforcement of the Federal Trade Commission Act is somewhat different from that of the rest of the legislation. It is not enforced through the courts but by administrative procedures. There are no criminal offences under the Act and, unlike the other Acts, private individuals may not bring civil actions for damages resultant on its infringement.

The Federal Trade Commission was set up as a government agency given joint responsibility with the Department of Justice for enforcing the Clayton Act through the criminal courts but also administrative powers to make 'cease and desist' orders including injunctions against persons where the Federal Trade Commission Act has been found to have been infringed. The Commission can also, having given notice to the Attorney General, issue an action for compliance or to obtain an injunction from a district court to give effect to any cease and desist order which it feels likely to be disregarded.

Once the time for appeal runs out the cease and desist order becomes final and thereafter any violation of it attracts a maximum \$10 000 fine. In addition, there is a somewhat draconian provision that there is a \$10 000 per day fine for failure to respect the order although this provision is often not enforced. A person receiving such an order may seek judicial review of it by the Circuit Court of Appeals (usually Washington DC Circuit) and ultimately to the Supreme Court. Appeals are on a point of law only. The conclusions of fact of the Commission, if supported by testimony, are to be conclusive and binding on the courts hearing any subsequent appeal.

Because of the courts' interpretation of the scope of the Federal Trade Commission Act, in many cases the Commission has the choice of pursuing someone under the Sherman or Clayton Acts through the criminal courts or through its own administrative procedures, and a number of cases which in form (i.e. appearance) are Federal Trade Commission Act cases, are in substance Sherman or Clayton Act cases, and instructive as such.

To proceed under s5 the Commission must have reason to believe that a proceeding by it in respect of the infringement would be in the public interest, so there is no obligation to take proceedings, but rather a discretion to do so.

On the other hand, the Act does oblige the Commission to issue a complaint in all cases where it has reason to believe that some infringement of the Act has taken place.

Mention should also be made at this point of the possibility of consent orders, which may often go very far but which are often favoured by potential defendants because they eliminate the expense and publicity of a trial. These consent decrees are settlements reached between the government and the defendants and sanctioned by the courts, which terminate the proceedings and under which the defendants usually undertake to cease the objectionable practice, divest themselves of the assets which have made them monopolistic, etc.

14.7 EXTRA-TERRITORIAL APPLICATION OF THE US LAWS

It is a feature of American antitrust provisions that they claim international application in as much as even those arrangements which affect solely trade with 'foreign nations' are covered by the Sherman Act.

The American legislation reflects to some degree the stance taken by the US courts in developing a doctrine under which foreign transactions which have a substantial and foreseeable effect on US commerce are rendered subject to US law regardless of where they take place and a concept of personal jurisdiction which may be exercised over persons transacting business in a certain place even if they are not 'in' that place in a traditional sense.

The effect of the case law is such that if, for instance, two British producers, or one British and one French producer competing in the US market, enter into an agreement to fix or stabilize their prices to the USA, such horizontal arrangements at least on their face constitute violation of the Sherman Act.

The reason for the arrangement, and the intentions of these foreign producers, is to restrain competition between themselves in the US market. The US jurisdiction will apply even more clearly when foreign producers operate through US subsidiaries or through US affiliates. Price fixing or other equally proscribed arrangements may be established or implemented between US affiliates of foreign producers. Thus arrangements solely between foreign producers or between their US affiliates can raise serious problems under the Sherman Act, and that is true if all discussions or communications take place abroad.

Although of less significance for this discussion, it should be mentioned also that foreign companies are subject to US antitrust laws other than the Sherman Act when they are doing business in the USA. Thus, a foreign company which charges different prices to competing US customers is subject to the Robinson Patman Act ban on price discrimination.

Nevertheless, the dangers of antitrust should not be exaggerated. US antitrust authorities regard imports as a good thing for competition in the domestic market and will only act if you are the sole or dominant suppliers in the US market. Customers are usually in a position to benefit from any unlawfulness. And US based competitors seem to prefer to act not under the antitrust law but under the antidumping law which, although it only tackles undercutting by importers, is quicker, does not make provision for antitrust counterclaims by the defendant foreign firms, does not involve such extensive rights of discovery against the plaintiff in favour of the importers and does not raise difficulties of jurisdiction raised by antitrust cases.

From the point of view of the defendant, such antidumping suits do not carry the treble damages clauses in antitrust actions which are sometimes available to plaintiffs and the discovery rights available against the importers are also more limited in the import cases than in antitrust cases. The balance seems to lean in favour of bringing antidumping actions as they are much less hazardous for the US plaintiff. So far as the activity of American firms operating outside the USA in a way which affects only non-USA markets is concerned, these American exports are to some degree exempt from the US antitrust considerations – but it must be stressed only in so far as they have no effect on the US market itself.

Closely linked with these provisions from the point of view of foreigners is US export administration legislation, one of the most controversial aspects of US technology policy. There is no general requirement for governmental consent to an international licensing agreement involving an American party or affecting the USA, but substantial regulation does exist of technology transfer to nations which the USA considers unfriendly to itself, where that transfer might lead to the hostile country receiving goods or technical information, directly or indirectly, which could contribute to economic or military development of that country. This is discussed further in Section 14.10.

14.8 ANTITRUST AND INTELLECTUAL PROPERTY RIGHTS

At a number of points, the case law on antitrust begins to merge with other doctrines developed in the courts to limit the scope for anticompetitive actions by intellectual property owners, those of fraudulent patent procurement and of misuse of patent rights. And where antitrust and misuse are concerned, sometimes the courts fail clearly to distinguish these two causes of action. How do these antitrust provisions interact with the patent and other monopoly rights created by the intellectual property statutes in the USA?

Under the USA Constitution it is provided that Congress has the power to promote the progress of science and useful arts by securing for limited times to authors and inventors the exclusive right to their respective writings and discoveries (Article 1 Section 8). Thus, the power to enact legislation granting patent monopoly and similar laws is embedded in the USA Constitution. However, the courts are anxious to confine such protection to its minimum.

14.8.1 Patent procurement fraud

The US courts have developed the rule that any applicant for a patent must disclose to the Patent Office all facts concerning possible invalidity or scope of the patents applications in issue. Any failure to provide all this information might of course lead to the invalidity and unenforceability of the patent, but the Supreme Court case of Walker Process Equipment Inc. v. Food Machinery and Chemical Corp. (1965) held that the fraudulently obtained patent rights could in themselves found an antitrust action under the Sherman Act s2 if the other requirements for such an action could be established.

In the result two lines of case law have developed, one looking at the requirements for a finding of invalidity through non-disclosure without any regard to possible antitrust consequences and the other looking at the problem in the light of antitrust liability too.

So far as the former category is concerned the position is now well summarized by the 1977-amended rules of the American Patent Office which (rule 56) require individuals to disclose all of the 'information they are aware of which is material to the examination of the application. Such information is material where there is a substantial likelihood that a reasonable examiner would consider it important in deciding whether to allow the application'. The actual application of this rule in the courts has not been too consistent, some tribunals adopting a broad, and some a narrow construction of the rule. This makes it quite essential to provide patent agents procuring patents in the USA with as much information as possible about the prior art of which you are aware at the time of making the application.

Any attempt to monopolize by use of such a patent procured through 'fraud' in the rule 56 sense is also antitrust violation under s2 of the Sherman Act. The only consolation for the patentee is that patent fraud must be distinctly proved as must some level of intent to defraud the Patent Office, although in a number of cases the courts have been willing to infer such fraud. Indeed it has been said on occasion that the court will infer the necessary intent wherever the material which should have been disclosed was prior art of which the counsel did know or should have known at the time of the application though other cases hold that mere negligence is insufficient and some cases indicate that a higher degree of fraud is required in an antitrust action than in a mere patent fraud case, negligence sufficing only for the latter, and the former requiring proof of intent. If the non-disclosure is in good faith however, no such antitrust consequences arise.

14.8.2 Patent misuse

When it comes to patent misuse, i.e. use of a valid patent to preserve or achieve unfair market advantages, the basic approach taken by the US courts has been that patents are privileges granted in the public interest and that where patent rights are used in a manner not in the public interest, or to an extent which goes beyond that necessary to protect the patent and for anticompetitive purposes, the exercise of the patent right in for instance a licence will not justify conduct which would otherwise be a clear infringement of general antitrust principles. In cases where the subject matter of an agreement does not have monopoly protection under law, such as know-how, the courts take an even more aggressive approach to anticompetitive provisions in licences.

Nevertheless, it is still true that the patent owner can to some degree act in a way which would at first sight appear to infringe the provisions of the antitrust legislation. The mere exercise of patent rights is not an infringement of antitrust laws and even misuse of the patent will not in all circumstances be so serious as to constitute an antitrust offence. But any attempt to extend the patent rights outside the legitimate scope of its grant will not benefit from any exemption from the antitrust legislation.

14.8.3 Use of improperly obtained trade secrets

A number of cases have found that the unlawful misappropriation of trade secrets by a dominant party or parties pursuant to an attempt to injure competition is a *per se* illegal act under the Sherman Act s1, with all the consequences that implies, quite apart from being actionable under the general trade secrets laws.

14.9 US ANTITRUST AND JOINT R AND D VENTURES

Recent amendments by the Reagan Administration to antitrust law make some specific provision relevant to intellectual property in dealing with joint ventures in research and development, many of which would otherwise fall firmly within the scope of the antitrust laws. The *National Co-operative Research Development Act 1984* says that joint research and development ventures are to be judged on the basis of reasonableness, and taking into account all relevant factors affecting competition, including, but not limited to, effects on competition in properly defined, relevant research and development markets. Any party to a joint research and development venture, acting on such venture's behalf, may, not later than ninety days after entering a written agreement to form such a venture, file simultaneously with the Attorney General and the Commission a written notification disclosing the parties and the nature and objectives of the venture. This is then published in a Federal Register. This has the effect of limiting to a claim for actual damages (and not treble damages) any plaintiff who wins an action under the antitrust laws in respect of the registered agreement. The scope of joint research and development venture agreement is fairly narrowly defined.

14.10 USA'S GENERAL APPROACH TO LICENCE AGREEMENTS

When we come to look at the attitude of the USA's authorities to licences we find that a number of terms have in the past clearly been thought to be illegal attempts to extend the scope of the patent etc. outside its grant.

Tying clauses have been held to be illegal, because they plainly seek to extend a monopoly position beyond the patented product or process to other products or processes.

Royalty clauses extending beyond the seventeen year term of a US patent also sought to extend the scope of the grant.

Patented products subject to price fixing between competitors and cross licensing agreements or agreements which exclude third parties from licences, patent pools (i.e. agreements to cross licence all patents owned by the members of the 'pool') which dominate the market were not exempt.

Only limited scope existed for grant-back clauses, and even excessive accumulation of patents might be unlawful under the Sherman Act.

Even fraud in front of the Patent Office in procuring patent rights might in addition to providing a defence to an alleged infringer also give him or her an action under the Sherman Act s2.

Other clauses in licences are more generously treated but are still subjected to close economic analysis and are viewed with suspicion by the judiciary.

In the area of know-how and trade secrets courts have taken a hard line with market restraints such as price control, market division, time leads, etc., although in most cases these decisions have been based on reasonableness – with the exception of tying clauses and royalties beyond term these agreements are not necessarily *per se* illegal, but depend on whether the restriction is merely reasonably related to the patent, trade mark, copyright interest being exercised by the owner-licensor, in other words whether they are reasonably ancillary to a legitimate commercial purpose, or whether they go beyond this and amount to attempts in essence to restrain competition.

In fact, the case law on trade secrets is relatively sparse and problems have arisen out of attempts to argue cases on trade secrets as if they were patent licences, when the different nature of the trade secret requires in some measure a different response.

Until 1981 the USA's Department of Justice had a list of nine clauses which were thought to be irredeemably bad. This list has now been abandoned but it remains of use as a danger list:

(1) Tie-ins
(2) Grant-backs
(3) Restrictions on resale
(4) Tie-outs i.e. undertakings not to buy other products
(5) Licensee veto of further licences
(6) Package licensing
(7) Licences conditional on taking other licences
(8) Minimum price for resale of patented goods
(9) Quotas and restriction on sales of licensed process products by licensee.

The latest *Department of Justice Antitrust Division Guidelines* have this to say about restrictions in intellectual property licences (paragraph 2.4):

> '. . . Such restrictions often are essential to ensure that new technology realizes its maximum legitimate return and benefits consumers as quickly and efficiently as possible. Moreover, intellectual property licenses often involve the coordination of complementary, not competing, inputs . . . Unless restrictions in intellectual property licenses involve naked restraints of trade unrelated to development of the intellectual property, or are used to coordinate a cartel among the owners of competing intellectual properties, or suppress the creation or development of competing intellectual properties, the restrictions should not be condemned.'

14.11 REMEDIES FOR INFRINGEMENTS OF US ANTITRUST LAW

In addition to the general points already made it should be mentioned that in the case of intellectual property rights, in particular patents, it has been commonplace for the antitrust authorities to order compulsory licences to be available to all-comers where a company has built up a dominant position through amassing a collection of the patents for any given market, even where those patents were not unlawfully acquired but were developed all in house by the company's own R and D team. Orders have extended not only to all the company's existing patents, but where its lead in research is such as to require that rivals be given time to catch up, even future improvements patents and new patents in the field may be subject to the compulsory licence regime, if that is necessary to open up the market sufficiently to make it truly competitive.

The courts apply the rule of reason test to determine how far they should go in making the order appropriate. Such orders may be backed up with orders which require the disclosure of the know-how necessary to make the patents work satisfactorily and competitively, usually in the form of technical operations manuals.

Even more severe from the point of view of the patentee, although usually the patents licences will be ordered at a reasonable royalty payment, dedication to the public – i.e. royalty-free – licences have at times been ordered by the courts, where such an extreme measure has been felt necessary to open up the market and prevent continued monopolistic practice. It is not a material factor to consider how the monopolist used or abused his position, but what is necessary to create competition.

14.12 THIRD WORLD RESTRAINTS ON LICENSING

A number of the Third World countries have very extensive provisions not only on what is to be required before government approval will be given for a licence transaction involving inward technology transfer but also as to what types of clause are enforceable or are even legal in their country. The range and implications of these provisions is too great for a book of this size, but care should always be taken when doing one's homework for the agreement to determine what restrictions if any there are. Poor experience in the past in which Western companies have, to put it bluntly, given poor value in technology transfer have left the Third World governments with a somewhat jaundiced view of the licence agreements accepted in the West.

In many of these countries provisions at least as strong as the antitrust of the USA will prevail, often backed up with laws requiring that, for instance, royalties may not be withdrawn from the country, requiring prior government approval for any transaction and its terms, or setting out maximum rate royalty percentages, the employment of local labour, local joint ventures in which the foreign company has no more than a 49% equity stake or perhaps requiring some minimum capital investment, and so on. Clauses requiring most favoured nation treatment and local purchase of raw materials or supply of materials at world market prices are again quite common.

14.13 EXPORT ADMINISTRATION CONTROL LAWS

A further practical restriction on the British exploitation of some technology is the recent export administration control legislation enacted in response to the legislation along similar lines of the USA. For under the original US legislation, and now under the UK law, severe criminal penalties may be invoked against businesspeople for even unintentional infringements of the law. Only a brief resume of this hopelessly complex subject can be given here, to alert the reader to the risks which may be run under these laws. Furthermore the regulations vary in their details with an alarming regularity and thus the up to date position has to be checked before any individual transaction: it is never safe to rely on what was the position the last time the particular transaction in question was contemplated.

Accordingly this section is deliberately vague and imprecise on this topic. You should never assume that it does not apply to you and never assume that even though it applies you are doing nothing wrong. There have already been certain prosecutions in relation to quite low technology products and know how. For a UK trader the closest sources of advice are the US Embassy and the UK's own Department of Trade and Industry.

14.13.1 The US position

The US *Export Administration Act 1979* as later renewed and amended by the new *Export Administration Amendments Act 1985* has provided that there should be some ten categories of products, covering most types of industrial goods, and technical data, all of which may not even be exported from the USA without a prior export licence.

All overseas countries with the exception of Canada are listed in one of several fixed categories of security risk nations. The list is in summary form as follows:

Category of country

Q Romania
S Libya
T North America, Central America, Bermuda and the Caribbean area, South America.

V All countries not in any other group except Canada (dealt with specially in the regulations)

W Hungary
Poland

Y Albania
Bulgaria
Czechoslovakia
Estonia
GDR
Laos
Latvia

Lithuania
Mongolian People's Republic
USSR

Z Cuba
Kampuchea
North Korea
Vietnam

In addition to these lists there are certain other special country policies relating for instance to South Africa and Namibia and also to the People's Republic of China, which are treated less favourably than their Class V rating might suggest. There are also special commodity policies each relating to armaments etc., and to crime control, antiterrorism and regional stability policies. It is very important to stress that technical data as well as the tangible property are covered by these laws and thus any licensing of know how, etc. may well be covered by some licence requirement. There are also certain scarce commodity regulations covering all countries including Canada.

14.13.2 The licensing procedures

(a) General licences

Under the US laws certain categories of goods now require only 'general licences' which will be granted automatically in respect of goods worth less than some $1000. For these there is now no need to apply for any licence: these general licences are published in the regulations and they apply generally to any such export transaction. Some other general licences relate to the type of country, e.g. CoCom member states, as well as to the type of commodity.

(b) Validated licences

But other transactions require a full validated licence, i.e. a licence must be applied for specifically. This licence will authorize the export of a particular commodity to a given place for a particular purpose by a particular applicant to a particular consignee.

(c) Other licence categories

There are also various intermediate positions, which involve simplified procedures for certain types of licence such as project licences (which are granted covering all items in a given large scale project, e.g. as in a major dam construction project); distribution licences (for a general authorization to an individual in respect of a variety of items); qualified general licences (for multiple exports of certain types of product by an individual to given countries); and the service supply licences (designed for unlimited export of spares or replacement parts for countries in, for example, the Groups T or V in respect of materials which are already authorized for export and for certain replacements but not for spare parts for countries Q, W and Y, and for

Afghanistan and the People's Republic of China, and in certain other cases).

Where a licence has to be applied for the body administering the scheme will usually be the office of Export Administration in the Department of Commerce but in respect of certain specialized items, such as those related to defence, energy, nuclear equipment, drugs, etc., different regulatory bodies have the powers given by the Act and these must be approached in the same fashion.

The result of any one application will depend on the view taken of the given technology in relation to the category of the export country involved, i.e. in effect, whether it is Communist, terrorist, NATO and/or friendly Western European, etc. as classified by the law. Licences can be revoked or suspended and goods exported recalled or intercepted.

14.13.3 The scope of the regulations

The regulations extend not only to direct exports from the USA to overseas countries including the UK but to re-export from those destination countries, exports from foreign countries of items containing US origin components, and exports of equipment making use of US technical data. The law sometimes requires the exporter – as a condition of being granted a licence – to impose a requirement of a written guarantee of compliance with the regulations by the purchaser or the licensee of the technology or data, in effect against re-export to the forbidden places. The exporter will usually add his or her own voluntary clause, requiring the purchaser to even indemnify him or her for any fines incurred as a result of any breach of the law.

14.13.4 Jurisdiction over UK exporters, licensees

The significant thing from the point of view of the UK company about the US laws is that they claim to have extra-territorial effect, so foreign subsidiaries of US companies and indeed even wholly unconnected foreign purchasers and licensees are purportedly also subject to the laws and therefore prohibited from re-exporting in breach of the regulations. Equally, the law purports to control non-US technology in the hands of companies based overseas owned or controlled by US companies. Certain other related transactions involving either US citizens or overseas companies controlled by US companies or citizens are also capable of being controlled by the US law, where destined for places such as Iran or countries in classes Q, W, Y, or Z or nationals of those countries anywhere. If the overseas exporter is not subject to the US jurisdiction the law requires him or her to appoint a US agent to make the application on his or her behalf.

14.13.5 Penalties for breach

The penalty for a knowing violation of the US law or a licence granted under it is up to five times the value of the goods exported or $50 000, whichever is greater, or imprisonment for up to 5 years or both, with civil penalties of up to $10 000 per violation of the law even if the violation was in pure ignorance of the law. Conspiracies to breach the laws are also dealt with severely. And the rights to continue exporting may be limited or revoked entirely for good. Much heavier penalties exist for more serious offences under the law with national security or foreign policy implications, including a failure to report certain uses by others of the technology of which you become aware. End users as well as agents abroad can be caught and there is now provision to restrict overseas offenders from importing into the USA as well as from receiving US exports.

The precise regulations are arcane and extremely complicated. They require you in effect to obtain both expert and official advice at every turn, and the grant of licences can be a protracted affair, occupying much time, up to 60 days in many cases, 240 days in others, and even longer where CoCom approval is required as it sometimes is; although much low technology equipment is often in the general licence regime or a licence is in fact obtainable with relative speed. Overseas companies do seem to wait longer than US companies, in practice. It must be said that the 1985 amending Act did attempt to streamline the licensing procedure somewhat, mostly in an attempt to counter criticisms that it had been rendering US exports less competitive. Basically, the procedures for exports to CoCom states are much more streamlined than elsewhere and may be obtained more quickly.

14.13.6 The United Kingdom law on technology export

The present UK law is much more limited and really designed to put the UK in a well favoured position with regard to US technology export arrangements. The UK law is contained in the *Import Export and Customs Powers (Defence) Act 1939* which has sprung to life in recent times because of the large powers which it has always contained now being exercised in earnest again.

The 1939 Act permitted the making of statutory instruments controlling the export of goods to certain places and the regulations enacted in 1985 under this legislation now prohibit the export of much high/new technology material without a prior licence from the Secretary of State on pain of criminal penalties up to £1000 or up to two years imprisonment (under s68 of the UK's *Customs and Excise Management Act 1979*). Any such decision by the Department of Trade to refuse a licence is unappealable.

The UK regulations contain a long list of goods and technological documents divided into various set classes which are subjected to restrictions as to the destination of export. These may, according to class, prevent export to any destination, to Eastern Bloc countries, or permit export only to the EEC states, etc., etc. In all such cases by way of exemption the Secretary of State may issue a licence to an applicant to export to the country concerned despite the ban.

14.1 CONCLUSIONS

Here as always the moral is to do you homework and to vet any proposed transaction with your lawyer to ascertain its legality and the consequences of its illegality; there are often criminal sanctions which involve more than mere fines. Where local foreign law is involved check with the local foreign lawyer too.

15
Enforcing rights in new technology and design

15.1 INTRODUCTION

So far we have been concerned with the means of protecting new technology and design by the acquisition of intellectual property rights. Now we turn to the question of enforcing such rights. This is not intended to be a textbook guide for lawyers but an indication to non-lawyers of some of the aspects of intellectual property litigation which require thought or action on their part.

15.2 THE DECISION TO BRING OR FIGHT AN ACTION

Even though only a small proportion of patents are actually used there are still in terms of numbers many in active use and probably an equally large number of technical infringements. The same must be true of the other intellectual property rights though the incidence of infringements is impossible to judge. For a variety of reasons not every infringement is detected and of those detected not every one is pursued, not least because a judgment will have to be exercised about the strength of the right concerned in each case. It may often be preferable to let an infringement pass than to subject what may be a relatively weak right to the glare of an infringement action in which it might be held invalid or narrower than was thought.

Even a weak patent has some deterrent effect on competitors and will therefore confer a competitive advantage on the right owner. The existence of a patent or other right may persuade a potential challenger to take out a licence at a reasonable rate and thereby become an ally.

The vast majority of incipient patent and other intellectual property right actions are nipped in the bud by out-of-court settlement. Legal costs can rise dramatically after the writ is served and it will be an exceptional right which is worth the expense of the litigation which follows. What factors should be taken into account in deciding whether to sue or defend?

Factors in favour or against an action include:

(1) The prospects of success
(2) The costs of the action for winner and loser
(3) The value to you of an injunction or damages
(4) The risks of trade secrets being exposed in court
(5) Financial strength of the opposition
(6) Alternatives available

15.2.1 The prospects of success

This does not just relate to the legal question of whether the patent or other right is a strong right. It requires the assessment of a number of other factors, such as the likely performance of your witnesses in the witness box and even whether they will still be alive when the case gets to trial! If the prospects for your success are not high it may be better to preserve some credibility for the technology by agreeing on a licence out of court and thereby terminating any such potential litigation with the rights apparently still intact.

15.2.2 The costs of the action to each side

Even if you win on all issues you will not recover all of your costs in the action, either overt or covert costs. Even if you win the costs come back later, and the cash flow problems may become acute for the smaller firm in the action.

It is impossible to give estimates of how much an action is going to cost when you get started. So much depends on the aggression of the other side, and their willingness to settle, the technology involved, the law and financial implications of the technology, and so on.

To a limited degree the cost of litigation can be minimized by both sides by merely confining the issues in dispute to only one or two of the claims in question in the patent specification and so on, but this requires the tacit co-operation of the parties which may not be forthcoming. The other hidden costs must not be disregarded either. Litigation will inevitably distract highly paid and valuable employees from their jobs of developing or selling technology or designs and into the action itself. These are a major disruption to the business of the company and they should be weighed up carefully before taking action.

15.2.3 The value of the injunction or damages

This brings up the next point, the publicity value of a victory or a defeat, and the financial benefits from being able to appropriate damages for lost sales, or recovering the infringing products themselves by an injunction and order for delivery up. In a market where there are many small retail suppliers the infringement problem can be more easily dealt with if the larger wholesale supplier is tackled and squeezed out of the market altogether so that the retail supply dries up.

15.2.4 The financial strength of the other party

This will be crucial in many cases, because weakness on the other side's part may enable you to extract a shotgun licence out of them as a potential infringer or to gain better terms from a potential licensee who just cannot afford to do without the product but who cannot afford to challenge the intellectual property rights.

Although courts now go some way to preventing abuse of legal procedure by large firms against alleged infringers by stating that they will not make an order for an interim injunction restraining production by the defendant which will close a factory, if the defendant undertakes to keep accounts of the products made and sold, it is still true to say that interim injunctions are powerful weapons in the hands of a plaintiff who is good for the money should he lose, for the practical effect of the present UK law is that whoever can pay later gets the interim order now.

15.2.5 Risk of trade secrets being exposed

Although there exist procedures for safeguarding the information which has to be disclosed to the other side in pre-trial proceedings and in court and in some extreme cases there may even be court hearings in camera to protect the integrity of such confidential information, it is nevertheless true that to some extent there may be a risk of uncontrollable disclosure of information: for example, it may be necessary to disclose some know-how which the other side previously did not have even though they were infringing your patent, and the proof of subsequent misuse of such know-how may be virtually impossible for the party compelled to disclose it.

15.2.6 Alternatives available

The viability of alternatives such as negotiated settlements and arbitration through mini-trials should also be considered. The merits and the disadvantages of these alternatives are considered later in the chapter.

15.3 UNJUSTIFIED THREATS OF LITIGATION

Despite all of the risks of intellectual property litigation from the point of view of the right holder it is still true to say that even the mere threat of litigation is a potent weapon which can be abused. It is possible to exercise considerable power by alleging infringement of rights, with much attendant bad publicity for the victim of the threat even where those rights are of doubtful or no real validity. Threats may be enough without actually putting a right to the test before the court especially if a weak victim is chosen.

The general law gives very little redress for such actions. Where litigation is actually brought, the jurisdiction to strike out vexatious and frivolous actions is narrow and the sanction of costs may never arise if in the meantime the defendant has given in. But many cases do not even reach litigation and these pose a serious threat. Accordingly there are provisions in the case of UK patents and registered designs for an action to be brought in respect of such unjustified threats of litigation which may be of help on occasion. By contrast there is no such legal provision under the Copyright or Trade Marks Acts and the victim of such threats will be helpless unless the law of slander or libel or injurious falsehood can be invoked, although it has been suggested that the court under its inherent jurisdiction might be able to grant injunctions to restrain threats of litigation. Any threats should be vetted before issue.

As a footnote to these matters it may be worth mentioning the criminal liability which exists for falsely representing that trade marks, designs or patents are registered. In the case of trade marks, the *Trade Marks Act 1938* provides that it is an offence to represent that a mark is registered when it is not. But it is not an offence to state that an unregistered mark is a trade mark whereas use of the word 'Registered' in relation to a mark which is not in fact registered as a trade mark is an offence under the Act. Nevertheless the false use of the claim that a mark is registered has never barred an action in passing off brought by the person making the false claim, so the offence has no civil law consequences for the protection of the unregistered trade mark against others.

Parallel provisions exist under the *Patents Act 1977* and the *Registered Designs Act 1949*. So a false representation that there has been an application for a patent is also an offence so that use of terms such as 'patent pending' or 'patent applied for' may also lead to trouble, if there has been no application or it was abandoned. Nevertheless, so far as patent and patent applications are concerned, due diligence in preventing the offence is a defence even if the prohibited act occurs notwithstanding this diligence. Further, in the case of a patent or patent application, a period of grace following expiry or revocation of the patent or withdrawal of the patent application is permitted to enable the patentee to take steps to ensure that the representation ceases.

EXAMPLE

Cassidy v. Eisenmann (1980)

A toy imported from Italy bore the marking 'Brevattato Italia, Espania, Great Britain, France, Deutschland' but at the time only a UK patent application existed. A fine was imposed.

Clearly the safest policy is to avoid liability altogether by careful drafting of letters. An example of such a letter is included in the Appendices.

15.4 WHERE IS THE INFRINGEMENT ACTION FOUGHT?

These infringement actions may involve a number of courts in England and Wales: the Patents Court or the Chancery Division of the High Court; the Comptroller General of Patents, Designs and Trade Marks (known as the Registrar in cases involving Registered Designs and Trade Marks); Criminal Courts in some copyright or trades description cases; the County Court in highly exceptional copyright cases, usually not industrial in nature.

In licensing matters you may be involved with the High Court in breach of contract cases governed by English law or sometimes foreign law or enforcement of overseas judgments. Theoretically, in exceptional cases the claim might be small enough for the County Court, but this is virtually inconceivable.

Occasionally, you may be concerned with overseas national courts in cases abroad and/or governed by foreign law and/or needing enforcement overseas following an

English judgment against an overseas licensee/licensor. Instead of these state tribunals, there may be occasion to use private arbitrators in the UK or overseas in some cases.

There may also be involvement with the European Commission/European Court of Justice EEC, the Office of Fair Trading in the UK, or Federal Trade Commission or Criminal Courts in USA, if you are unlucky enough to become caught up in a competition law problem.

This chapter is concerned only with litigation in England and Wales in so far as it raises questions related to intellectual propety infringements. In a number of cases you will have a choice of forum. Which one you choose will depend on a number of factors, such as cost, remedies sought, etc. Again these are not just lawyer's questions. They require management decisions.

15.4.1 Patents

Although the *Patents Act 1977* applies to the whole of the UK, Scottish cases are heard in the Court of Session, not in the Patents Court. Only the English procedure will be considered here and in view of the earlier comment Patent Office procedure in infringement actions will be ignored. In all cases of patent infringement in England and Wales there is a choice of two fora: the Patents Court or before the Comptroller General of the Patent Office. The Patent Office may hear disputes over a number of matters:

(1) Amendment of specifications or claims
(2) Disputes over entitlement to the patent
(3) Employee compensation claims
(4) Licence of right and compulsory licence disputes
(5) Revocation claims
(6) Declarations of non-infringement; and
(7) To a limited extent infringement claims

The 1977 Patents Act considerably extended the jurisdiction of the Comptroller to hear and decide the infringement actions but only where the parties by an agreement refer such disputes to him or her. The Comptroller may decide that the case is more appropriately dealt with by the Patents Court and refer the matter there.

The procedures for proceedings before the Patent Office are set out in the Patents Rules and this is not the place to go into them, though in general terms they are more informal, cheaper, the Hearing Officer is an expert in the field, written rather than oral evidence plays a greater role, and there is greater scope for representation by patent agents. On the other hand, the Patent Office seems reluctant to engage in complex legal arguments and tends to favour the patentee more than the Patents Court does.

Moreover it is generally not a good idea to proceed before the Comptroller even if he or she consents, because although he or she has the power to grant a declaration or an award of damages he or she cannot grant the injunction, which is usually desired by the patentee and in any case the loser can appeal to the Patents Court, thereby merely adding another level to the action and increasing the costs of it. Furthermore, the Comptroller lacks certain powers of discovery and cross examination which the Patents Court enjoys and which makes the oral evidence before it of more value, and will decide issues which the Patent Office would defer.

There are proposals to change the balance of such litigation work in the case of patents by giving more work in the Patent Office than at present, and to much restrict appeals from the Patent Office to the Court. This will be considered in Chapter 16. For now we will examine only the present procedures.

In most cases it is cheaper to go direct to the Patents Court. But on occasion the initial procedure in the Patents Office may be of benefit because even if there is an appeal it will ensure that the technical side of your case has been reviewed by a Hearing Officer who is an expert in the field and whose findings will be very persuasive before the Patents Court judge on appeal. Where this course of action is taken the Patents Act provides that a decision of the Patent Office does not estop a party counterclaiming for revocation on either the same or different grounds in any subsequent infringement action.

The willingness of the Patents Court to let the Patents Office do some of this work is clear from the decision in *Hawker Sidley Dynamics Engineering Ltd v. Real Time Developments Ltd (1983)* where Patents Court proceedings for infringement between the parties were stayed pending the resolution of Patent Office proceedings for a declaration of non-infringement between the same parties.

The Patents Court was set up in 1977 under the *Patents Act 1977* as part of the Chancery Division of the High Court, and is staffed by Patent Judges who are all judges of the High Court. The present patent judges are Whitford J. and Falconer J. Clearly patent cases may involve the complex technical matters and judges are legal not technical experts. To avoid the dangers of misunderstanding of the technology involved in such complex cases, it is possible for the court to appoint its own scientific advisers in appropriate cases to assist the judges. These scientific advisers can only give assistance and report on matters of pure fact or opinion, not questions of law or construction of the patent specification, and the court need not accept their reports. They can assist either by being asked to inquire and report or by sitting with the judge at trial. The parties will usually prefer the former, because it gives them a greater opportunity to challenge the findings. Where the expert adviser does inquire and report he or she may ask the parties to arrange experiments or tests which he or she thinks necessary to enable him or her to make a satisfactory report, making arrangements with them about expenses, the parties to attend, etc., and if the parties cannot agree the court will settle the matter.

15.4.2 Registered designs

For registered designs, although technically no rule specifically assigns the matter to the Patents Court unless there is a counterclaim for rectification of the Register, in practice all infringement cases are assigned to that court and not to an ordinary judge.

15.4.3 Copyright and trade marks

For copyright actions involving no patent or registered design, the action for infringement will usually be brought in the Chancery Division of the High Court as will cases of infringement of trade marks. This is also true of passing-off actions.

15.5 WHO CAN SUE FOR INFRINGEMENT?

In sum, the registered proprietor for the time being of the intellectual property right alleged to have been infringed, where it is a patent design or trade mark, or the owner for the time being of the copyright or common law reputation, may sue. An exclusive licensee of a patentee may sue in certain circumstances. An assignee or exclusive licensee may sue in respect of infringements committed before the licence or assignment was granted only if the licence or assignment expressly confers the right to sue for past infringements. Patentees can sue for infringements before the grant but after publication of the final specification. Owners of a territorially divided patent may sue individually, whereas otherwise co-owners must sue joining the other co-owners as parties to the action.

Where an exclusive licensee or co-owner sues and the owner or co-owners as the case may be are unwilling to join the action they are joined as co-defendants but if the owner or co-owner as the case may be does not both acknowledge service and take part in the proceedings he or she will not be liable for any costs of the action in any event.

The position of co-owners of copyright, of registered designs and of registered trade marks is basically the same as that for patents, as is that of licensees of right in patent cases but in the case of registered users of trade marks there has to be a period of two months following a request to take action which the owner refuses before a registered user can sue for an infringement, subject to any agreement to the contrary between the licensor and licensee which in fact is quite common in trade mark licences. Indeed many licences of all kinds make express provision for assistance in litigation. An assignee of a trade mark may apparently sue in respect of infringements before the assignment was registered, although a proprietor could not commence action until the mark had been registered, even though that registration operates retrospectively to the date of the application.

The registered proprietor or exclusive licensee is presumed to be such by virtue of the fact of registration of the right in his or her name. In the case of a patent no damages or account of profits can be obtained by a plaintiff in respect of infringements occurring before the plaintiff's right was registered unless he or she registers it within six months of the date of acquiring that right or if that was not in the eyes of the court reasonably practicable then as soon as practicable thereafter.

15.6 WHO CAN BE SUED FOR INFRINGEMENT?

Obviously, infringers. But consideration should be given to the status of the employer, agent, partner, director, employee, promoter, of infringers, who may in certain circumstances be joined quite properly as co-defendants. The question of pursuing a sample defendant when there are many defendants should also be considered.

- Employers, in accordance with general principle, may be vicariously liable for the acts of an employee and should generally be sued rather than the employee, although it may be proper to sue the employee instead or or in addition to the company for whom he or she works where he or she is a senior employee substantially in charge of the employer's business alone.
- Agents, because such infringement actions are tortious (i.e. wrongful), acting on behalf of their prinicipals are liable, unlike in cases of contract, as well as their principals.

- Partners who were unaware of any infringement by their co-partners and who repudiate the act as soon as they learn of it will not be liable.
- Directors who personally committed or authorized the infringement may be liable as agents of the company, and where they are in substance the business infringing behind the corporate veil, it may be desirable to join them as co-defendants. This may be a serious question where trade marks or copyright are involved and the infringer is an importer of little substance of infringing goods manufactured overseas, for in such cases the companies as such will often have few assets with which to meet any judgments, let alone costs, whereas those behind the companies may be relatively wealthy.
- The costs of litigation dictate that in some cases pursuit of all possible defendants is not going to be feasible or desirable. Clearly the pursuit of some selected infringers will have greater impact than others in terms of publicity and in cutting off infringing articles from the market. Pursuit against the small-fry trader is expensive without being effective. In practice the right holder will seek to cut off supplies at the top of the tree, the manufacturer or importer. Where there are a number of similarly placed defendants it is common to select one as a test action and for the court to stay actions against the others pending the outcome of the test action, sometimes upon the others undertaking to submit to judgment in like terms should the plaintiff win. This saves time and money. The publicity value of a victory is such that the court will even grant an injunction in open court when the defendant consents.

15.7 INFRINGEMENT ACTION PROCEDURES

This is not a lawyer's textbook and only certain aspects of the intellectual property infringement litigation procedures are going to be looked at. How does a typical patent infringement action unfold?

Often the action begins with the plaintiff seeking to prevent the defendant from infringing his or her rights in the period before the full trial hearing. These will take the form of so-called interlocutory proceedings for an injunction or an order to admit the plaintiff's representatives to seek evidence of infringement in order to prevent further acts alleged to constitute infringements pending the outcome of the trial.

In general anything from eighteen months to two years elapse until the trial of action, i.e. the court hearing. There is then a judgment from the court either at once or reserved in a difficult case for the judge to think about it. After the judgment is given the loser has a period of six weeks maximum in which to bring notice of appeal.

15.7.1 Beginning the infringement action

It is not necessary to give any notice of one's intention to sue in respect of an alleged infringement and in some cases it will be essential in practical terms not to do so as for instance where there is to be an application for an interlocutory or Anton Piller order or a Mareva injunction, matters which will be considered below.

In all cases, however, although it has been argued that it is good practice to give notice (because if before the action has been heard the defendants give satisfactory undertakings – which they may well do if they were innocent infringers – a plaintiff who has begun the action without warning may be denied his or her costs) because of the danger of threats actions being brought against the patentee etc. in practice you often do not give notice. But where the infringement alleged is one which cannot be proved without proof of knowledge, and such proof may be difficult, notice should be given of the claim asking for an understanding that no further infringements will take place, and if this is refused the writ is then issued seeking an injunction restraining future infringements.

15.7.2 Interlocutory injunctions

A trial takes a long time to be heard and judgment given. In the course of that time the plaintiff's own business can be ruined. He or she wants immediate action. In an attempt to meet that need it is possible for the court to grant an interlocutory injunction pending trial restraining the defendant from doing the thing which it is alleged is infringing the plaintiff's rights. On what principles will such an interlocutory injunction be granted? The guidelines in reaching this decision are that:

- The court must be satisfied that the plaintiff's case is not vexatious or frivolous: he or she must have a serious issue to be tried but he or she need no longer prove a strong prima facie case. The court will not try to sort out the conflicting evidence.
- If there is a serious question to be tried, then the court should go on to decide whether the balance of convenience lies in granting or refusing the injunction that is sought.
- Would the plaintiff be adequately compensated by damages for the intervening period between this application and trial if he or she eventually won and would the defendant be able to pay them or would only an injunction now protect him adequately? If damages would be adequate there is no reason to grant the injunction. The same question must be asked on the opposite hypothesis, i.e. would the defendant be adequately compensated under the plaintiff's cross undertaking as to damages if an injunction were to be granted? If yes, there is no reason not to grant an injunction.
- If the answer for or against an injunction is uncertain, then the balance of convenience must be considered. What matters should be considered will vary from case to case.
- If the matter is still balanced evenly then preference should be given to preserving the status quo and only in extreme cases should attention finally be given to the strength of the parties' cases.

The main question is what factors do the courts take into account when determining the balance of convenience. The fact that the defendant had taken great care to avoid infringing valid patents is an immaterial factor in deciding whether to exercise the discretion in his or her favour. But equally the courts also indicate that the relative size of the plaintiff and defendant may be of relevance especially if it appears that the larger defendant has exercised his or her economic power to intrude into the legitimate scope of the plaintiff's monopoly with a view to driving him or her out of the market he or she has built up through cut-throat tactics such as loss leading or undercutting, or conversely if the continued presence of a small infringer can do little more than dent slightly the profits of the larger plaintiff.

The courts will look to see if the decision one way or the other will shut factories or lead to unemployment, whether it will nullify any investment programmes undertaken in good faith by one or other of the parties, and so on.

And non-commerical factors may influence the court too. The fact that the patented drug alone under dispute is a potential life-saving drug and a reasonably accurate assessment of loss could be made assists in a decision ro refuse an injunction.

The form of undertaking which the defendants are prepared to give may be particularly influential. For instance, the undertakings of the defendants to pay profits from the product allegedly sold in breach of confidence into an account on trust to protect the plaintiff should the defendants become insolvent and to deliver a monthly record of sales figures to the plaintiff's solicitors. Likewise an undertaking to pay a reasonable royalty into an account until the trial of the action for infringement of a larger company's patent.

15.7.3 Determining the conduct of the trial

The court will make certain orders as to how the parties are to conduct themselves in the trial. The matters which are dealt with include whether further details should be given of the parties' claims and defences in the case, and whether the parties should be permitted to inspect each other's equipment and under what conditions. This may be restricted to certain qualified persons.

EXAMPLE

Warner Lambert v. Glaxo Laboratories (1975)

The ordered disclosure was restricted to the plaintiff's counsel, solicitors, patent agent and his expert, and their chief executive in the USA, subject to an undertaking not to disclose to others. It was denied to the legal counsel of the plaintiffs in the USA, to their Italian scientists, and to their US Patent Counsel.

The court will also determine whether certain documents should be subject to discovery, i.e. revealed to the other side or what questions about them may be put, whether either side should make any admissions of facts alleged and whether any experiments should be conducted relating to the trial issues.

In the case of experiments the court will usually require the party conducting them to demonstrate them to the other side or where they are relying on third party experiments, to reveal the results to the other side, before consent to admit the evidence is given, how many expert witnesses each side should be permitted, and for taking affidavits on matters of expert evidence and filing them and sending copies to all the parties.

The court may order the disclosure of photographs and models relied on as evidence of prior art in challenging novelty etc. and decide whether a particular matter should be heard as a preliminary issue, including any question as to the construction or specification of documents.

The issue of whether there is to be a preliminary hearing on the construction of the specification should be given careful thought, because it may be that consideration of this matter will determine whether or not the experiments to be ordered might not be necessary. These orders save money in the longer run because they prevent one side from being taken by surprise and having to seek an adjournment to consider the evidence, cut down on battles of experts and organize the trial so that a single crucial matter which could determine the whole issue is disposed of first.

In cases which involve any particularly difficult questions of technical fact, it is now common practice for the parties to the action to prepare an agreed account of the technical background to the case, which includes what has become known as the 'Simkins list', a list of previous unsuccessful attempts to solve the problem which the patent in dispute solves, and also for the parties to make written summaries of their submissions in difficult cases even where they cannot reach agreement on the facts.

15.7.4 Delaying infringement proceedings

It may be the case that the same invention is protected by a number of patents throughout the world all of which are under attack by the same party. The expense of multiple world wide actions involving the fate of the same invention cannot easily be avoided, even though the outcome of one case might in effect decide the rest too.

EXAMPLE

Western Electric v. Racal Milgo (1979)

The defendants in an infringements action applied for the action to be stayed, i.e. postponed pending the outcome of the corresponding action in the USA. This stay of proceedings was refused by the court because it was felt it would prejudice the plaintiff unduly. In the USA action the defendant through an associate company was seeking an injunction restraining the plaintiffs from continuing with their UK action on the basis of antitrust law, but the court felt that the UK action should proceed, because although the patented invention was the same, the patents were not and both different issues and different claims might emerge. Practice in the USA seems to be identical to UK practice on this kind of case.

Equally, the defendant might seek the stay of an infringement action on the basis that he or she is to allege an infringement of Articles 85 or 86 of the Treaty of Rome which requires a reference to the European court of Justice at Luxembourg.

The Court of Appeal has said that an action should not be stayed on this ground and that the reference should only be made after the full trial, which seems sensible because of the risks of time-wasting abuse by a defendant seeking to use this ploy. Indeed, the court should consider whether the reference is a bona fide one or merely one designed to obstruct or delay proceedings which will otherwise almost inevitably lead to a finding against the party seeking the reference, so as to deny the probable winner his remedy, although any potentially important and not obviously hopeless argument suffices to require the consideration of giving reference to the court.

Nevertheless on giving of a suitable undertaking by the defendant such as paying money into court to make provision for future costs and damages and to pursue the Treaty of Rome issue before the Commission with all due diligence, the court may grant such a stay of proceedings.

15.7.5 Discovery of documents

The right of discovery, i.e. inspection of one party's documents may be limited to the other party's legal advisers and expert advisers so as to protect the integrity of the trade secrets which might turn up in the process. There is in any case also an implied legal restriction on the uses to which the material revealed may be put, limited to the purposes of the action, and an implied undertaking to that effect.

In trade secrets disputes the court will often exercise their inherent jurisdiction to permit access to information by the adversary or only by an independent expert. Success in restricting access to an independent expert is something of a double-edged sword because failure to indicate all the relevant details (of what the expert inspects) that are alleged to be confidential will prevent the party in default from adducing evidence of those matters in the subsequent trial of the action. Express undertakings to use the materials thereby disclosed only in particular ways and not to disclose them to persons other than those listed in the order or authorized by counsel may and often will be made. Likewise, the inspection of machines is often extremely important. Again certain restrictions to protect the integrity of the other party's trade secrets will usually be made on similar terms.

In very extreme cases plaintiffs have been able to obtain at least interlocutory protection by injunction without disclosing in their particulars of claim all of the relevant secret material to the court, but the majority of decisions adhere firmly to the principle that sufficient details of what the plaintiff seeks to protect must be adduced before the court and generally a failure to particularize the trade secrets complained of with sufficient certainty will lead to the action being lost or at the very least deny the plaintiff any interlocutory relief before trial, when full disclosure will be insisted upon by the court.

15.7.6 Privilege in intellectual property actions

The privilege enjoyed by legal advisers from the other side being able to inspect any documents passing between them and their clients about the litigation is extended by the *Patents Act 1977* to communications in pending or contemplated patent proceedings before the Comptroller under the Act and before Convention courts under the European Patent Convention (EPC) and Patent Co-operation Treaty (PCT) in connection with such patent applications as well as granted patents. The Act makes provision for conferring privilege on patent agents giving opinions about patentability, and on any infringement etc., in connection with contemplated proceedings for patent infringement.

The patent agent is in a completely different position from the solicitor or counsel outside the scope of this, however, as was recently pointed out by the court in the case of Wilden Pump Co. v. Fusfeld (1985), where it was emphasized that the patent agent is not a kind of lawyer and does not enjoy even common law privilege for his or her communications with clients, but only that provided by the Act in relation to patents, and therefore communications relevant to copyright and designs are not protected.

The case law on the range of documents thereby privileged is substantial and this is not the place to go into it in depth but one case which should be mentioned is that of Halcon v. Shell (1979) in which discovery of documents was ordered which extended to those documents in the possession of the defendant concerning the research and development leading up to the invention, which the court felt would assist it in deciding whether the invention was the result of an inventive step or a mere routine development, and experiments which had failed, and also the papers held by the patent agents on the application for the patent. This was a complete break from past practice in which such documents had been considered irrelevant to the trial.

15.7.7 Hearings in camera

The general principle is that all proceedings take place in public. But clearly the blanket application of this rule would prejudice the integrity of the trade secrets or other confidential information which it was sought to protect, especially in light of the fact that the details of this information must generally be pleaded in full, as has just been mentioned.

Accordingly where the court feels it proper to do so it will hear cases involving trade secrets in camera in so far as it is necessary in order to preserve the integrity of those trade secrets, although this will be only in rare cases and will be restricted to those parts of the evidence or proceedings essential to preserve confidentially. Where such a step is taken to protect the secret process, discovery or invention in issue before the court, the *Administration of Justice Act 1960* provides that disclosure of the information so heard shall be of itself a contempt of court. The court will back up such measures by ensuring that on the court record details of the process, etc. are concealed or represented by letters or simply omitted from the judgment, kept separate or sealed or both, and may make similar orders in respect of the documentary evidence furnished to the court by the parties pending any appeal, before ultimately having them destroyed.

15.8 COLLECTING EVIDENCE OF INFRINGEMENTS

15.8.1 Evidence through trap orders

In many cases of passing off or of patent, trade mark, or design infringement, it will be necessary to acquire evidence of a specimen infringement through a sale. This is often achieved through a 'trap order', i.e. by asking for the trade marked product and being served with a counterfeit or a different product passed off as the trade marked product, or requesting an example of the patented product being produced by the infringer.

The courts have shown themselves willing to accept such evidence but have required high standards of execution of the traps on the part of the conditions which would prevail if, instead of being circumstances where the victim of the trap is given a fair opportunity to contest the evidence on an equal footing by providing either written evidence of the trap order or by alerting him or her to an oral trap as soon as practicable after it has been sprung, when the incident is fresh in his or her mind. Specialist advice must always be taken in advance of trying to trap defendants in this way to ensure that the evidence so obtained will be admissible in the court.

15.8.2 Evidence through surveys of market opinion

Market research evidence can be particularly useful in trade mark infringement cases and in passing off claims to establish the presence of a reputation in a product or service associated with a particular trader, or to establish the similarity and confusing nature of two marks. But again the courts have been suspicious of it and have required it to be as free as possible from bias in the design of the market survey questions posed. The courts require that the questions asked in the survey and the answers given be clearly and carefully recorded and that the person conducting the survey be available for questioning in court. The survey must be formulated in such a way as to preclude a weighted or conditioned response and there must be clear proof that the questions were faithfully and accurately recorded and drawn from a true cross section of the public or tradespeople whose impression or opinion is relevant in the matter in issue. A reputable and experienced market research organization familiar with the court's practices, and its requirements, likes and dislikes, should thus always be employed, with carefully constructed questionnaires.

15.8.3 Evidence of expert witnesses at trial

We have already mentioned that the court's own advisers may be appointed to report on certain matters. The parties too may be permitted to call witnesses. An expert witness can give evidence on the significance of the specification to a technician skilled in the art, on the meaning of a technical term and on the state of the art at the relevant time; but not on whether the specification is adequate, whether the defendant has infringed and whether the invention is novel or obvious. A similar approach is taken in respect of registered designs.

Similarly, where trade marks are concerned, although the persons deceived are to be the ultimate purchasers, expert evidence will be admissible as to the trading environment, the features which sell goods of this type, and so on, but beyond that the best that can be done is to bring full market research evidence, the vulnerability of which has been considered already.

15.8.4 Evidence from use of Anton Piller orders

One of the most recent and the most controversial means of acquiring evidence in intellectual property cases is the Anton Piller order, named after the legal dispute in which it was first granted, Anton Piller KG v. Manufacturing Processes Ltd (1976). An Anton Piller order is an order which may in appropriate cases be granted in camera and ex parte – i.e. without the defendant being heard or even notified of the proposed order in advance – to order the defendants to permit the plaintiff's solicitors to inspect the premises in question, to search for any relevant documents or infringing items and to seize them.

It has even been found possible to order that a defendant resident in the jurisdiction should disclose the contents of his flat outside the jurisdiction (in this case in France) and permit it to be searched by a French advocate who was also an English barrister! By contrast English courts will not order a defendant to permit searches in Scotland under an Anton Piller order; you must go to the Scottish courts, and searches are usually for defined areas – because the courts are very unwilling to permit searches for any premises under the control of the defendant, and restrict them to specific addresses. Three conditions have to be satisfied for the Anton Piller order to be made.

(1) There must be extremely strong prima facie evidence against the defendants.
(2) Damage, actual or potential, arising from an infringement must be clearly very serious.
(3) There must be clear evidence that the defendants are in possession of incriminating evidence and there is a real possibility of its destruction if an application is made to inspect of which they have notice.

The court will require proof that the plaintiff seeking the order is good for the damages and costs he or she will have to pay if the cross undertakings which are usually extracted have to be enforced by the defendant. In practice the sort of evidence which will usually satisfy the judge of a prima facie case of infringement sufficient to grant the order is a trap order. In some cases evidence of general dishonesty on the part of the trader concerned has been held sufficient.

Clearly the whole system depends heavily on the integrity of the lawyers acting for the plaintiff and presenting the case and for this reason the evidence that you do have should be reviewed very carefully to make sure that counsel can support his or her claims and that you are not merely launching yourself on a fishing expedition in the hope of finding some incriminating evidence to justify retrospectively your obtaining the order. Plaintiffs must not use the Anton Piller order as a short cut to determine what allegations they might bring. Indeed it has been said that even if the Anton Piller search is successful in that it turns up useful evidence, if it is made upon insufficient inquiries so that the evidence could in any case have been obtained otherwise, the court will disallow the cost of getting the order and executing it.

In seeking an order counsel for the plaintiff must give full disclosure of relevant material. Failure to do so will lead to the immediate discharge of the order even without its merits being considered, though since it is rare to have an inter partes hearing this is unlikely to come to the attention of the judge in time.

The courts have emphasized a need for disclosure not only of those facts which were already known to the plaintiff, but also of those facts which ought to be in the purview of the court and enquiries as to which the plaintiff ought reasonably to make. Once a failure to present full evidence as to the credit rating of the defendant in the possession of the plaintiff was fatal.

When executing the order it is often wise to consider asking the police to attend if you fear any violence. The police cannot execute the order for you although their very presence may of itself induce the defendant to admit you. The solicitors are usually required to explain clearly and accurately what the order is and what it requires of the defendant, and give him or her a reasonable opportunity to take legal advice. Generally the plaintiff is also required to serve the order by a solicitor – who is himself or herself an officer of the court – and to serve copies of the evidence together with copies of the copiable exhibits though a legal executive may sometimes to permitted to serve such an order. The plaintiff himself or herself cannot serve the order.

The effect of the order is such that the order has sometimes been called a civil search warrant but it differs from a search warrant in that in a search the police can ignore the person whose premises are searched whereas this is not so in the case of an Anton Piller order although the defendant who fails to comply with it will be guilty of a contempt of court and in any case a refusal to admit the plaintiff's solicitors does look fairly incriminating.

Although a defendant may delay admitting the plaintiff armed with an order, so as to attempt to have it set aside, the fact that this was the reason for their refusal will not exempt them from punishment for contempt of court if it is found that the order was properly granted. If they use the time thereby won to destroy the evidence the penalties will be particularly severe.

Nevertheless the courts are anxious to separate the two procedures, one civil and the other criminal, and to that end, where in some cases it may be that the police will at the same time be executing a warrant to search pursuant to some criminal matter, the court will take steps to ensure that the solicitors executing the Anton Piller order are not seen to be connected with the police activity.

15.8.5 Discovery against third parties

A further extremely useful order obtainable in intellectual property actions is that of discovery against a third party, not a defendant in the actual infringement action. This will enable the plaintiff to discover the identity or whereabouts of suppliers of infringing items. It may extend beyond documents in the possession of the defendant to names and addresses etc. known to him or her.

EXAMPLE
Norwich Pharmacol v. Commissioners of Customs and Excise (1974)
Goods imported into the country on their face clearly infringed the plaintiff's patent. These passed through the hands of the Commissioners of Customs and Excise to whom the importers paid excise duty but whose officials clearly were not party to any infringement. Nevertheless, an order was obtained to disclose to the plaintiffs the names and addresses of the importers.

A fortiori, where the person who knows of these names and addresses is one who has knowingly assisted in the infringement the courts will oblige him or her to disclose the name of the infringer sought by the plaintiff.

Where the innocent defendant such as that in the Norwich Pharmacol case is sued only for discovery the plaintiff will bear the costs of the action but can recover those costs from infringers thereby detected on winning the infringement action against them.

15.9 REMEDIES FOR INFRINGEMENT OF RIGHTS

Several remedies for infringement of intellectual property rights may be considered by a court: they are damages; account of profits; injunctions and orders to destroy or hand over infringing copies.

15.9.1 Damages for infringement

The object of damages is of course to compensate the plaintiff for the loss he or she has suffered through the infringement of his or her intellectual property right. The mere fact of an infringement will entitle you to some nominal damages but to obtain substantial damages you must establish the amount of loss suffered and you will not be awarded more than those losses, apart from some exceptional cases in which exemplary damages may be given: this is considered below.

The measure of damages in intellectual property actions is based on the damage done to the right itself. But this basic principle does not mean that there is a single, rigid legal rule for assessing these damages; they may be based on one of a number of methods of calculation. The point to stress is that the profit or benefit enjoyed by the defendant is irrelevant to the calculation of damages: if these are substantially more than the damages suffered by the plaintiff then the plaintiff should consider instead of damages seeking the award of an account of profits.

Generally, exemplary damages, that is damages for particularly gross and fraudulent infringements, and parasitic damages, that is for example damages for the loss of non-patented product sales as well as those of the patented product, are not awarded. There are some exceptions to this general principle of the measure of damages to be applied both in favour of the defendant and in certain cases in favour of the plaintiff. These are matters for your advisers and the specialist books.

15.9.2 Account of profits

In practice an account of profits is not always a popular remedy, because of the expense and difficulty in determining them. The measure of compensation is somewhat different from that of damages (in which the profit made by the defendant is wholly irrelevant, the question being one of assessing the loss suffered by the plaintiff) and whereas awards of damages are made by estimation and without a great deal of precision, accounts of profits are just what they are called and require a detailed review of the defendant's accounts to assess his or her gain which must be surrendered to the plaintiff.

The cost of the exercise rarely justifies the advantages it may bring to the plaintiff. In cases of simple appropriation of a plaintiff's design or some other right and sale of entire infringing copies the taking of accounts may be relatively easy but where the right infringed is, for instance, some confidential information relating to one component in a machine which makes that machine cheaper to produce, the courts have to ask a much more difficult question, the extent to which the defendant's profits have been increased by infringement or breach of confidence, and some exercise in apportionment of the profit has to be undertaken.

The account will look to the profits which can be regarded as improperly made by the defendant and the decision as to who is to bear the costs is deferred until the outcome of the inquiry is known. Where the profits are made out of items of which only a component is an infringing item, there is some uncertainty as to whether there should be apportionment of the profits to take account of the non-infringing part or whether the profit should be awarded on the whole of the item sold.

Accounts of profits are available not as of right but only in the discretion of the court. A plaintiff cannot have both damages and account of profits against the same defendant and the court will at most give him or her the option of which he or she wants. To award both account of profits and damages for the wrong to the intellectual property right would be to risk awarding a remedy twice over. It is usual to ask for the two remedies in the alternative in the statement of claim. Accounts against innocent infringers are usually refused because of the equitable nature of the remedy which is to be awarded only against unconscionable acts, as will they when the sums are small, or the cost of the exercise is wholly disproportionate.

15.9.3 Final (perpetual) injunctions

The remedy of damages or an account of profits for past infringements is often insufficient, not least because it provides no guarantee that the defendant will not risk infringing the plaintiff's right again. To avoid persistent actions against the same defendants the courts will in appropriate cases on giving judgment for the plaintiff either in addition to or instead of damages for breaches proven to have occurred in the past and in some exceptional cases where there has been no breach but merely a threatened future breach, grant an injunction restraining the defendant from infringing the plaintiff's rights, the disobedience of which is a contempt of court. This is so for all the intellectual property rights and extends to actions for passing off the trade libel, threatened or actual, and breach of confidence. Wholly innocent defendants may be made the subject of an injunction.

In practice such injunctions will be granted almost as of course where there is some threat or probability that the infringement will be repeated by the defendant, but it is generally not necessary to show repeated infringements in the past before being granted the remedy, one infringement being shown to have occurred is usually enough unless it is an exceptional and accidental case which the court is confident will not be repeated and which the defendant undertakes to the court not to repeat. It is usually quite difficult for the defendant to argue that an undertaking to refrain from doing something will suffice and that no injunction should issue, for if he or she intends to keep to his or her undertaking what objections can he or she have to an injunction being granted ordering him or her to refrain from doing it?

However, although the existence of a proven breach is helpful, even in the absence of any existing infringement it is possible to gain an injunction to restrain a threatened infringement, although it is often said that the burden of proof on the plaintiff of matters necessary to establish his or her case for such an injunction will be higher.

Any breach of an injunction or an undertaking made to the court is a contempt punishable by fine or even imprisonment and costs. Any contempt committed by a company may be punished by sequestration of its assets or by committal of its directors. And a director of a company may be committed for a breach by his or her company even if he or she did not know of the breach although to be committed it must be shown that the person concerned was aware of the injunction or undertaking.

Notice is achieved by serving the order on him or her but where one is committing officers of a company which is in contempt, the court can exercise its discretion to dispense with service of a copy of the order where it thinks it just and it suffices that the director or the other officers concerned were aware of it in some way. Furthermore, in cases of injunctions restraining a party from doing something, telephone or other notice will suffice, and this will of course be essential if the interlocutory injunction is to be of any force against a defendant, where it is granted urgently as in the case of an Anton Piller order.

Generally it is better to make sure that all of the directors themselves are also defendants to the action and not merely in charge of a defendant company, especially where they are more substantial parties and the company is only nominally the defendant because they operated behind the shield of the company.

Because of the seriousness of the consequences of being held in contempt of court there must have been a clear breach of the injunction for this to follow. Accordingly if the injunction is drawn up in broad and insufficiently precise terms it will be almost impossible to find a contempt if any possible doubt arises. Equally if the injunction is drawn up in very specific terms – which the courts prefer because it is then clear to the defendant exactly what he or she must not and may do – then any act which does not infringe the letter of that undertaking or injunction, whatever its spirit, will not lead to committal.

15.9.4 Delivery up/destruction on oath

Another remedy related to injunctions is the order compelling the defendant to deliver up articles which infringe the rights of the plaintiff. The underlying justification for an order for delivery up is that it is the only way in which the plaintiff can be protected from the defendant's infringement.

Such an order will not be made if the articles infringing a patent can easily be rendered unobjectionable simply by alteration or removal of the infringing component, and merely the necessary alteration will alone be ordered.

The actual delivery up of those machines and moulds used to make the infringing items may be ordered in the appropriate cases. So far as goods which infringe a trade mark are concerned it is sometimes sufficient to order the mere erasure of the trade mark from the items, without going any further. The defendant may and usually will then be given the option of delivery or destruction by himself or herself and where this is so the plaintiff cannot demand one or the other, though he or she may demand to be present at any destruction if the order is for destruction. On occasion and if the court simply does not trust the defendant it may order actual delivery rather than destruction on oath of the infringing goods.

15.10 AWARD OF COSTS IN INFRINGEMENT ACTIONS

Costs lie in the discretion of the court, though normally the unsuccessful party pays the successful party his or her costs. In some cases of course the successful party overall will have lost on some of the issues. The practice of the courts varies on such occasions.

Sometimes although costs have been awarded to the overall winner he or she has been denied the costs of an issue which he or she lost and which substantially increased the cost of the trial and those costs have been awarded to the other side or more simply not awarded either way. In some cases there has been apportionment, in others not.

As in normal actions, security for costs can be demanded of a foreign party who chooses to bring an action in the English courts. From the point of view of the UK firm suing a foreign firm with assets in the jurisdiction, the possible question of seeking a Mareva injunction, considered below, should be borne in mind.

It may be feared that the defendant is likely to dissipate his or her assets or remove them abroad or conceal them in the UK, especially where a counterfeiter of little substance and much mobility is involved or where he or she is a foreigner importing into the UK. The *Supreme Court Act 1981* now gives recognition to a practice which the courts developed known as the Mareva injunction, after the case in which it was first ordered, designed to prevent this frustration of the plaintiff's award, and confirmed its extension beyond the early cases which were restricted to foreigners likely to remove assets from the country to all types of defendant.

The Act gives the courts the power to in effect freeze the assets of the defendant if he or she is likely otherwise to remove them from the country. The courts have interpreted this widely so that the order prevents selling, charging or otherwise disposing of assets and not merely removing them, and extends to all assets including jewellery, vehicles, *objets d'art*, etc. Indeed, where appropriate the court will also order the defendant to disclose the nature and location and value of these assets and accounts and even require delivery of them to be kept by the plaintiff's solicitors until the trial.

Again, because of the need for speed and secrecy this order is granted by the court ex parte, without notice to the defendant. Like the Anton Piller order this injunction can do enormous damage to a defendant and so has to be used with prudence; but it can be of great assistance to a plaintiff who might otherwise be deprived of his or her just deserts. Accordingly there are limitations on the exercise of this jurisdiction. The order does not extend to assets already outside the jurisdiction. It is not to be used as a weapon to put pressure on the defendant to settle or to give the plaintiff priority over other creditors: it is purely to prevent the plaintiff from being cheated of his or her award by the defendant. It should also permit the drawing of everyday living expenses including medical and legal fees and proper business expenses in so far as that does not conflict with the underlying purposes of the Mareva injunction. The orders may also make more detailed provision to safeguard legitimate interests of innocent third parties who may be affected by these proceedings.

There must be shown some ground for believing that the defendant has assets here and that there is good reason to believe that he or she will dispose of or conceal them before judgment against him or her can be enforced. The order takes effect the moment it is pronounced by the judge and it covers everyone who has notice of it and those persons will be in contempt of court if they do not try to ensure it is complied with. Accordingly the plaintiff's solicitors will give notice as soon as is possible to banks etc. in charge of the defendant's assets. In practice the order will be granted initially for a short time only to give the plaintiff a chance to give notice of the injunction and the defendant a chance to contest it later.

15.11 USING PUBLIC AUTHORITIES TO ENFORCE RIGHTS

15.11.1 Customs seizure for trade marked goods

Under the *Trade Marks Act 1938*, a trade mark owner may procure the seizure of goods imported bearing a disputed trade mark by giving notice to the Customs and Excise that he or she is the proprietor or registered owner of the trade mark, that goods bearing that trade mark are expected to arrive in the UK at a specified place and time in a specified consignment, that the use of the mark in the UK will infringe his or her rights, and that he or she requests them to treat the goods as prohibited goods. The named goods are then treated as prohibited goods unless they are imported for the private and domestic use of the person importing the goods.

The trade mark owner has to pay a fee of £5 in respect of the notice and give security to the Customs and Excise against any liability or expense incurred in relation to the seizure or detention of any item to such value as the Customs and Excise think fit. Within seven days or such other period as the Customs and Excise allow the person giving the notice must furnish them with a certificate from the Registrar of Trade

Marks as to the ownership and infringement of the trade mark etc., otherwise the goods are released and the notice is of no effect except in so far as it obliges the person giving it to indemnify them.

Very similar provisions exist in relation to printed matter imported in contravention of copyright, but other types of copyright material e.g. manufactured items protected by artistic copyright are not within the scope of those provisions: only literary or dramatic work is covered.

15.11.2 Criminal proceedings in the Copyright Act

There are criminal offences under the *Copyright Act 1956* for manufacture for sale or hire, or the sale or hire, or trade display for sale or hire, or the importation other than for private and domestic use, or for distribution either by way of trade or so as to affect prejudicially the owner of the copyright of articles which one knows to be infringing copies of copyright works, or for the possession infringing plates for the making of infringing copies in the UK.

Apart from the penalties which conviction under this section may impose on the defendant, the Act provides that the court before which a person is charged with an offence under this section may, whether he or she is convicted of the offence or not, order that any article in his or her possession which appears to the court to be an infringing copy, or to be a plate knowing that it is a plate intended to be used for making infringing copies, shall be destroyed or delivered to the owner of the copyright in question or otherwise dealt with as the court may think fit. The mere possession of infringing copies is not an offence. These are summary offences and dealt with therefore in the Magistrates Court, though an appeal against seizure lies to the Crown Court.

The Copyright Act makes no provision for searches and seizure of any infringing copies. This and the difficulty of proving knowledge beyond all reasonable doubt – the standard of proof in criminal proceedings – has meant that in practice this section has been little used by copyright owners. Any more serious dishonest activity by two or more persons may justify prosecution for conspiracy to contravene the *Copyright Act 1956* or for conspiracy to defraud.

15.11.3 Trade Descriptions Act 1968

So far as smaller counterfeiters and traders are concerned, who will rarely be good for large damages, it may be worthwhile cutting down on the expense of pursuing them in large numbers by enlisting the support of the Trading Standards Officers (TSOs) in exercising their powers of search and prosecution under the *Trade Descriptions Act 1968*. False trade descriptions extend to statements as to the persons by whom the goods are manufactured, produced, processed or reconditioned.

The TSOs are given fairly extensive enforcement powers, in addition to their capacity to prosecute detected offences, to find offenders and search their premises and inspect goods. These powers may be backed by a warrant where a magistrate is satisfied that entry is likely to be refused or that an application to enter would defeat the objects of inspection and this may permit entry by force. There is some support for the view that TSOs are better off using the criminal provisions of the *Copyright Act 1956* where possible since these are easier to enforce through the provision for forfeiture of goods.

15.12 INSURING AGAINST LITIGATION

The costs of bringing and defending intellectual property litigation are now very high indeed. In the government Green Paper of 1983 the costs of a simple patent case were estimated at £50 000 to £60 000. This may be potentially destructive for a smaller firm faced with infringements of its rights by a larger business.

At one time insurance against litigation was unlawful but now it is legally possible to take out litigation expenses insurance for just about anything, including intellectual property litigation. Thus, at the time of writing a standard premium for up to £100 000 cover per UK patent or patent application might be just under say £100, with additional premiums of a slightly lesser amount for each further state in which a corresponding patent is taken out. Alternatively blanket protection for intellectual property portfolios as part of a general commercial legal proceedings policy might be arranged. This can extend to the insurance of licensor–licensee disputes.

The advantages of insurance in this field are those of any other field: the benefits of a very strong financial backer which small firms might not otherwise enjoy, namely greater negotiating power in any later settlement discussions. But the advantages are not of course unqualified.

As with any insurance policy the level of profit and the range of exclusions favours the insurance company above everyone. You may well find that the policy on offer pays out 100% when you win – when the costs actually to be paid by you are relatively small, because most will be paid by the opposition – but still requires you to make a substantial contribution when you lose, to deter hopeless claims!

The policy may also carefully exclude insurance for liability on your part for any infringements arising out of your own attempts to design around another's patent or copyright. Thus as in all insurance policy documents you have to read the small print very very carefully indeed.

15.13 PLANNING A LITIGATION CAMPAIGN

How you plan your litigation campaign will depend on the nature of the opponent(s). You may be faced with a situation in which you are chasing a number of small-time plain counterfeiters, or a bona fide manufacturer, or a manufacturer who is deliberately designing around your product but you feel sailing too close to the wind with his design. The source of the infringing goods may be the UK or imports from another country. The rights affected may be being challenged at home and abroad. It requires in each case a different approach from your legal advisers, in consultation with you: again these are business decisions, not, ultimately, legal ones, albeit taken on legal advice. What factors will you be taking into account?

15.13.1 The preliminary decision to take action

In formulating the decision whether or not to take action the following questions should be asked:

(1) Which legal system are you going to be under?
(2) What are the prospects of enforcing a decision?
(3) How strong is the right in question?
(4) What evidence in favour and against is there?
(5) How reliable/how old are the witnesses(!)?
(6) How important is it to your business?
(7) How much money have you got to spend?
(8) Who is the easiest target for the most publicity?
(9) How strong is the evidence of infringement?
(10) Will they settle out of court? For how much?
(11) What alternatives are there to litigation?
(12) How dependent on goodwill of the company are you?
(13) What non-legal strengths have you, for example withheld fees?

15.13.2 The tactics of the case

These depend on the size and finances of your opposition as well as yourself and will determine the approach you take and targets you choose. The actions taken will vary according to whether the opposition is:

(1) Small trader/distributor: will they have money?
(2) Big manufacturer/importer: can they outspend you?
(3) Home or overseas opposition: enforcement problems?
(4) Deliberate counterfeiter/designer around/bona fide
(5) What alternatives there are to litigation

One tactical advantage which it may be possible to employ if you do have a number of patents covering the same basic product is to use them selectively when you are pursuing infringers. The device of such multiple patent protection – in other words a series of patents covering different aspects of the product – may be more expensive than a single patent but it has the advantage of creating an umbrella of protection which enables one or two of the patents in question to be exposed to some challenge at any one time in litigation by suing for an infringement of only those and keeping the others back. This minimizes the risk of losing all protection at one time through an adverse decision.

15.14 ALTERNATIVES TO LITIGATION

One possibility to resolve a licence dispute or an infringement dispute in which both parties are in good faith and there is some genuine disagreement as to the scope of the rights alleged to be infringed is to seek arbitration. The formats taken by arbitration differ widely in different countries: some arbitrators are required to give a decision according to the law, others are permitted to make compromise decisions not strictly in accordance with the legal position.

The alleged advantages of operating under such a private arbitration are that it will be private, i.e. a secret affair; that it will be quicker, cheaper, and more informal than litigation. But arbitration is not necessarily cheaper or quicker in the long run than out-and-out litigation. The procedures of arbitration can be used by one party to delay the inevitable and the enforcement of an arbitration award is not always as easy as it might be. The possibilities of gaining the information necessary to establish your case to the satisfaction of the arbitrator are wholly inadequate, there being no equivalent to the power of the court to order discovery, so the arbitration is only possible where there is no substantial difference between the parties as to the facts (as against the interpretation of those facts or the law applicable).

The tendency is for firms to have resort to arbitration only when the sums involved are so small as not to justify all the expense and risk of a full-blown court action. Where the stakes are high firms tend to prefer to go direct to court instead in most cases. And, furthermore, the inability of the arbitrator to decide on the validity or invalidity of a patent etc. makes the usefulness of the proceedings only very limited for the parties to an infringement dispute for whom there is little point in invoking such a procedure.

One more recent possibility is the so-called mini-trial in which no binding award is made but in which the parties receive a very clear impression of the strengths and weakneses of their case from an experienced arbitrator in proceedings which may be very much cheaper than conventional arbitrations and taking place both quicker and over a shorter period of time. The philosophy underlying the mini-trial is rather different from that of arbitration. It is for the business people to get an idea of their prospects of success at trial, to narrow down the issues through argument and to rehearse the arguments in an informal confidential atmosphere. This may enable some sort of dialogue to begin where a way out of the dispute, often grant of a licence or other compromise, is agreed. Sometimes Chambers of Commerce or their equivalents, and in the USA even some private so-called dispute resolution companies, may operate as go-betweens or intermediaries in a dispute settlement negotiation.

These may lead to the grant of a licence in the rights subject to the dispute as a means of escaping an otherwise unavoidable conflict. But three things should be borne in mind. Firstly, it is often only the actual issue of legal proceedings which induces the grant or acceptance of a licence in the technology or design in dispute. Secondly, the grant of a shotgun licence of this kind is not best calculated to promote co-operation and goodwill between the parties which we have already characterized as essential to a successful licensing operation. Thirdly, the resulting agreement may even be subjected to analysis for possible antitrust implications.

16
Prospects for reform

16.1 INTRODUCTION

In 1983 a Green Paper *Intellectual Property Rights and Innovation* (Cmnd 9117) was published advocating major changes in the British intellectual property system. Many of its proposals were met with hostility and it was thought to be a rather hastily put together paper for discussion. After the typescript for the present book was completed the British Government's long-awaited White Paper on *Intellectual Property and Innovation* (Cmnd 9712) was at last published, responding to the Green Paper on a number of points and certainly not accepting all of its recommendations. This makes a number of interesting new proposals directed at the reform of copyright and its related rights in the field of entertainment and music which will not be looked at here, but some of the most important new proposals for reforms and rejections of calls for reforms in the technical and design field may be outlined.

16.2 TRADE SECRETS (CHAPTER 2)

The Government has rejected calls to make it possible to order compulsory licences of know-how associated with patents under s48 of the *Patents Act 1977*. This is probably inevitable given the practical difficulties which such a power would give rise to. The questions of proof and compliance would be considerable and prejudice the effectiveness of the regime.

16.3 PATENTS (CHAPTER 3)

The Government has rejected calls for a so-called second level or petty patent regime as unsatisfactory in its operation, despite the considerable support it received from the author of the Green Paper to which this White Paper is a response.

The Government is however to press for the rapid implementation of the Community Patent Convention, and if needs be a smaller Convention omitting those

EEC Member States who for constitutional and other reasons cannot at present ratify, which is preventing the new Convention from coming into force. This seems to be a somewhat difficult proposal to follow up politically.

16.4 DESIGN COPYRIGHT (CHAPTER 4)

Facing the dissatisfaction with the present long-term copyright regime, the Government now proposes to introduce a modified short-term copyright for purely functional designs which will last for only ten years and which will be subject to a licence of right after five years. The Secretary of State will be able to order that licences of right should be available at any time if the Monopolies and Mergers Commission concludes that the right is being exercised contrary to the public interest.

The present spare parts defence will not apply to the new unregistered right but a more limited right of repair by the owner or on his or her behalf will be enacted. Licence of rights disputes as to terms will be resolved by the Comptroller General of Patents. The non-expert defence will disappear too. A Crown user right will exist.

The new unregistered and modified copyright will be available not only for functional designs but also for designs which are registrable under the *Registered Designs Act 1949* and the protection of that Act with its monopoly will be extended to a twenty-five year term. The definition of a registered design will be amended to make the exclusion of functional designs more clear.

When a design has been registered the modified copyright will not apply to protect it: there will be no duplicate protection. Full-term copyright will be retained for works which have value as artistic works.

Protection of this kind will be restricted to UK designs (i.e. designs originally marketed in the UK or designed by a British national or by a person domiciled or resident in the UK or in a country which protects such UK designs) and to those designs originating overseas in a country which gives UK designs corresponding protection.

This is arguably the most important proposal of the whole White paper, an attempt to remove the uncertainty surrounding industrial designs which has prevailed for many years.

16.5 TRADE MARKS (CHAPTER 5)

The Government proposes no further changes in the trade mark regime. This means that certain anomalies which have become apparent in the new legislation on service marks will go unchanged for some time.

16.6 SPECIAL PROTECTION PROBLEMS (CHAPTER 6)

The Government proposes to consider whether to support reforms to deal with the changing relationship between patent and plant breeders' rights in the light of recent developments in the science of biotechnology. Likewise a possible variation in patent term in respect of certain types of invention which take a long time to gain official approval such as pharmaceuticals and agrochemicals. Generally it seems that arguments that current law is too restrictive towards biotechnical inventions is meeting with some support in the British Government. The Government is also going to review the position on deposits of samples of micro-organisms.

There does seem to be some prospect too for action to further clarify the position on computer software and databases, to the extent that the new legislation will make it clear that copyright subsists in works fixed in any form in which they can in principle be reproduced and it will amplify the protection of copyright against modern forms of piracy. No further clarification on the ownership of computer-produced texts using compiling programs will be given in the absence of any true consensus on the matter.

16.7 EMPLOYEE RIGHTS (CHAPTER 7)

The Government Paper was uncommitted to the case for enhancing rights for employee inventors to take their inventions if these were not being exploited by their employer and ordered a further enquiry, the results of which are to be studied. The earlier Green Paper had been rather more enthusiastic about this type of approach. Some attempts to tidy up the legislation on ownership of copyright in photographs is to be made to give copyright to the photographer.

16.8 PATENT AGENTS (CHAPTER 8)

The Government is continuing its investigation of the profession of patent agent with a view to seeing whether the monopoly is to be preserved and if so on what terms. The Office of Trading has recommended abolition of the monopoly.

16.9 PATENT LITIGATION ETC. (CHAPTER 15)

The Government proposes to change the pattern of patent litigation quite markedly with the object of making it both cheaper and simpler, by moving to a less adversarial and more inquisitorial regime before the Patent Office rather than before the Patents Court with limited appeal rights only.

Under the proposed regime where the validity of a patent may be put in question and the Comptroller and Patents Court at present share jurisdiction, parties will be obliged to commence proceedings before the Comptroller and only in exceptional cases will the case be transferred to the Patents Court. The Comptroller will have additional powers, such as to award interlocutory injunctions, which previously only the Patents Court enjoyed, and much more written evidence will be used. Appeals will no longer be as easy to obtain as at present.

The Government proposes to abolish the anomalous 'conversion damages' remedy for the breach of design copyright but the power of the courts to award certain additional (i.e. punitive or exemplary) damages will be enhanced to deal with deliberate counterfeiters. The power to order delivery up for destruction of the machinery making the copies will only be available in respect of machinery designed specifically for this purpose and not for machinery which may be used for legitimate purposes too.

No financial assistance is to be given towards the pursuit of counterfeiters, and particularly of foreign counterfeiters, as was suggested in the Green Paper, but the Government intends to assist in the collection of intelligence about counterfeiting.

16.10 INFORMATION SOURCES (CHAPTER 17)

The present position is considered in Chapter 17. The Government proposes to give a higher priority and profile to the publicity of the value of an awareness of intellectual property rights and the Patent Office, which is to be hived off as a non-governmental public body will be making greater efforts to communicate and educate. As part of this it seems that the Patent Office will be able to duplicate some private sector information database services, but without being able to charge at levels which will be unfairly competitive to the public sector. There will also be a review of ways of increasing government-funded aid to assist and advise users of the intellectual property system.

17

Information sources about technology, design and licensing opportunities

17.1 INTRODUCTION

The object of this chapter is to survey many of the various sources of written and on-line information available to help those who are either seeking some new technology or design to license in or a medium through which to advertise technology or design which they wish to license out; or more simply to examine by legitimate means what their competitors are doing in the field of technology or design in which they compete before their new products actually reach the market place. Clearly much information comes to you informally through person-to-person contact, but if this fails to produce the information you need and want, where else can you look?

17.2 PATENTS AS A SOURCE OF TECHNICAL INFORMATION

Patents are extremely useful sources of technical information which can be used to:

(1) Spot licensing opportunities to fill product gaps
(2) Avoid unnecessary duplication in research work
(3) Find solutions to specific practical problems
(4) Provide a state-of-the-art review in a given area
(5) Update the reader on new technical developments
(6) Keep an eye on one's competitor's research effort

A good example of the last activity, which is quite legitimate through using published sources of information, is the speculation over the shape of the keel of the America's Cup yacht *Australia II*. This was in fact available for all to see even before the yacht appeared in public with its keel concealed behind a protective cladding, having been published in a patent specification some weeks earlier!

Less obvious are the use of patents to assist in your own research programme or to spot opportunities to complement your existing product line by licensing in technology as a short cut to creating your own version of the product or process.

Figure 7 Patents information network. Patents are a vital source of detailed information on the latest technology. A complete holding of British and foreign patent publications can be seen at the Science Reference Library in London. For users outside London there are selected holdings of patents and related indexes at libraries in 25 provincial towns – from Aberdeen to Plymouth, Belfast to Norwich – and at the National Library of Wales. The supply of most of these publications is financed by the British Library and, with the Science Reference Library, these libraries from the UK's patents information network. Each library gives public access to its own holdings of patent publications, helps enquirers in using patents information, can provide photocopies of patents, and will quickly locate patent publications held elsewhere in the network. Reproduced with permission of the Science Reference Library.

Patents are a greatly under-used resource. A survey conducted in 1982 by the British Library found that the patent information services available in the UK were virtually unknown by those UK companies and research groups investigated. Yet once this knowledge was made available to potential users in the Newcastle-upon-Tyne area it resulted in a threefold increase in the use of the local patent collection by local industry.

So the information, if properly understood and used, is of some considerable potential benefit to the user. Where, then, can you find such patent information and what can it tell you which is of any use? How do you use it?

17.2.1 Patent Office documentation

Apart from the various private collections of such patent specifications of various patent agents and some companies, there is in the UK a public national patents information network financed by the British Library and available free of charge to any member of the public, as well as photocopy and on-line services, plus a broad range of abstracting services to make the huge tasks of searching through and using these patent specifications that much easier. These services must be examined in a little more detail.

The map (Fig. 7) and the table of library addresses (Table 11) indicate the location of the public patents information network in the UK.

Table 11 *Libraries in the patents information network**

ABERDEEN
Central Library
Rosemount Viaduct
Aberdeen
AB9 1GU

0224–634622

ABERYSTWYTH
National Library of Wales
Aberystwyth
Dyfed
SY23 3BU

0970–3816/7

BELFAST
Belfast Public Libraries
Royal Avenue
Belfast
BT1 1EA

0232–243233

BIRMINGHAM
Patent Department
Paradise Circus
Birmingham
B3 3HQ

021–235 4537/8

BRADFORD
Central Library
Prince's Way
Bradford
West Yorkshire
BD1 1NN

0274–733081

BRISTOL
Library of Commerce and Technology
College Green
Bristol
BS1 5TL

0272–276121

COVENTRY
Central Library
Bayley Lane
Coventry
CV1 5RG

0203–25555

EDINBURGH
Central Library
Reference Department
Edinburgh
EH1 1EG

031–225 5584

GLASGOW
Science & Technology Dept
Mitchell Library
North Street
Glasgow
G3 7DN

041–221 7030

HUDDERSFIELD
Kirklees Public Libraries
Princess Alexandra Walk
Huddersfield
West Yorkshire
HD1 2SU

0484–21356

HULL
Technical Library
Central Library
Albion Street
Hull
HU1 3TF

0482–224040

LEEDS
Patents Informatin Unit
Leeds Public Libraries
32 York Road
Leeds
LS9 8TD

0532–488747

LEICESTER
Central Library
Bishop Street
Leicester
LE1 6AA

0533–556699

LIVERPOOL
Patent Library
William Brown Street
Liverpool
L3 8EW

051–207 2147

LONDON
Science Reference Library
25 Southampton Buildings
Chancery Lane
London
WC2A 1AW

01–405 8721

MANCHESTER
Patents Library
St Peter's Square
Manchester
M2 5PD

061–236 9422

MIDDLESBROUGH
Central Library
Victoria Square
Middlesbrough
Clevelend
TS1 2AY

0642–248155

NEWCASTLE-UPON-TYNE
Central Library
Princess Square
Newcastle-upon-Tyne
NE99 1MC

0632–617339

NORWICH
Central Library
Bethel Street
Norwich
NR2 1NJ

0603–611277

NOTTINGHAM
Country Library
Angel Row
Nottingham
NG1 6HP

0602–412121

PLYMOUTH
Central Library
Reference Department
Drake Circus
Plymouth
PL4 8AL

0752–264675

PONTYPRIDD
The Polytechnic Library
Llantwit Road
Pontypridd
Mid Glamorgan CF37 1DL

0443–405133

PORTSMOUTH
Central Library
Guildhall Square
Portsmouth
PO1 2DX

0705–819311

PRESTON
Reference and Information
Service
Haris Library, Market Square
Preston
PR1 2PP

0772–53191

SHEFFIELD
Commerce and Technology Library
Surrey Street
Sheffield
S1 1XZ

0742–734742

SWINDON
Swindon Divisional Library
Regent Circus
Swindon
SN1 1QG

0793–27211

WOLVERHAMPTON
Central Library
Reference Department
Snow Hill
Wolverhampton
WV1 3AX

0902–773824

* Reproduced with permission of the Science Reference Library.

17.2.2 Types of information available

The first point to be stressed is that information is available not only in respect of the UK but also for virtually all of the world's patent systems, apart from some countries where for political reasons records are not being sent to the UK. The UK's Official Journal of the Patents Office is published weekly but the Patent Office library keeps a vast range of foreign patent and trade journal material too. And the point of practical importance which this leads to is that not only does it provide a vast reservoir of knowledge about what is going on elsewhere in the world outside the UK, but also the patents taken out elsewhere may well not have been taken out in the UK: this is true of many US patents to give but one example. And what has not been patented in the UK is open for use in the UK without royalty, all other things being equal. So even relatively new patents may provide free information to the careful reader.

(a) Technical information

The evidence is that large amounts of commercially very valuable technical information are simply never published anywhere else but in patent specifications, (some estimates put the figure of patent specification material not published elsewhere as high as 90%) or if they are, that these patent specifications appear much earlier. This will be inevitable given the very strict novelty requirements of UK and European patent law. And this makes patent specifications a valuable source of technical information in their own right.

Such patent specifications provide both detailed information about problems to be solved and how to solve them as well as a large amount of background information about previous attempts and failures, all of which can save time and money for the research worker. A quick review of the existing patent specifications may reveal whether a particular line of research will lead to a dead end, whether it will lead to success but has been done before and so cannot be exploited, or has been done before and is subject to an expired patent so that the technology may be exploited without charge and saving on the time it would have taken to reinvent it. Such literature can provide a valuable starting-off point for much further research, and as the patent literature itself shows the great bulk of commercially successful R and D results is simply improving on the earlier known techniques and products.

The three major criticisms of the present patent literature as a source of information – and the three criticisms which lead it to be given less importance by scientists than it deserves – should be mentioned.

The first is that the literature is old. This is not true for most systems. Although in some countries the patent literature does not get published until after grant, some three years or more from the date of application, in an increasing number of cases initial publication arises much earlier than this at around eighteen months at most from the initial application.

The second is that the literature does not tell the whole story. Some people look on them as wholly negative, telling you what you cannot do rather than giving you ideas as to what you could do. It is true that a patent specification will take established knowledge for granted and not repeat it: for instance, in a patent for making a new form of tool, it will not tell you how to make the raw materials from which the new components are made if the raw materials are common knowledge. But every patent specification does have to provide by law to be valid enough information for a person skilled in the art, i.e. a technician working in the field, to be able to repeat it. Clearly an inventor or his company seeking the patent will try to reveal as little as possible but there are certain legal minima. It is also true that there are variations in the amount of information and type of information contained in the world's various patent specifications. As T.A. Blanco White has put it, 'specifications of United States origin tend to be exceptionally full and clear in detail but weak when it comes to wider ideas and implications; those from Germany tend to be vague as to practical detail but tend to contain asides that are often illuminating'.

As for UK patent documents, it is sometimes said that the legal requirements of patent specifications under the 1977 Act which now no longer require the patentee to disclose the best method of performing the invention (unlike the 1949 UK patents legislation and the present US law) mean that UK patent specifications are not as informative as they might be, and in particular not as informative as true scientific papers because inventors or their companies, through their patent agents, sail as close to the wind as they dare in not revealing all they know and still thereby preserving some degree of trade secrecy. Patents still provide most of the information you need and must be able to show you at least one way actually to perform the whole invention. What they will often not reveal is the detailed technical know-how which really makes that invention work profitably.

The third is that the language is incomprehensible legal jargon. The apparent legalese of specifications is sometimes criticized as obfuscating the issues for a scientific reader. There is only a limited degree of truth in this criticism. The patent specification will in general be very clear and straightforward, and the abstract even more so. And one thing you can be sure of, one or two rogue patents apart, is that they will provide a highly practical answer to solving a highly practical problem. The claims i.e. the part of the patent which claims the legal monopoly do tend to be very precise and therefore sometimes wordy as they must be if they are to protect the patentee adequately. But the claims are of much less importance to the scientist than the specification and abstract itself.

(b) Using the patent documentation literature

It is sometimes said that patent documents are now simply too long and too difficult to handle. It is true that it needs some training to conduct searches through the vast amount of patent material which at first sight is unwieldy. In fact the greater problem now is simply the sheer numbers of these patent specifications, about one million patent documents being published each year in the world, which makes undisciplined, indiscriminate searching a painful experience. But the material once found is as easily read as any other scientific papers once you are familiar with its style.

Just as it is common to make use of abstracting journals to find original scientific papers in journals or in surveying them before reading on, so in the case of most patent literature you do not have to plough through the full patent specification of every patent published when making an initial search. It is possible to search for patents by the use of their abstracts and their other searching and classification aids. And you can search for relevant patent information in a number of ways. A search may be by:

(1) Subject matter using the scientific or technical classification of the information with which they deal; by a catchword index to their subject matter; or by a search through related 'families' of patents.
(2) Name, using the name of a specific inventor or a company owning the patented invention.
(3) Number searches to find specific patents cited in the trade or scientific and technical literature, in other patent specifications or on patented products.
(4) Sometimes it is possible to make a patent family search, which will often be undertaken in order to find equivalent patents to an overseas patent in which you are interested which is available in an easier language so as to eliminate expensive translation fees, but because of the different legal requirements of the specifications in different countries or simply the additional research done between the inventor's initial application and its overseas equivalents, these other patents may prove to supply additional information that makes a comparison worthwhile.

A typical patent document in most countries will contain the following sorts of information about the technical content of the invention. There will be information about the inventor and the owner of the patent application or granted patent, which may of course be different, and about the date when the invention had reached a point where it was felt to be patentable and a patent therefore sought. The patent will usually contain a brief discussion of the so-called prior art, that is the technical developments prior to the application and from which the invention advanced, as well as indicating the nature of the technical problems which the patented invention sets out to overcome and possible applications it was sought to meet. Where the invention involves some new process or idea at least one and sometimes more worked examples of it in operation will be given and there will be some diagrams and illustrations to clarify the points made.

Furthermore, when you have identified a specific patent which looks to be interesting you do not have to read through it all to get an idea of what it is all about. This is because every UK specification under the 1977 Act contains an abstract summarizing the specification, and for the pre-1978 UK patents there is an equivalent 'examiner's abridgement' performing the same task. Other patent laws require similar abstracts abroad. A short list to cut the searching down may be arrived at through use of the classification key which classifies all the patent specifications into set technical fields.

In many cases these searches no longer have to be done manually or by post but can be undertaken on line from you own desk or via your patent agent through the Telecom Gold or similar systems keyed into a database to which you have subscribed (see Fig. 8). In certain places you can use other people's equipment to undertake searches in this way, e.g. through the Science Reference Library, and where searches can be undertaken on your behalf by trained searching staff who may be employed at a fee to do such work. The question of the methodology and searching techniques to be employed in searching for information from patents is too large a subject for the purposes of this book. But for those interested there are a number of guides available of varying degrees of usefulness.

Even if you intend to employ a searcher rather than undertaking the search yourself it is a good idea first of all to familiarize yourself with these basic techniques in order to be able to give the searcher adequate instructions, asking for the right kind of search and explaining its precise purpose which will be treated in confidence by the professional searcher. A list of searchers may be obtained from the Patent and Trade Mark Searchers Group, Institute of Information Scientists, Harvest House, 62 London Road, Reading RG 5AS. Alternatively consult your patent agent.

(c) Business information

It is possible to search for all the patents and applications held by a particular individual or firm, to determine what their specialisms or current research trends are, who of their research staff is responsible for their main inventions and which overseas markets they have decided to patent the new technology in, whether they have been diversifying or moving into new market areas, who new leading research staff are, and so on. This simply involves searching the name index which will provide a list of all the patents held under that name, a brief title describing what they are about, and the patent specification numbers so that you can search more deeply.

Alternatively you might want to keep abreast of all recent developments in the technical field in which you work by searching the standardized classification of subject matter which fits your particular interest. There are slight complications in this because each country has its own slightly different classification system and these are revised from time to time, so selecting the classifications in which you search may require a degree of care. Again your patent agents can monitor these for you.

(d) Searching in overseas material

To overcome the language barriers in reading some foreign patents some countries have adopted an agreed international coding system on the front page of their patent documents to indicate at a glance some of the bare essentials about the patent, called the INID code. This makes it a little easier to decide whether or not to seek a

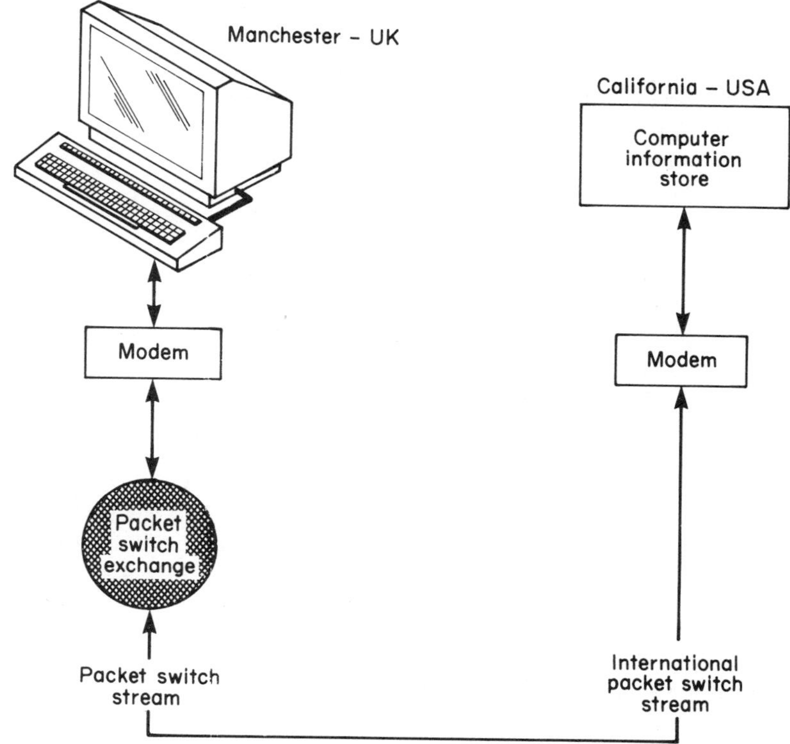

Figure 8 Making the online connection.

translation of the patent. Another short cut is as was stated above sometimes to trace the overseas patent to another one based on it or one from which it takes its priority in another language which you can understand, which will then give you much more starting information on which to decide whether to obtain a full translation.

In addition there is at the Science Reference Library some limited free help in translating the content of such scientific or technical matter, though the staff will not sit down and give a professional written translation. Numerous translation services and abstracting services which will produce respectively full and summarized translations are to be found which perform this service for a fee.

(e) Legal information

Finally you might be interested in a particular patented invention because it relates to something you would like to be able to license in, rather than spend money duplicating or designing around; or perhaps even because you think it infringes your rights or that you might be infringing it. You can find out both the owner and more details about the product or invention by referring to the specification numbers and tracing him or her from those.

Information sources about technology, design and licensing opportunities

A major use of patent documentation is of course for legal purposes, to determine through examination of the claims the precise scope and the validity of the legal monopoly conferred by the patent grant acquired or now sought, whether these be domestic or foreign patents – for instance where a manufacturer wishes to begin exploring into a new territory where there might be a patentee with local protection who did not bother to patent the invention in the UK, he or she may wish to check out his or her legal position there in advance.

Alternatively the patent literature may reveal that an invention it was desired to patent has already been anticipated and so is either not novel or not sufficiently inventive to merit the expense of what may turn out an unsuccessful application. The very high level of patent applications rejected by the Patent Offices of Europe on these grounds indicates that perhaps not enough use of the patent literature for this purpose is made by inventors and their advisers.

An offshoot of this official literature is the now wide range of commercial summaries and patent abstracts available with its own classification and presentation, designed to be more reader friendly. These are dealt with in more detail below but include not only the mere reproductions of patent specifications in abstract but also such selected publications as the list of *Expired British Patents and Licences of Right* which is derived from the Official Journal, published by Scientific and Medical Information Services. This may provide some examples of technology available which are of interest, although in general technology available on a licence of right has been too unattractive for others to take it up, and expired patents are with rare exceptions not of enormous interest to a company seeking innovative technology!

No one is suggesting that the patent literature, useful as it is, is the only answer to a researcher's needs for information. It does of course have certain limitations: it does not generally discuss any basic or fundamental theoretical research. It is not always the most suitable form of literature for research and the other traditional sources of the technical trade and business information can never be ignored. An initial review of the scientific literature may well be better sought in a scientific journal, where the bulk of the most useful literature is already brought together for you, which may need updating but provides a very good jumping-off point for further study, perhaps through a state-of-the-art survey of recent patent literature. It is merely intended to make out the case for including patents more often as one of the important sources of information to which a researcher or business person will have recourse in surveying the technical or business environment in which he or she is operating.

17.3 DESIGNS AS A SOURCE OF INFORMATION

Parallel to (though much less useful than) the patent material is the register of designs held at the Patent Office. This will provide some information about registered designs in much the same way as the patent information discussed above, but because of the much lower usage of design registration and the great legal protection afforded in the UK by design copyright it is not possible to use the register with confidence so as to clear your own design or to seek out licensing opportunities.

17.4 TRADE MARKS AS A SOURCE OF INFORMATION

For similar reasons, in the UK at least, the Trade Marks Journal is not a reliable guide to the freedom to use a trade mark because of the English common law's protection afforded to unregistered trade marks, and for this purpose it is always necessary to use one of the trade journals to identify whether anyone owns and uses a particular trade mark which does not necessarily appear on the register. In exceptional cases it may be a way of tracing the importer or the manufacturer of an interesting product which is not otherwise identified from its pack or accompanying literature.

17.5 OTHER SOURCES OF TECHNOLOGY AND BUSINESS DATA

Outside of the official publications of patents designs and business statistics, there is a wealth of trade and business literature which can be used by the company or by the individual seeking information about opportunities to license in or out, or to monitor the activity of his business competitors or simply to keep abreast of the new technologies and designs appearing in the UK or overseas markets.

17.5.1 Abstracting journals

A huge number of specialist and general abstract services exist which record in summary form the recent patents throughout the world in specific fields, or by country or industry; abstracts which deal both with the patent and with non-patent disclosures such as research papers published in the world's scientific journals or technical reports and trade exhibition publications. In almost any research or technical field one can find a specialist publication, from textiles to tobacco, food to footwear, biotechnology to information technology, and so on. These may be used as a short cut to further investigation of those entries which appear interesting on first reading. Again, searching can be by subject or name etc. There are too many to mention them all in this book but there are a number of directories of such abstracting and similar indexing journals. The UK's own Science Reference Library provides a free list of these journals (ask for the SRL Aids to Readers No. 30). But a note of warning should be sounded. Abstract journals contain the abstracts of patent specifications often presented in a modified form to conform to their house style. They do vary in their quality between each other and internally depending on the field and on individual abstractors or technical writers concerned, etc.

Two of the most useful databases of this type are the Derwent Publications and the INPADOC databases, which provide very good illustrations of the potential uses to which these may be put by the searcher.

Derwent Publications Ltd provide patent documents from most of the important patent offices in the world; and abstracts from a smaller range of countries which appear in hard copy as well as in machine-readable form so that a subscriber may make up his or her own database derived from the Derwent data. The World Patents Index weekly service updates subscribers on new patents in the classifications, General, Electrical and Mechanical along with patent family lists. Their Central Patents Index also provides an abstracting service in chemical patents. These services are available not only in hard copy but on line, though searching on line though very quick when done properly does appear to be a skilled task and beginners' attempts may be costly. There are of course limitations on the scope of older material covered by these databases, which generally go back to the late 1960s only.

INPADOC was set up by the Austrian Government in 1972 under a contract with WIPO (World Intellectual Property Organization). Its data are directly derived from the data provided by the national patent offices of the 45 or so countries which it covers and there is much less direct input by INPADOC itself when compared with Derwent and other commercial databases of this kind. It provides information in some categories only to patent offices and not to public libraries, e.g. such as the patent family services. There is a weekly fiche service but the weekly tape service is now only available to patent offices and similar official state agencies, but it should also be stated that Derwent itself impose certain restrictions on the type of clients which it will permit to become subscribers of certain services. INPADOC is also available on line and its range of coverage is greater than Derwents.

Further services which may be particularly useful include the Chemical Abstracts series which covers not just chemicals patents but also many other abstracts of relevance to chemicals searching in much the same way as in the other services we have already mentioned, a feature of the Chemical Abstracts material being a more sophisticated subject matter description in addition to the basic classification system of identifying patents by subject matter, an English language material only request and a patent only or non-patent or both search request. It covers 26 countries' patent documents. It is of more science interest than scope of legal protection interest to the searcher in practice. It has an update service issued weekly and is available on line. A very similar service is produced by INSPEC for abstracts in physics, electronics, computer and control subjects.

17.5.2 Using databases on line

As we have just mentioned the Derwent database is now available on line as is, for instance, Chemical Abstracts and its offshoot services. Such on-linel data facilities are an increasingly comprehensive and quick but also expensive searching method. The many on-line patents directories and abstracting services which are growing up alongside the more traditional hard copy directories and services may be of great use. But the unit costs of accessing this information can be very considerable. Most have print-out facilities, but these of course cost extra. Unless you are intending to use the service regularly it is often cheaper to make use of a patent agent with his or her own subscription or go through the UK Science Reference Library (SRL) or some other public source which has reached agreement with the database's owner over the use of the subscription service for the client in this way. A list of some of the databases on line can be obtained free of charge from the SRL (ask for Aids to Readers No. 20).

17.5.3 Monitoring services

A number of firms exist duplicating the function which many patent agents will undertake of monitoring patent applications throughout the world, not just for patents which you may wish to challenge, but for those which are of specific or more general interest in the technical field in which you operate, either for your scientific research or for the purpose of keeping an eye on your competitors, in which case a company name monitoring service will be provided too. The evidence is that a number of companies through their information officer as a matter of policy pass these on to R and D staff without in fact bothering to use such an outside monitoring agency, though the take-up rate of the staff to whom these are passed on seems to be variable from what little statistical research has been done. There are of course the more general monitoring services provided by newsletters and other updating sources described in the paragraphs below, which do not look specifically to the client's own requests, but serve a general subscription market or can be consulted in libraries. Similar services exist for trade marks.

17.5.4 Company reports

In theory the Companies Registry should contain a good deal of useful information for those examining the viability of a potential licensing partner, though in practice and despite the sanctions for non-compliance with the obligations of the *Companies Act 1985* the Register tends to be rather out of date and incomplete. A company can however hardly object to any requests for accurate and up-to-date information on matters which it is obliged by law to supply to a public register. This register should be able to give you details of company accounts, date of incorporation, changes of names, etc. The *London Gazette* contains information about directors' ownership, reports of non-case assets, and the like.

Some reports to shareholders occasionally contain very useful information about new products and their R and D. But often more useful are the various updating company monitoring services such as the Extel and the McCarthy card series which summarize company activities from press cuttings classified by name and by industry.

There are a number of other well-known sources of information like the *Jordan's* and *Kelly's* directories and on-line services with information about the financial status and similar business information on companies which may be useful in the preliminary evaluation of potential licensing partners.

17.5.5 Scientific and technical journals

These are probably the most commonly read papers by a scientist or technical problem solver when seeking some assistance in his or her work, but just as this type of person should not neglect the much-neglected UK patent literature, which is at least as likely to turn up the help needed, so neither should the business person ignore the purely scientific press any more than the business and product directories, patent literature and the financial journals. These science journals often do contain papers or notes or reviews about new technical developments. They are often either authorized for such publication after the patent priority dates etc. have been established, but before production has been begun, or are even published deliberately to deny anyone at all from ever gaining patent protection and therefore dedicating the new invention or discovery to the public whether or not patentable previously: and at least one journal in the UK exists specifically for this purpose. Many potentially useful commercial ideas can be found, albeit usually in a fairly raw state by mere searching through some of these scientific journals.

One should also consider use of lists of research projects and the resultant published theses in certain of the areas sponsored by the research councils, which are published and available for consideration. Although sometimes too abstract and pure science oriented these occasionally may turn up the kernel of a sound working solution to a real practical technical problem. There are examples of such theses being found to be possible sources of invalidity of UK patents through the prior disclosure of the subject matter.

17.5.6 Conference proceedings

These usually provide very similar information to the science journals just referred to, though there are instances of very useful information being picked up which would not otherwise be published, arising out of the personal contacts between scientists attending such events and from unauthorized asides after papers in the question times which follow, or simply over dinner (and after dinner!). Special colloquia proceedings may be published *ad hoc* or as part of an on-going series in the journal of the society concerned. One disadvantage of these is the time lapse between the conference and publication of its proceedings: it pays to be there in person, rather than reading it long after the event.

17.5.7 News journals, letters, forecasting notes

A number of business news journals exist all of which regularly carry updates on changing commercial and industrial environments in other countries, changes in laws and taxation, political stability or the lack of it, and so on, which can be invaluable in assessing the market prospects for export overseas to particular countries. Some more ambitious forecasting services are also available predicting market trends in the short to mid-term future and analysing other existing data.

Many overseas statistics which are not otherwise easily available and studies of particular industries can be found in these journals. These often give very valuable insights into trading practices, government requirements and even social customs in most of the UK's overseas markets. A selection of some of the many report services dealing with overseas markets can be found in the SRL Aids to Readers No. 32.

A publicly funded service of some use for similar purposes is the Department of Trade and Industry (DTI) Overseas Technical Information Unit. This provides a series of newsletters and reports available to UK companies and R and D organizations on the recent development overseas in certain technical fields for which these organizations have subscribed. The service also provides training seminars and answers enquiries on overseas development or puts the enquirer in touch with the source of the development. It also provides a more general advice service on technological development in specific countries. Similar services exist in the British Overseas Trade Board directed at exports in general rather than technology transfer in particular and much useful information and contacts in overseas markets can be obtained through this route.

17.5.8 In-house trade literature

The search for a product to license in or out may take you to the various specialist trade directories, many of which exist. These will give information as to the products supplied by various companies listed, data on the companies themselves and their activities. There are even directories of trade directories.

Unfortunately some of these directories tend to be indexed by name and not by product which makes searching much more difficult if you are seeking a particular line of new technology. In these cases the patents registers may often be a quicker route to the product you want. These directories may also provide a useful source of information about one's own business competitors.

In-house publicity handouts and staff newspapers reviewing new products are another fruitful source of information about one's rivals. It was information of this sort being published which in part deprived G.D. Searle and Co of any action at law when Celltech poached a number of their employees (see Chapter 2). There are some services enabling you to make use of this in-house material more easily, such as those which are published by Technical Indexes Ltd which microfilm manufacturers' catalogues and similar publications.

17.5.9 Market research reports

Market research can be a very important business information service. But the cost of market research, whether done for one single client or for a number who club together from an industry, is often phenomenal and the quality of these market research reports does vary enormously, depending on whether they are the result of an expensive original survey or of often cheaper secondary market research using published material. One possible option in cutting the cost of your market research is to read about other people's research as a substitute for doing your own. A wide range of publications exist with the final results of such research. These range from specialist reports for their individual clients through market research periodicals, emphasizing new consumer trends, to trade and business journals which as part of their more general treatment of an industry include many relevant results, and even abstracting journals specializing in summaries of research results, which may be used as a starting point for further, and more detailed, in-depth investigation by the reader of the fuller original data sources.

Of course, many such market research organizations do actively prevent the acquisition of their research results by UK libraries where they can be consulted by the public and in any case the mere cost of copies of those which are now available generally make it quite impossible for libraries to purchase more than a very few out of the many prepared. A selection of these market research sources can be found listed in the SRL Aids to Readers No. 29.

Closely related to these sources of information are the official and the unofficial collections of production figures produced by the DTI and by trade associations throughout the world which can indicate what level of supply is maintained in certain products and services. These can all be used in the search for market niches and weaknesses. It is worth consulting publications of the DTI such as *British Business* and the relevant trade associations to determine what information is available to members of this kind. One very useful source is the DTI's Statistics and Market Intelligence Library, at 1 Victoria Street, London SW1H 0ET, which has a wealth of material on overseas markets and development plans, directories, catalogues, etc.

17.5.10 Technology/product licence directories

A number of technology directories exist which are used by inventors or their agents as advertising papers for products seeking a licensee or some venture capital backing. The degree of usefulness of these journals is debatable but they should never be ignored either as a reader or advertiser. They do tend to lead a short and chequered career though there are some notable useful exceptions in both the private sector and more usually the public sector where governments maintain a regular output of licensable technology from their state-run laboratories. Of this latter kind of publicly sponsored directory an example is the Techalert scheme, which has provided a list of about 500 items annually of the most interesting technical output from publicly funded R and D in the UK and overseas. The list is published in ten technical journals in the form of short summaries, and there is a follow-up service to go into greater depth and to provide a source for contact with a potential licensor or other supplier of the technology and also to make available a list of related research activity.

Of the private sector journals they generally will advertise a new product, patented etc. or not, and state that it is available for licensing or that it seeks a distributor and so on. The general standard is not very high. These publications often are classified according to the technological field of the product but appear as a general list, often randomly distributing the entries and usually both sides of the licensing transaction pay for their entry or subscription as the case may be. Apart from the growing number of commercially operated databases, there is a growing tendency for the UK's universities and similar institutions to create their own hard copy and now on-line databases of their technology available for licensing or assignment. One older example is the Manchester University hard copy technology directory *The Manchester Partners*; a new example is the on-line database run by BEST (British Expertise in Science and Technology) universities.

Another possible source of ideas and technical opportunities outside the main-stream of in-house or research institutional inventions may be found in the Institute of Patentees and Inventors which has a list of individual inventors and firms wishing to sell or license products. These too are generally of very low quality.

17.5.11 Technology brokers' search facilities

A number of technology brokerage services possess their own closed lists and databases for use only by their own clients, backed up by personal contacts and often extending beyond mere product lists and lists of those companies willing to license in to market surveys and feasibility studies, marketing packages preparation etc.

These will undertake searches on behalf of buyers and sellers of technology of their own databases and of the other available sources, including the periodicals and other trade literature, taking much of the legwork out of the process and coming up with lists of products matched to the technology requirements of the client or potential licensees for their client's products, often advertising his or her needs in their own technology transfer directories and newsletters, holding stalls at trade fairs and using personal contacts for direct marketing. After such initial stages and of course after further payment more financial reports, technology evaluation and advice on licence fees and other lump-sum payments may be made for the client.

Such services are generally available on a one-off basis or as an on-going annual service which is paid for on retainer. In some cases the broker will charge for preliminary work only by way of a flat royalty on any resulting licence deal struck, i.e. on a contingency basis rather than through a single up-front payment which may be attractive (at least in the short term) to the small company or individual inventor who at first cannot afford substantial expense in seeking the partner with whom to exploit the technology or design.

The service offered by these brokers is extremely variable in quality and cost and it is advisable to seek references from satisfied clients the company is prepared to name and to make your own enquiries about their skills and honesty. Where services are charged of course you should always seek a quote for the fees to be charged in advance of authorizing any further search or other activity especially involving on-line database searching, the cost of which can mount up alarmingly. It should also be borne in mind that the cost of entry in many of the technology directories is high and the results of rather dubious value to the company. These brokers advertise regularly in trade journals and their brochures will give you a good idea of the range of services each will provide, though charges are always to be found on application!

If you still fail to find suitable technology at home you can contact research institutes such as Arthur D. Little, PA Technology and Fulmer Research Institute, to identify overseas firms working in a similar area and which may be interested in joint research. In the USA technology brokers such as the Dworkowitz group are now common, though these are expensive and once again of debatable and variable value in much the same way as the UK groups, some of whom are merely subsidiaries or associates of these US based organizations.

More open register facilities which are of some possible use are those of the Trade Openings Bureau of the Confederation of British Industry which has various information services for trading and commercial matters and covers new and existing products, suppliers of raw materials, journals, directories, information services, exhibitions and trade fair details and so on.

17.5.12 Patent agents, banks, accountants, etc.

One should not forget the potential of one's own patent agent as a source of information about any new products, which he or she may have worked on as a professional adviser for another client drafting the now published patent paper. He or she can direct you to such products and other new devices without any breach of professional ethics or breach of confidence, the information being publicly available, and may have a good local knowledge or knowledge of a specific technical field arising out of patent searches etc.

Patent agents can and do undertake the searches of many of these patent databases for any of the various purposes mentioned above in Section 17.2, though such a task is time consuming and any professional time comes expensive. Their advice should certainly be sought in respect of the various facilities on offer, since they will have experience in their usefulness for the type of client in question and the type of technology sought or to be advertised.

Similarly, some accountants will now have a good idea of sources of money to put into new ventures and will maintain Business Expansion Scheme registers and the like which can be of considerable assistance in finding at least relatively small levels of investment. These too can be a good source of advice as well as helping you to put together the marketing package with which to sell the new product or design.

17.5.13 Culture collections

Increasing numbers of culture collection databases are being set up around the world, both because of the compulsory regimes of the patents legislation and for the more general research and commercial purposes. The current position is in a state of flux. There now exist directories of the world's culture collections and the list of current directories of such culture collections can be obtained from the Science Reference Library.

17.6 INFORMATION ON VENTURE CAPITAL SOURCES

The nature of the various sources of UK venture capital for innovative projects is beyond the scope of this book. The Appendices do no more than list some sources of information about these venture capital opportunities; not sources of venture capital themselves. Apart from those sources of information listed there divided into the private and public sector publications some local authorities also publish advice booklets and pamphlets about which local enquiries may be made.

Further information can again be obtained from banks and accountants. Occasionally even patent agents and solicitors will have encountered some contacts with enough available capital to merit an enquiry.

17.7 SUMMARY OF INFORMATION SOURCES AND ADDRESSES

17.7.1 Finding where to find information: the SRL

There are vast amounts of written material of many varying degrees of usefulness for various needs of the reader. It is beyond the scope of this book to give any comprehensive list of the material, let alone to review its use for different purposes. The best way of getting to know the literature and its uses is perhaps to make use of the UK's Science Reference Library bibliographic guides to it. Throughout the book reference has been made to the many SRL publications of relevance to this field, most of which are freely available without any charge at or from the SRL itself. A list of the most relevant publications is set out in the Appendices in one place for easy reference.

17.7.2 Finding out yourself: courses and training

There are a number of services apart from books or other teaching literature available to the scientist or business people seeking further information about the use of intellectual property. These take the form of seminars, training courses, lectures and video films by both the public and private sectors. In the public sector, the Small Firms Service of the DTI Factsheet 5/85 states:

> 'The Department of Industry's Patent Office can help explain the benefits to individuals and businesses of intellectual property protection (patents, trade marks, designs and copyright). Speakers are available to attend seminars arranged for small business advisers of the Small Firms Services, local enterprise agencies and Chambers of Commerce, or events arranged for representatives of small firms themselves. The Office also has a short explanatory video film (U-matic, VHS and Betamax formats) available for loan or for showing at the above events.'

In addition the universities and polytechnics have run awareness courses in some parts of the country. In both Manchester University and London University for instance there are from time to time basic courses on intellectual property law and practice for business people and scientists as well as regular updating seminars on licensing and recent legal developments, often held in conjunction with the professional bodies. These may be more widespread in the future. The Science Reference Library in London itself has run successful training courses in Patents, Scientific and Business Information usage. The SRL also has a training room for individuals or small groups to learn how to use its British patent publications as a source of information. The room is equipped with a sample set of material, including some British specifications, abstracts and official journals' indexes and classification keys. A training package comprising a guide to the literature and practical work is available to enable participants to work at their own pace. The Newcastle-upon-Tyne Polytechnic has made a video recording for sale which introduces the patents system to the beginner. And an extensive local patents information publicity campaign has also been undertaken in Newcastle-upon-Tyne by the local Patents Library and polytechnic.

In the private sector there are a small number of private run for profit business seminar companies which run introductory and advanced level course both on the law of intellectual property and licensing as part of their programmes. These include companies such as Oyez, European Study Conferences, and others, who also have corresponding publishing activities. These conferences tend to be based in either London or Paris, Brussels or Amsterdam and their organizers charge accordingly. The Institute of Patentees and Inventors also runs courses and seminars for its members as well as providing newsletters and similar services. Some patent and trade mark agents will provide informal seminars and talk-ins.

18

Conclusions about intellectual property

18.1 THE MERIT OF THE INTELLECTUAL PROPERTY SYSTEM

Particularly to the scientist, the patent system and perhaps by implication the rest of the intellectual property system often seems a hindrance to research and the dissemination of knowledge which is not justified by its results and which leads others – especially the dreaded lawyer – to impose restraints on the way in which he or she works. To the accountant it often seems to be a substantial drain on company finances with no evident guarantee of a significant return at the end of the day.

The commercial significance of the intellectual property system to the individual firm has already been discussed. It is clear that, as a matter of business, given that the system exists, it is foolish to ignore it. But is the system itself a good thing? Does the system merely hinder the development of science and technology? Should we abandon it altogether? Several arguments are often advanced in favour of the present intellectual property system. It is argued that the system:

(1) Encourages dissemination of knowledge
(2) Encourages technology transfer
(3) Encourages research and development
(4) Encourages new investment in production
(5) Is morally justifiable in rewarding inventiveness
(6) Restrains abuse of technological dominance

Equally, however, there are those who object to the intellectual property system as a concept. They often argue against the intellectual property system that it:

(1) Is inimical to pure research and science
(2) Costs too much to use
(3) Takes too much time to use
(4) Is a political weapon of the West
(5) Is morally unjust and permits abuse
(6) Is economically inefficient and wasteful

(7) Is anticompetitive in a free market economy

Some of these arguments should be looked at quite briefly to consider their validity. Is the system worth having? Or are the sceptics right?

18.2 ARGUMENTS FOR INTELLECTUAL PROPERTY RIGHTS

18.2.1 The system encourages dissemination of knowledge

It is often said that without the intellectual property system people who wanted to protect their ideas from commercial exploitation by others would be afraid to publish them. Secrecy would be the order of the day and the dissemination of vital information would be seriously hindered.

It does seem to be true that patents and similar rights promote disclosure. In the very short term the demands of the patent system require that the inventor refrains from disclosure of the invention until the patent is applied for. But after that stage there is no hindrance to disclosure.

No patent lasts for ever. After a relatively short time the rest of the world is free to use the invention and in return for a relatively short period of monopoly the inventor has permanently given the world his or her knowledge in a readily accessible form: he or she was not compelled to disclose the invention, merely encouraged to do so. Nor was he or she ever obliged to use the intellectual property system and he or she could have disclosed the invention to the world for free and let everyone use it at once: many academic scientists do. But he or she chose to use the system, or the person who employed him or her and therefore spent money enabling him or her to do research chose to do so.

In practice, of course, the merits of this argument are qualified. The patent offices of the world contain a vast treasure house of technical information. But they are under-used as a means of acquiring knowledge. And the efficiency of the patent system has not gone unchallenged as a conduit for information. Disclosure in the specification of a patent of the working of an invention though it varies in style from country to country is not always – indeed is rarely – as full as it might be. The barest minimum to comply with legal requirements is disclosed by the inventor and his or her legal advisers and many inventions simply cannot be exploited profitably on the basis of what is disclosed in the patent itself. Furthermore, the organization of patent offices and their resources, together with public ignorance of their facilities, results in the under-use of this information.

Nevertheless, the potential exists and substantial efforts are now being made to improve the role of the patent in the dissemination of technical and scientific information. In the UK Government Green Paper of 1983, *Intellectual Property Rights and Innovation* (Cmnd 9117) one of the main recommendations was for an awareness campaign to be conducted to be aimed at increasing the usefulness of the patents information to be found in the national database. Limited attempts have been made by the British Library's Science Reference Library staff to this end but information services of this kind are still relatively under-developed as well as being very under-utilized. With a view to acting on this the Government's White Paper of 1986 *Intellectual Property and Innovation* (Cmnd 9712) proposed hiving off the Patent Office as a non-governmental body whose function as a storehouse of information could be expanded in competition with private sources of information about new technology. The range of information available was considered in an earlier chapter.

18.2.2 Intellectual property helps technology transfer

The commercial aspect of the dissemination of information is the transfer of technology from those who have it to those who do not – at a price. The transfer of technology can be effected in one of a number of ways: by setting up production and training facilities in the destination location; by licensing the technical know-how for others to produce there; and by exporting the goods and other results of the know-how to the destination location. The intellectual property system acts as a lubricant in the technology transfer machine in all of these. It reduces risk for the owner of the idea, and without the protection of the law for their inventions, traders in developed countries would often simply not embark on transactions which would weaken their grasp on their ideas. The intellectual property statutes, in creating property rights, reasonably closely defined, out of inventions and the like, facilitate the dealing in and exploitation of these developments in the commercial world – in short, they enhance the free marketability of the inventions with as few obstacles in the way as possible. To a certain extent this applies not only to those inventions protected by the patent system but also to the non-patentable know-how associated with and often essential for the exploitation of the patent.

In a curious way this was one motivation behind the early monopoly grants made by the medieval monarchs in England which were the forerunners of the modern patent system; a bait with which to induce foreign craftsmen to introduce their skills into fifteenth century England, and to overcome the protectionist guide which had provided a barrier to these entries.

The underlying policy was also reflected in the grant of patent rights to the first importer into the country of an invention, irrespective of whether the importer was the original inventor. It was a means of protecting embryonic industries for a short time in order to encourage in the longer term the development of those industries on a larger scale and to defend them from powerful industrial interest groups who stood to lose by the introduction of competitive new technologies and skills. Later, when the English industrial power had developed to make England one of the industrial giants, at the beginning of the industrial revolution, inventors such as Arkwright found the State equally willing to impose restrictions designed to prevent technology flowing out of the country, however willing it was to see it flowing in.

It is also true to say that much patent licensing takes place on a tit-for-tat basis where barter rather than money payments is agreed and where patents may be cross-licensed with their associated know-how between market competitors. This practice, which can sometimes develop into the establishment of a formal or informal patent pool, can have most serious implications for the competitiveness of those companies outside what may become something of a 'magic circle' of technological development, and its implications have attracted the attentions of both the European Commission and the USA antitrust authorities. Equally it can lead to useful co-operation in research and eliminate duplication of effort.

18.2.3 The system encourages research and development

It is easy to say that people should not make money from inventions when they have not put up the money which led to them or took any financial risks. The intellectual property system provides a means of helping to ensure an opportunity, not a guarantee, of a return on investment adequate to recoup one's on-going research costs. This is the classic argument of the large pharmaceutical drugs companies. The high cost of research in the field demands that the company be able to recoup its expenses. Without the world intellectual property system to help it would not be possible for them to devote such resources to research projects, to employ so many research workers, in a field in which so many projects fail and only a few succeed. There would be less research and far fewer employed researchers!

This argument has taken on an truly international dimension. Nowadays, where export markets are of such size that they affect investment decisions by large multinational corporations, the knowledge that foreign patents are likely to be available may provide that essential degree of protection required to induce these particular types of innovation and investment by the firm involved in a large-scale project. Economies of scale are now developing to the point where genuinely international protection may be necessary to justify initial research and development costs. Furthermore the patent system is said to reflect market needs far more accurately than for instance government subsidy, since what determines the success and therefore the price of patent products is the level of public demand for them. If demand is high enough, others will pay sufficiently to reward the inventor and risk taker, and their desire to get 'a slice of the action' will induce any rivals to seek to make similar technological strides and improvements.

It has been argued that the patent system is no longer the incentive that it was to the large corporate research and development departments when supported by sufficient market power to lead an industry even with free competition. But this argument fails to take full account of the fact that even with the larger research corporations, the attitude to patents may well differ depending on the area of activity in which they are trading and that a company which is confident of its ability to lead the market even without the assistance of patents in a field in which it has long played a leading role may nevertheless feel the need of the assistance of intellectual property rights in later bolstering its diversification into new areas of trade activity. Thus F.M. Scherer argued in *Patents and the Corporation* (1959) (F.M. Scherer *et al.*) Galvin, Boston, USA:

> 'The value of patents as a stimulus to technical investment in large and well-established corporations is similar to their value to the independent inventor or to the small and struggling firm. As long as other factors such as distribution channels, relative costs of production, brand preference, and engineering know-how are well-established, patents are relegated to an unimportant niche in the decision-making process. But when corporations contemplate moving into areas where they have very little experience or market following, where they must in effect begin all over again just as the small company must begin, then patents can become an important factor. The security of good patent protection makes up for the lack of security regarding those other factors upon which the company's day-to-day

business success is based.'

In some ways it might be argued that a far greater undermining of the classical patent system has arisen out of the transfer of so much research and development investment, particularly in the case of fundamental science and in the major capital intensive areas of research, from private industry to government funded institutions – anything from 50–60% of expenditure on R and D being funded by government in Western Europe. Where is the incentive to innovate if all the profits and the expenditure are removed from those who undertake the research, all the risks and all profits passing to the funding provider – the state?

18.2.4 The system encourages investment in production

It is often argued that the intellectual property system encourages investment in new plant and in other production facilities to produce new lines which would not be profitable if other manufacturers could learn from the mistakes of the initial producer to set up more cheaply and produce cheaper rivals to the new product. In other words the market lead justifies otherwise unjustifiable investment in new plant and equipment. There has been much scepticism about this traditional argument in favour of patent and similar protection and no empirical research has really been possible. The theoretical economic analysis has been equally inconclusive.

It may be true that patent protection is far more significant to a smaller and economically weaker firm than to a large firm in this regard. This is because the technical significance of the patented product may overcome the disadvantages of small-scale production and of the high initial costs.

It has been argued that the intellectual property system is thus no longer the incentive that it once was to larger corporate research and development departments supported by sufficient market power and by finance to lead an industry even without patent or some other protection.

But this argument fails to take into account the fact that even with larger companies the attitude will vary according to the technical or the commercial field in which it is operating. And it is true if at all only in respect of patents with wide valid claims to monopoly which place the company well ahead of its competitors in the same field. So similar considerations to those mentioned by F.M. Scherer apply here too.

18.2.5 The system is moral in rewarding inventiveness

It is often argued that the system encourages the orderly and fair competition between companies in the market place, especially in the field of trade marks and designs, and discourages industrial espionage or at least helps to fight it by imposing obstacles in the way of exploiting the results of espionage. It rewards human ingenuity. We reward and applaud talent in many fields and this is merely one of them. It is just to reward the intellectuals for the creative contribution they have made to the human race and to compensate them for their past activity. This has always been one of the weakest arguments in favour of the system. The fact that a patent has been granted does not in itself give the inventor any financial reward, indeed it costs him or her money. As such it does not assist him or her in the task of bringing the patent or other right to any profitable use, and employee compensation schemes apart, the firm employee gets no special bonus for inventing something which by law belongs to the employer: though of course as one employed to invent he or she was in receipt of a salary which may have been paid through years of failed but expensive research. A socialist scheme of inventors' certificates with financial and other state-provided rewards could perform this task of rewarding a man's or woman's contribution to humanity just as well and indeed in the past Parliament has gone so far as to vote a special award to an inventor for the contribution the inventor has made to the industrial development of the kingdom and who is otherwise inadequately rewarded. Nevertheless, the pure emotional appeal of this justification for the patent system is strong.

18.2.6 The system limits abuse of technical dominance

Some people argue that although the intellectual property system does confer considerable rights on the owners of intellectual property these rights are always tempered by restraints on the exercise of the rights which would be lacking if the owner had simply relied on trade secret protection. If a monopoly right is not being adequately exploited or is being deliberately suppressed, compulsory licence provisions exist so as to enable others to exploit the invention or design even without the consent of the owner. Certain obvious pernicious practices such as requiring a licensee to buy non-patented goods at a high price as a condition of gaining a patented product licence are outlawed. The use of the intellectual property system, which is still sufficiently attractive to induce registration, brings the new technology out into the open and subjects it to legal control which would otherwise be lacking.

18.3 ARGUMENTS AGAINST INTELLECTUAL PROPERTY RIGHTS

18.3.1 The system is inimical to 'pure' scientific work

It is often argued, with some justification, that the commercial orientation of the intellectual property system diverts money away from fundamental research with no immediately apparent commercial application into more pragmatic short-term commercial research, leaving government if any one to pick up the bill for research into projects such as CERN. But it is far from clear that without the intellectual property system things would be any better in this regard, indeed it might exacerbate the trend. One might respond by saying that the demands of the intellectual property system do impose a discipline on applied research which directs the mind to the practical applications of what he or she has been working on when contemplating patent or other protection.

It is also interesting to observe that the patent system expressly excludes from its ambit 'pure' science. You cannot monopolize any idea as such. Only the practical manifestation of that idea in the form of an industrial application can be protected from duplication by the use of a patent. The idea itself remains – or rather becomes – free for the world to use.

Furthermore, no patent system prevents the conduct of genuine research into the patented invention. Only the commercial applications of that research can be prevented by the patentee. The same is true of the other intellectual property rights. It is quite open to a research worker to reproduce an invention protected by a patent for the purposes of conducting further pure research into it or even into ways or circumventing it or improving on it so as to make the original patented invention redundant or commercially worthless. Indeed, many valuable ideas come from the necessity of working around patented technology and designs; the patent or other protection is then in itself a stimulus to the further research work.

18.3.2 Patents etc. take up a researcher's time

This is a criticism of the working of the system not of the concept of intellectual property. Scientists sometimes object to the amount of working time that a patent application seems to require of them, not only in drafting it but in reacting to the Patent Office's objections to the application based on obviousness or lack of novelty – an implied affront to their inventive capacity or their cherished idea which a Patent Office examiner thinks is old hat or too imprecise.

This is partly a reaction of vanity and partly a failure to appreciate the commercial significance of what they are seeking. If patents and other rights are to have any credibility they must be thoroughly examined to prove their validity. If the applicant is to be given these extensive legal rights against others it must be shown that he or she has done what Parliament has said must be done to qualify.

The issue of time and complexity of the research is sometimes used as an excuse for not bothering with patents or other intellectual property. Sometimes this is simply indisciplined thought or an unwillingness to apply the mind to the practical application and precise workings of the invention or design involved. And often these difficulties can be overcome simply by the better organization of the research and development procedures within a company and by a closer liaison between the staff and the patent agent or other legal adviser. This was considered in an earlier chapter.

Similar problems arise in the vexed question of in-house security and in the restraints which the intellectual property system – especially in the UK – places on publication of result by fame-hungry research workers, whose own professional standing may be considerably enhanced by what they have discovered but whose firm refuses publication for good commercial reasons. The idea will usually be published after relatively short delay. Fame may be nice but fame and money is nicer.

18.3.3 The system costs too much money

This again is a criticism of the workings of the system rather than of its existence. It is a particular problem for the individual inventor and the small firm, especially if the costs are multiplied by seeking more protection overseas. Most of the costs of acquiring any intellectual property rights relate to the professional charges involved, which can be daunting. Professional fees in many field of activity are high. The average cost of acquiring a low to medium technology patent in the UK alone might well be around £1000 at present rates, though it is impossible to generalize about such matters. The other rights cost much less to acquire but of course the costs of obtaining foreign protection all multiply. The costs of defending, licensing and perhaps exercising these rights in court may also proliferate quite alarmingly if you are unlucky. But five points should be made.

Firstly, a very high proportion of all potential litigation is settled by agreement between the parties outside court, so that these high costs often fail to materialize.

Secondly, the value of patents and other rights in themselves as deterrents to copying and their influence on the decision of an imitator to withdraw his or her products from the market should not be underestimated.

Thirdly, the cost of obtaining and defending your rights should be set against the cost of lost sales and royalties which a failure to take out legal protection may involve you in, not to mention the lost reputation which inferior imitations may expose you to.

Fourthly, so far as such intellectual property litigation is concerned, as with all areas of activity, insurance cover is available for such contingencies, at a price.

Fifthly, it is of course possible to cut down the costs of an intellectual property system in the short term. For instance one can virtually eliminate the costs of maintaining a strict examination system, which is kept to ensure that rights are not granted wrongly; some countries do, either because they are unwilling or unable to bear the cost of the full examination system. But the less pre-grant examination, the more likely expensive litigation is to follow when the rights are challenged, for then the validity of the rights will have to be examined in full and that tends to result in a system which is cheap only for the state and not the individual concerned.

In any case, inventing and exploiting one's new technology and designs is, like it or not, a classically entrepreneurial activity, and the fruits of success in it can be great. Looking at the type of applicant for intellectual property protection, it is evident from the statistics that the individual applicant maintains a high profile seeking the benefit of intellectual property rights, which would indicate that he or she feels that such protection is worth paying for.

18.3.4 The system is a political weapon of the West

The principal critics of the intellectual property system as a whole and the patents system in particular have been the Third World states and some international organizations such as UNCTAD (United Nations Conference on Trade and Development), which view the classical patent system as instrumental in the continued economic and technical subjugation of the developing countries by Europe and the USA, and a legacy of colonial rule. Critics argue that patents are used not as a means of protecting the development of industry and training of personnel in the host nation but simply as a means of protecting the West's exports markets, to the detriment of foreign exchange balances, exploitation of domestic resources and to education and employment in the host nation.

These critics argue that of the patents granted by the host nation 85–90% are granted to foreigners of which only 5–10% are exploited locally other than by imports. In fact, few countries administer systems of patent protection which are used more by their domestic applicants than by foreigners, and even the most highly developed nations have the highest numbers of foreign patents – including the highest developed of the Third World states. Although only 5–10% of the patents may be exploited domestically, the vast bulk of the remainder are not exploited at all, whether by imported goods or otherwise. Many products would not be viable if they had to be produced in small quantities throughout the world, and a central or limited number of centres of production is essential to keep costs down. Trading conditions in the host countries themselves may be the major reason for failure to produce locally and in some cases it may well be cheaper for that country to import than to set up local production. And often the goods imported are themselves essential prerequisites to local production of other goods.

It is probable that the economic dominance of the foreign concerns and their much much greater investment power, production base and know-how combined with their prestige is a much greater factor in the establishment and perpetuation of their dominance in Third World countries than the patent system. Nevertheless, the number of patents held in the Third World by larger European or American companies is pretty small, and the economic arguments advanced by these underdeveloped countries have been questioned even by their own economists.

In fact few Third World countries have felt able to do without at least formal protection and have directed their policies to greater governmental control over-exploitation through the introduction of sterner legal provisions providing for revocation of patent and other rights for failure to work the invention in the country, compulsory licences, most favoured licensee provisions and the like. This is sometimes known as patent erosion, and patent erosion in the Third World (though not exclusively in the Third World) has been particularly marked in the field of pharmaceutical drugs.

However, any attempts to weaken patent protection or to impose working requirements can be dangerous for a smaller host nation, since often the company will simply withdraw from the market, being able to survive without it, leaving the nation wholly without the technology it so badly wanted.

It is a curious fact that the arguments advanced by these Third World countries and rejected by many advocates of the patent system in Europe, and some of the measures that Third World states have taken, bear a striking resemblance in many cases to the arguments and measures which the courtier of Tudor or Elizabethan England would have been familiar with. They reflect the same economic problems – an absence of technical infrastructure in the country concerned. Attempts to achieve some compromise between protagonists and antagonists of the patent system in the form of the UNCTAD draft codes of conduct for the transfer of technology have so far made little headway; as have the wholly uncompromising UNIDO draft guidelines for pharmaceutical licences. The WIPO Model Patent Law proposal was much more acceptable to potential licensors than the proposals of either the UNCTAD or UNIDO (United Nations Industrial Development Organization) but with the Third World countries now in the ascendancy in WIPO too, the tide there is also turning in favour of more extreme technology transfer proposals which seem likely to lead to a slowing down rather than a speeding up of the desired flow of design and technology into the Third World at a price it can afford.

Of course, alternatives to patents and to similar forms of protection do exist. The patent system is a market economy phenomenon, devised and perfected in the West. In the context of a wholly state-planned and 100% controlled industry it has been questioned whether the patent system has any role to play and in response to arguments that it did not a number of socialist states have followed the model of the Soviet Union's law on inventors' certificates. This is a regime in which the incentive to invent was preserved by means of state compensation awards and privileges rather than by means of market monopolies, the right to exploit vesting in the state.

However, in more recent times since the Second World War, even in the USSR a regime of patent protection which more or less closely follows Western market economy systems has again been gaining ground and the patent rights of overseas traders respected in preference to mere compensation for use granted under the inventor's certificate scheme with the result that there are now two forms of protection available in a number of socialist countries, the patent and the old inventor's certificate. Patents too are available to the Soviet inventor: they confer an exclusive right and his or her consent must be obtained to exploit, license and assign them, but the formalities for acquisition are stricter and overseas exploitation needs State consent.

The inventor's certificate confers on its holder no monopoly rights – the right of exploitation goes to the State which is under no obligation to exploit – but earns him or her awards and privileges such as better housing, further education and the like, based on an assessment of the benefits and savings to the national economy, and is available for a number of inventions which do not qualify for patent protection under the Soviet law. Because of the nature of the awards – a patent is no use to one who has no access to means of production or exploitation – domestic inventors effectively always obtain an inventor's certificate, foreigners now seek protection through the Soviet patent system. The USSR has also come round to the view that it needs patents to protect its potential exploitation overseas of its home-grown ideas. Hence the USSR has been party to the Paris Convention of 1883 since 1965. China has recently enacted an intellectual property law too, for the same reasons: the necessity to attract foreign technology.

Other COMECON (Council for Mutual Economic Assistance) countries too have moved away from the Soviet inventor's certificate system and closer to what one might term a classical patent system: some of them have dual systems, whereas others have abandoned such inventors' certificates altogether. Of course, the level of protection in practice rather than theory may be doubtful and patent enforcement practically impossible in many of these countries as – probably – in the USSR itself for a foreign concern. In practice patents to foreigners tend to be taken out for purely strategic reasons as a gesture of co-operation and faith in a joint venture with a state concern.

18.3.5 The system is morally unjust

Much of the Third World hostility to patents is founded on the argument that it is contrary to a basic morality, that the fruits of human ingenuity should not be kept out of reach of those who most need but can least afford to pay for them. This moral view has been accepted to a small degree in many of the modern patent systems and is reflected most clearly in the attitude taken for many years by certain states with regard to pharmaceutical and medical products. This debate has taken on a new lease of life with the emergence of much more advanced techniques of genetic engineering and other aspects of biotechnology. How this feeling has been put into effect has varied between states and products.

In some countries, such as Italy, protection was entirely denied to pharmaceutical products for many years. In the UK under the 1949 Patent Act, there were once provisions for compulsory licences for food and medicines. As a result of much lobbying by the industry such provisions have now disappeared, though today the largest UK consumer, the Department of Health and Social Security can still in effect make use of such pharmaceutical products by means of the wider provisions enacted in the 1977 Act for compulsory licences to be granted to the Crown in certain cases. This makes use of an entirely different power – that of the sovereign to demand the use of a patented invention or process for the good of the State, and fits in more closely with those provisions which enable the Crown to prohibit publication of inventions for which patent applications have been made and which appear to have implications for State security. Immoral and antisocial inventions and designs and trade marks may be denied protection under the UK law and in most other systems.

18.3.6 The system is economically inefficient

Many of the basic premises on which the case for intellectual property rights and in particular patent rights are founded have been challenged by economists in the past and in the present: in the nineteenth century there was a strong anti-patent movement based on arguments of free trade.

Nowadays, even though some arguments about the compatibility of patents with market competition theory remain, attention has perhaps focused more onto the efficiency of the patent system. Some see the system as wasteful of scarce resources – thus it is said that the patent monopoly encourages attempts to work round other patent technology and the duplication of work which involves the wasteful deployment of research.

In the absence of available empirical evidence, these views are as unprovable as those in favour of the patent system's alleged economic benefits. It is undoubtedly true that some patents are an obstruction to the natural development of a rival's product line, which that rival would have reached in time, or to the improvement by others of the original idea. 'Blocking' and 'fencing-in' patents may be used as a means of preserving a market lead by denying any progress to others rather than taking a corresponding step ahead of them, and such patents may give rise to considerable litigation. Some commentators feel that the initial market lead attained by being first in the field is sufficient to induce extensive research and development expenditure by the producer if he or she is really efficient.

Ultimately a nation's patent system is the product of a political judgement. To weight up in the balance the social technical and economic pros and cons of a patent system is impossible since no statistical analysis possible can measure present progress and performance with patents against a hypothetical model of which there is no real modern equivalent, that of a modern highly industrialized capitalist society without some form of patent system.

18.3.7 The system is anti-competitive in free markets

Closely intertwined with the question of economic efficiency is the question of competition. It was for many years argued that patents and other allied rights stimulated competition where it matters, in the rapid development of technically more accomplished products and processes which operate efficiently and cost less to produce, in better, safer, more attractive designs, and in ensuring that better informed consumers can buy the products and exercise a choice between them without deception as to their source.

There has however been a growing body of opinion which takes a diametrically opposed view and which regards the way in which the intellectual property system permits such technological advance to be used as profoundly anticompetitive in effect. Many critics have argued that patents and similar rights, especially trade marks, have at times been used by their owners to inhibit international trade. This is because, by using the intellectual property system, goods which would otherwise be imported into a country can be kept out by the right of a patentee or trade mark owner to permit the manufacturing and sale of his or her patented or trade marked article only by the person he or she chooses to license. In many markets a trade mark is as powerful a right as a patent: take the soft drinks market as an example.

This power may be used just like a tariff barrier to protect otherwise inefficient domestic production. And in an economic climate in which there has been progressive lowering of tariffs by agreement between trading nations, the importance of non-tariff barriers to imports, including those arising out of the national intellectual property systems, has become more and more apparent. Indeed, whereas a tariff barrier may be an economic obstacle to trade which any very efficient foreign producer or importer could still overcome, a patent or similar right may impose a total legal ban on any rival goods being imported, which no degree of efficiency on the past of foreign producers and importers can ever overcome.

In the EEC this view has influenced the provisions of the Treaty of Rome on free movement of goods and the competition laws of the common market. These now impose significant restraints on the way in which intellectual property rights can be used and licensed and may answer many of the criticisms levied against the intellectual property system.

18.4 SUMMARY

As in all matters the reader must make up his or her own mind about these arguments. The stance taken by the author is that intellectual property rights, whatever their merits in the wider economic context, and if used with care, are well worth having and keeping, not only from the point of view of the individual but also in the UK's national interest. Given that they exist and that they are unlikely to disappear, it is incumbent on all those in UK industry, research staff, managers and lawyers alike, to exploit their potential for the nation's benefit to the greatest possible degree.

Appendices

Appendix 1
Useful addresses

(a) SOME PROFESSIONAL ORGANIZATIONS

The Law Society
113 Chancery Lane
London WC2
01 242 1222

The Senate of the Inns of Court and the Bar
11 South Square
Gray's Inn
London WC1R 5EL
01 242 0082

The Chartered Institute of Patent Agents
Staple Inn Buildings
London WC1V 7PZ
01 405 9450

The Institute of Trade Mark Agents
69 Cannon Street
London EC4
01 248 4444

The Licensing Executives' Society (Britain and Ireland)
33–34 Chancery Lane
London WC2A 1EW

Patent and Trade Mark Searchers' Group
Institute of Information Scientists
Harvest House
62 London Road
Reading RG1 5AS
0734 861345

(b) SOME PUBLIC INFORMATION SOURCES

Patent Office and Trade Marks Registry
25 Southampton Buildings
Chancery Lane
London WC2A 1AY
01 405 8721
and State House
High Holborn
London WC1R 4TP
01 831 2525

Patent Office and Trade Marks Registry
Designs Registry Branch
State House
High Holborn
London WC1R 4TP
01 831 2525

Science Reference Library
25 Southampton Buildings
Chancery Lane
London WC2A 1AW
01 405 8721

European Patent Office
Erhardtstrasse 27
D 8000 Munchen 2
Federal Republic of Germany

(Most publications available via UK Patent Office)

UK Regional Patents Information Network: See list of addresses on pp. 345–7

Department of Trade and Industry
Statistics and Market Research Library
1 Victoria Street
London SW1H 0ET
01 215 8444/5

(Material on overseas markets and development plans, directories, catalogues, etc).

Technology Advisory Point
Ebury Bridge House
2–18 Ebury Bridge Road
London SW1 8QD
01 730 9678

(Contact service for technical enquiries, referring the enquirer to the appropriate public sector laboratories)

Small Firms Advisory Service
Ebury Bridge House
2–18 Ebury Bridge Road
London SW1 8QD
01 730 8451

British Overseas Trade Board
1–19 Victoria Street
London SW1
01 215 7877

Money For Business
Industrial Finance Division
Bank of England
Threadneedle Street
London EC2R 8AH
01 601 4444

(Guide to types and sources of finance)

(c) SOME PRIVATE SECTOR INFORMATION SOURCES

(This list does not pretend to be comprehensive and no representations are made as to the quality of service of those included or of those not included in the list)

Centre for Innovation in Industry
London Science Centre
18 Adam Street
London WC2N 6AH
01 930 3258

(Contact service which puts inventors in touch with sources of technical and financial assistance)

Trade Openings Bureau
Confederation of British Industry
Centre Point
103 New Oxford Street
London WC1A 1DU
01 379 7400

(Information service for trading and commercial matters covering new and existing products, supplies of raw materials, existing journals, directories, information services, exhibitions and trade fairs)

Ideas and Resources Exchange Ltd
Snow House
Southwark Street
London SE1 0JF
01 633 0424

(Clearing house for technology licensing opportunities)

Some private sector information sources

British Venture Capital Association
1 Surrey Street
London WC2R 2PS
01 836 5702

(*Publishes directory of venture capital in the UK which lists the members and some detail on investment policy*)

Venture Economics Ltd
37 Thames Road
London W4 3PF
01 995 7619

(*Maintains a confidential database on venture capital from British Venture Capital Association and publishes a journal on venture capital in the UK, called: UK Venture Capital Journal (bimonthly)*)

Investors' Chronicle (weekly)
Investors' Chronicle
Greystone Place
London EC4
01 405 8969

(*General investment magazine which regularly publishes details of new venture capital funding sources in the UK with details of preferences and policy on funding*)

Sources of Venture Capital
Stoy Hayward (Chartered Accountants)
8 Baker Street
London W1
01 486 5888

Venture UK (monthly)
Redwood Publishing Ltd
68 Long Acre
London WC2 9JH
01 836 2441

Venture Capital Report (monthly)
Venture Capital Report
20 Baldwin Street
Bristol BS1 1SE
0272 272250

Outline Guide to Business Expansion Funds (bimonthly)
Investment and Tax Planning Services
7 Regal Lane
London NW1 7TH
01 267 0133

Peat Marwick Mitchell Prestel on-line database
1 Puddle Dock
London EC4V 3PD
01 236 8000

ICC database
ICC House
81 City Road
London EC1Y 1BD
01 250 3922

(*On-line database on companies and financial status and industrial sectors market size trends and prospects*)

Dataline, Finsbury Data Services
68–74 Carter Lane
London EC4V 5EA
01 248 9828

(*On-line financial and company accounts reports*)

Appendix 2
Countries party to international conventions

(a) PARTIES TO THE PATENT CO-OPERATION TREATY

Australia
Austria
Barbados
Belgium
Brazil
Bulgaria
Cameroon
Central African Republic
Chad
Congo
Denmark
Federal Republic of Germany
Finland
France
Gabon
Hungary
Italy
Japan
Democratic People's Republic of Korea
Republic of Korea
Liechtenstein
Luxembourg
Madagascar
Malawi
Mali
Mauritania
Monaco
Netherlands
Norway
Romania
Senegal
Sri Lanka
Sudan
Sweden
Switzerland
Togo
Union of Soviet Socialist Republics
United Kingdom
United States of America

(b) PARTIES TO THE EUROPEAN PATENT CONVENTION

Austria
Belgium
Federal Republic of Germany
France
Italy
Liechtenstein
Luxembourg
Netherlands
Spain
Sweden
Switzerland
United Kingdom

(c) MEMBERS OF THE PARIS CONVENTION

Algeria
Argentina
Australia
Austria
Bahamas
Barbados
Belgium
Benin
Bourkina Fasso
Brazil
Bulgaria
Burundi
Cameroon
Canada
Central African Republic
Chad
China
Congo
Cuba
Cyprus
Czechoslovakia
Denmark and Faroe Islands
Dominican Republic
Egypt
Finland
France including Overseas Depts and Territories
Gabon
German Fed. Rep.
German Dem. Rep.
Ghana
Greece
Guinea
Haiti
Holy See
Hungary
Iceland
Indonesia
Iran
Iraq
Ireland
Israel
Italy
Ivory Coast
Japan
Jordan
Kenya
Korea (North)
Korea (South)
Lebanon
Libya
Liechtenstein
Luxembourg
Madagascar
Malawi
Mali
Malta
Mauritania
Mauritius
Mexico
Monaco
Mongolia
Morocco
Netherlands
Netherlands Antilles
New Zealand
Niger
Nigeria
Norway
Philippines
Poland
Portugal with Azores and Madeira
Romania
Rwanda
San Marino
Senegal
South Africa
Spain
Sri Lanka
Sudan
Surinam
Sweden
Switzerland
Syria
Tanzania
Togo
Trinidad and Tobago
Tunisia
Turkey
Uganda
Union of Soviet Socialist Republics
United Kingdom
Hong Kong
Isle of Man
United States of America including Guam, Puerto Rico, American Samoa, Virgin Islands
Uruguay
Vietnam
Yugoslavia
Zaire
Zambia
Zimbabwe

Note: not all states have adhered to the same text of the Paris Convention. The position for each country must be checked in every case.

Appendix 3
Sample and draft letters, contracts, record forms

(a) DRAFT CONTRACT OF EMPLOYMENT TERM ON INVENTIONS

1. The employee will not without the prior written consent of the company use or disclose except for the benefit of the company during and for a period of one year after termination of employment with the company any information which he or she may acquire during his or her employment which is of a secret or confidential nature relating to the business, products, processes, procedures, financial affairs, future plans, contracts or terms of contracts, technical data, equipment, staff or requirements of the company, whether or not that information was originated by the employee or disclosed to him or her intentionally or unintentionally by the company in the course of his or her employment or otherwise and whether or not he or she was authorized to receive that information.

2.1. Any invention, design or improvement made or discovered by the employee either by himself or herself or jointly with any other person (whether or not also employed by the company) shall as against the employee be the sole property of the company if that invention, design or improvement is connected with or is capable of being used in connection with or in the course of the business of the company, its associated companies or subsidiaries, and was made or discovered in the course of his or her employment or by reason of or with the assistance of facilities enjoyed by virtue of his or her employment, and whether or not on the premises of the company.

2.2. The employee shall promptly notify the company in confidence of every such invention, design or improvement referred to in Clause 2.1. Where a formal employee suggestion or invention record scheme operates at the premises of the company at which the employee is employed, the employee shall fulfil his or her duty of notification in accordance with that procedure.

2.3. The employee will during and after his or her employment with the company at its request and expense but without charge to the company assist the company or its nominees to obtain and vest in it or them full legal title to inventions, designs or improvements belonging to the company in accordance with Clause 2.1 above and in any resulting patents, designs, trade marks, copyrights or applications for the same in all countries, by executing all necessary and proper documents.

3.1. In respect of any invention, design or improvement which belongs to the company by law or under the terms of Clause 2.1 above or which is assigned to the company by the employee and which is patented, the company shall at its discretion reward the employee in accordance with what the company considers to be fair and reasonable having regard to the terms of the *Patents Act 1977* ss39–43. Any reward paid or offered by the company under this contract of employment shall not prejudice the right of the employee under the Patents Act.

3.2. In respect of any invention, design or improvement which is not patentable or which being patentable is not patented and which belongs to the company by law or under the terms of Clause 2.1 above or which is assigned to the company by the employee, the company shall reward the inventor to such extent if any as the company in its sole discretion considers to be fair and reasonable.

(b) SAMPLE COLLECTIVE AGREEMENT: UNIVERSITY OF SUSSEX

The following procedures relating to the proceeds of inventions by teaching faculty are hereby published.

The proceeds of an invention shall be dealt with as follows:

(a)(i) Any imputed and marginal costs identified by the finance officer as having been generated by work leading up to the granting of the patent shall be a first charge on such proceeds.

(ii) The remainder of such proceeds shall be shared between the University and the member in such manner as shall be agreed by the finance officer and the member in each case prior to the granting of a patent, provided that the University will not be entitled to more than 50% of such remainder.

(iii) Any part of such remainder accruing to the University will be placed in a University research fund to be administered in a manner decided by the appropriate committee.

(iv) In the case of an invention arising from work supported by research contract with, and/or a research grant to, the University, a condition of which is that the rights to any such invention shall belong to the awarding body and/or the grantor, these provisions will only apply to such part (if any) of the proceeds as are payable to the University.

(b) Any dispute arising out of the provisions of paragraph (a) shall be referred to an independent expert who shall:

(i) be appointed by the Registrar and Secretary and the member or in default of agreement by the senior pro chancellor of the University and

(ii) be deemed to act as an expert and not as an arbitrator and

(iii) make a determination which shall be final and binding on the University and on the member and which shall include the apportionment of the costs and expenses therefore payable by each of them.

(c) SAMPLE COLLECTIVE AGREEMENT: BRITISH RAILWAYS BOARD

INVENTIONS AND NEW DESIGNS

To accord with the coming into force of the Patents Act 1977 fresh Regulations have been prepared. These revised Regulations, which are effective from 1.6.78, cover the patenting of inventions made by members of the staff of the Board which relate to or are connected with or are capable of being used in the course of the business of the Board and were made or discovered in the course of their employment or by reason of facilities enjoyed by virtue of that employment arises out of or can be related to their employment.

These regulations, which have been agreed with the Trade Union representatives, are applicable to all staff employed by the British Railways Board (which shall include, where the context so requires, subsidiaries of the Board in the case of employees of those subsidiaries).

Inventions submitted will be considered and a decision reached as expeditiously as possible so that any necessary patents protection may be secured quickly.

It is essential that no disclosure of inventions should be made until an application has been filed at the Patent Office, otherwise the application may be invalidated.

It has been agreed with the Trade Union representatives that, if a member of the staff is dissatisfied with the award for an invention, he can ask his Trade Union to take the matter up with the British Railways Board or Subsidiary of the Board.

The arrangements apply with the necessary adaptations to any new or original design of any kind.

PATENTING OF INVENTIONS: REGULATIONS

1. Every officer and member of the staff of the British Railways Board (hereinafter called 'the Board' which shall include where the context so requires subsidiaries of the Board in the case of employees of those subsidiaries) shall at once advise his employer through his immediate superior of every invention or improvement of any kind which he may make or discover either by himself or jointly with some other person if it relates to or is connected with or is capable of being used in the course of the business of the Board or in any subsidiary thereof and was made or discovered in the course of his employment or by reason of facilities enjoyed by virtue of that employment.
2. In respect of any such invention or improvement which is capable of patent protection and which belongs to the Board by operation of the Patents Act 1977 ('the Act') then
   (a) if a patent is granted in respect of such invention or improvement the Board shall at its discretion reward the inventor in accordance with Clause 3(a) below

(b) if a patent is not applied for or granted in respect of such invention or improvement the Board shall at its discretion reward the inventor in accordance with Clause 3(b) below

3. (a) The Board shall reward the inventor to such extent (if any) as the Board in their discretion consider to be fair and reasonable having regard to Sections 39 to 43 of the Act.

   (b) The Board shall reward the inventor to such extent (if any) as the Board in their discretion consider to be fair and reasonable as if the provision of Sections 39 to 43 of the Act applied to such invention or improvement

   (c) Any reward offered or paid by the Board under these Regulations shall not prejudice the right of an employee under the Act

4. In respect of any such invention or improvement as described in Clause 1 above which, whilst not capable of patent protection belongs to the Board either by operation of the Act or in accordance with the Inventor's Contract of Employment with the Board, the Board shall reward the Inventor to such extent (if any) as the Board in their discretion consider to be fair and reasonable.

(d) SAMPLE EMPLOYEE INVENTION RECORD FORM

Reference? .... Bloggs & Co.
Date received?....................... R & D Division
Action taken?.........................

It is important in order for the company to determine the facts relating to an invention, design or other copyrightable material (e.g. computer software) and to protect the rights of all members of staff who may be involved, that a statement of inventorship is lodged with the company as soon as possible. The information provided on this form may be used to assist the company's patent agent and as a basis of a statement of inventorship in respect of a patent application or other measure designed to protect the invention. In some instances incorrect or incomplete information may lead to the patent or other form of protection not being granted or being declared invalid.

When you have completed this form please return it to: ....
If you have any query do not hesitate to telephone him on ....

Inventor(s) name(s)
Nationality
Department
Title/Subject of invention

Other staff involved (as team members or otherwise)
Names
Nationality
Department

FUNDING OF RESEARCH OR DEVELOPMENT WORK.
Please give ALL external funding which has been used in connection with this or related work, including grants, contracts, studentships, etc.

TURN OVER

Brief Description of Invention (circa 150 words)

Any earlier work of the same kind of which you know

Date and place of invention and by whom the invention was:
Conceived:
Reduced to practice:
Please give a full description of the invention. This should be attached using 2–3 A4 pages. The purpose of this is to help in briefing a patent agent to determine whether your invention is patentable.
Other relevant information

This form should be signed and dated by those named as inventors.
Signed

Dated

(e) SAMPLE LETTER TO RESEARCH EMPLOYEE LEAVING THE EMPLOYMENT OF A FIRM

Bloggs & Co.
R & D Department

Dear ....,
I understand that you will soon be leaving the company. I am writing to remind you of your position in relation to the company's trade secrets and any inventions which you may have made or been associated with during your employment by the company.

Should you contemplate an appointment with another company or starting up your own business during the next few years you may not use or disclose to others information relating to the trade secrets of this company. If you should find yourself in a position where a risk of inadvertent disclosure may occur e.g. if you were to take up a position with a competitor of the company, it would be advisable for you to let me know (preferably in writing) so that we can together take such steps to protect all parties concerned from any possibility of future conflict and embarrassment.

Under the contract of employment which you signed on joining the company, you undertook to disclose to the company any invention or discovery arising in any way from or during your employment with us and relating to the duties which you were asked to undertake. I should be glad if you would kindly confirm that there are no ideas or inventions of which you are aware which you should have but have not yet disclosed. Also, to avoid any embarrassment in the future, would you please let me know in confidence of any idea or invention which you conceived or made during your employment with the company which you consider to be outside the areas connected with your employment by us. If there are any doubts we can clear them up quickly.

I should also be glad if you would confirm that you have returned to us all confidential or restricted documents and copies of such documents or parts thereof to which you have access during the period of your employment with us. In this context I should be glad if you would check at home to ensure that you have not accidentally left any such document there and to return all such documents before you leave the company to prevent any inadvertent breach of secrecy procedures.

In order to clarify your past responsibilities within the company while these are still fresh in your mind and ours, I should be glad if you would list below the positions which you have held during your employment with us where you may have had access to any trade secrets or other confidential information of the company. Please return this to me so that a copy can be included in the documents which you will collect on leaving the company.

Finally, I should like to take this opportunity to extend my best wishes for your future success and happiness and to thank you for your co-operation.

Yours sincerely,

Dear ....,

I have noted the contents of the above letter.

I confirm that I have returned to the company all items of property belonging to the company and that I do not have in my possession any documents or parts thereof relating to the trade secrets or other confidential information belonging to the company.

I also confirm that I have passed to the company the details of all ideas and inventions made by me or with which I have been involved while in the employment of the company, namely: (use an attached sheet of paper if necessary)

The following ideas or inventions have been made by me either before or during my employment with the company but are in my opinion not related to my duties with the company (this information will be treated in the strictest confidence):

The positions which I have held while employed by the company and in which I had or possibly had access to the trade secrets or confidential information of the company are:

Signed ....
Date ....
Received ....

(f) DRAFT LETTER OFFERING TO MAKE CONFIDENTIAL SUBMISSION

Dear Sirs,

I have invented a piece of agricultural machinery very suitable for use in tilling land on smallholdings which I believe will be of considerable interest to your company since it fits in with your present product range but represents a considerable advance on existing machines in terms of cost and efficiency. You will understand that this piece of machinery involves technical information of a highly confidential nature. I am willing to submit to you copies of the full technical data, drawings etc. of this machinery for your review and consideration if you will accept the materials submitted in confidence and agree not to copy or use the materials or disclose any information in them without my prior written consent, and will return the materials in full after thirty days. If you agree to this arrangement please sign the original of this letter and return it to me, upon receipt of which I will at once dispatch to you the materials referred to above. I enclose a copy of this letter for your files.

Yours sincerely,

Mr J Bloggs

We accept the terms stated in this letter

For Agrimechanicals Ltd

(g) SAMPLE DRAFT CONFIDENTIAL DISCLOSURE AGREEMENT

This agreement is made the .... day of ...., nineteen hundred and , between:

.... Hereinafter referred to as the Company, and

.... Hereinafter referred to as the Client.

WHEREAS the Company has carried out experimental work into techniques concerning .... (hereinafter referred to as the Techniques) and WHEREAS the Client has expressed interest in examining and evaluating information relating to the Techniques (hereinafter referred to as the Information);

NOW, therefore, the parties hereto agree the following.

1. The Company shall disclose the Information to the Client in writing, orally or by demonstration in its discretion, in sufficient detail to enable the Client to fully evaluate the same.

2. In consideration of the disclosure of the Information by the Company, the Client agrees at all times until such time as the Information becomes part of the public domain through no fault of the Client that it will treat the Information with all reasonable and practicable care to avoid disclosure of the same to any other person or organization and the Client shall be liable for unauthorized disclosure or failure to exercise reasonable and practicable care with respect to the information.
3. The Client shall have no obligation with respect to the Information or any part thereof which
(a) is at the time of disclosure already known to the Client from a source other than the Company under this agreement
(b) becomes known to the Client from third parties having lawful title to the same
(c) is approved for release from the provisions of this agreement by written authorization from the Company.
4. The Client will at the written request of the Company promptly return all papers, documents, photographic materials, specimens, models, computer storage media or other materials supplied on which the information has been recorded together with any copies or extracts thereof which have been made.
5. Without prejudice to its obligations under paragraph two above, the Client shall be entitled to subject the Information to such tests, analyses, experiments, or clinical studies as are warranted in its judgement or of interest in it and to disclose the information only in confidence to its employees and through a signed non-disclosure agreement to any consultant or agent not an employee of the Client for evaluation purposes only.
6. Without prejudice to its obligations under paragraph two above, the Client will use its best endeavours to ensure that the Information is treated as restricted information within the Company and that the Information will not be disclosed to any third party not directly concerned with the evaluation of the Techniques. The Client will take all reasonable steps to ensure that persons to whom the Information is disclosed will at all times keep the same secret and confidential and will not use the same other than for the purpose of evaluating the Techniques.
7. This agreement confers no rights on the Client to use the Information other than for evaluation purposes stated herein set out, such evaluation to be completed within sixty (60) days of the date hereof.

SIGNED
for and on behalf of (the Client)

duly authorized

dated

SIGNED
for and on behalf of (the Company)

duly authorized

dated

(h) SAMPLE DRAFT RESEARCH OR DEVELOPMENT CONSULTANCY AGREEMENT

This agreement is made the .... day of ....,
nineteen hundred and , between:
.... Hereinafter referred to as the Company,
and
.... Hereinafter referred to as the Client.

WHEREAS the Client has requested the Company to provide consultancy services in connection with ....
in such manner and at such times during the currency of this agreement as shall be requested by the Client and which the Company has agreed to provide in accordance with the terms herein set forth;

NOW it is hereby agreed as follows

1. Definitions..

1.1. The Technology ....

1.2. The Patent Applications..

1.3. The Know-How ....

1.4. The Technical Field ....

2. The Consultancy Services

2.1. The Company shall provide Consultancy Services to the Client during the term of this agreement in connection with the Technology which shall include but not be limited to

(a) research and development activity on the part of the Company directed towards the extension and development and application of the Technology; and

(b) the effective transfer of the Technology from the Company to the Client together with the results of any activity relating to the particular requirements of the application of the Technology.

2.2 Such services will be performed by such suitable employee or employees designated by the Company as shall from time to time be approved by the Client, such approval not to be unreasonably withheld.

2.3 Such services will be provided by the Company at such time and place as shall be mutually agreed provided that the Client shall in no event be entitled to more than thirty (30) person days per annum consultancy services from the Company under the terms of this agreement. Nothing in this agreement shall prevent the parties from entering into further agreements for the Company to provide the Client with Consultancy Services in excess of thirty person days per annum.

3. Payment

3.1. The Client shall pay to the Company

(a) The sum of .... for all the Consultancy Services provided under Clause 2.1(a) of this agreement.

(b) The sum of .... per person day of consultancy services provided under Clause 2.1(b) at the Company's premises; save that in the first twelve months of this agreement the Company shall provide to the Client up to a maximum of five person days of consultancy services under Clause 2.1(b) of this agreement at the Company's premises and in the second twelve months of this agreement the Company shall provide up to a maximum of three person days of consultancy services under Clause 2.1(b) of this agreement at the Company's premises in each case without any charge under this clause. All consultancy services to be performed by the Company under Clause 2.1(b) other than at the Company's premises shall be chargeable at the rate of .... per person day whenever performed.

(c) All the reasonable and properly incurred travelling accommodation, and out of pocket expenses incurred by the Company's employees in performance of the agreed Consultancy Services, upon confirmation by the Company that these expenses have been properly incurred.

3.2. The Company shall as soon as possible after the last day of March, June, September and December furnish the Client with a statement of account in respect of consultancy services supplied under this agreement and at the same time supply an invoice for the same. The Client shall pay all such sums due to the Company within thirty (30) days of the date of the invoice.

3.3. The sum of .... payable under Clause 3.1(b) shall be increased on each anniversary of the date hereof in accordance with any increase in the United Kingdom Retail Price Index from the figure at which the Index stands as at the date hereof provided that in the event of the Index ceasing to exist the parties agree to use the nearest equivalent index available for this purpose. The Company shall notify the Client in writing of any such increase together with the first statement of account and invoice to be supplied under Clause 3.2 after each anniversary of the date hereof.

4. Intellectual Property Rights

All intellectual property such as patentable inventions, non-patentable processes, know-how, registrable designs, copyright and the like arising from or created by the Company in the course or in connection with the provision of consultancy services hereunder shall belong to and be the absolute property of the Company but shall be made available to the Client in accordance with the terms and conditions set out in the Option Agreement dated the .... day of ....,
nineteen hundred and eighty six, made between the Company and the Client. (For sample see Appendix 3(i).)

5. Confidentiality

5.1

(a) Each party shall keep confidential all technical, commercial or other information obtained from the other party under this agreement or any other confidential disclosure agreement signed between the parties.

(b) Each party agrees and undertakes that it will only divulge such information obtained from the other as aforesaid to those of its employees agents and subcontractors who are directly involved or engaged in the use of the technology pursuant to this agreement or the Option and who need to know the same in order to carry out their duties.

(c) Each party agrees and undertakes that it will ensure that all employees agent and subcontractors to whom such information is divulged are made aware of and undertake to comply with and do comply with the obligations as to confidentiality herein contained.

5.2. The provisions of Clause 5.1 shall survive the expiration or earlier termination of this Agreement by a period of five years.

5.3. The provisions of Clause 5.1 shall not apply to:

(a) any information other than any source code which shall come into the public domain otherwise than by breach of this agreement, or

(b) any information other than any source code obtained free from restriction on disclosure or use from a third party or reasonably believed to be so, or

(c) any information other than any source code which it is necessary to disclose for the purpose of utilization and exploitation of the Technology under the terms of this Agreement or the Option.

6. Term of the Agreement

6.1. Subject to Clause 7 below, this Agreement shall continue in force for a period of five years from the date hereof and thereafter until terminated by one party serving upon the other party three (3) months' prior written notice.

7. Termination

7.1. The Company shall have the right to terminate this agreement and all rights hereby granted upon the happening of one or more of the following events:

7.1.1. If the Client shall make default in the payment of any of the payments payable hereunder as and when they shall become due.

7.1.2. If the Client shall have a receiver of the whole or any part of its assets or if an order is made or a resolution passed for the winding up of the Client except if such winding up is for the purpose of amalgamation or reconstruction and in such manner that any limited company or other business resulting from such amalgamation or reconstruction shall (if it is a different legal entity) effectively agree to be bound by and assume the terms hereof or such other terms as may be acceptable to the Company.

7.2. Either party shall have the right to terminate this agreement in the event of the other committing a breach of this agreement or any undertakings on its part continued herein and failing to remedy such breach within thirty (30) days after written notice thereof to the defaulting party specifying the nature of the breach.

7.3. Should suitable employees acceptable to the Client cease to hold appointments with the Company and the Company is unable to supply the Consultancy Services provided for in this agreement, the Company shall not be in breach of this agreement but either party may terminate this agreement by one month's prior written notice.

7.4. Should the Client not exercise the option to enter a suitable licence agreement in accordance with the provisions of the Option Agreement referred to in Clause 4, the Client may terminate this agreement upon giving six months' prior written notice to the Company.

7.5. Any termination of this Agreement under Clauses 7.1, 7.2, 7.3 or 7.4 shall not prejudice the rights of the Company to recover payments due under this Agreement up to the termination date.

8. Interpretation of this Agreement

8.1. The Terms and provisions contained in this agreement and in the Option Agreement referred to in Clause 4 constitute the entire agreement between the parties and shall supersede all previous communications representations agreements or understandings either oral or written between the parties hereto with respect to the subject matter hereof. No agreement or understanding varying or amending or extending this agreement will be binding upon either party hereto unless set forth in a written instrument specifically referring to this agreement and signed by duly authorized officers of the respective parties. There are no warranties, covenants, terms or conditions with respect to the subject matter of this Agreement other than those which are expressly set forth herein.

8.2. The headings to the clauses of this agreement are for convenience only and shall be of no force or effect whatsoever in construing this Agreement.

8.3. No waiver of a breach by either party of any covenant or condition of this agreement shall be deemed to constitute a waiver or any other breach of the same or any covenant or condition.

9. Assignment

9.1. This agreement shall not be assignable by either party without the prior written consent of the other.

10. Notices

All notices required or permitted to be given under the terms hereof must be in writing. They shall be given by posting the same postage prepaid by recorded delivery or registered post.

Those to be sent to the Company shall be addressed to:

....

....

or such other addresses as the Company may designate in writing from time to time

Those to be sent to the Client shall be addressed to:

....

....

or such other addresses as the Client may designate in writing from time to time

11. Law and Date

11.1. This agreement shall be governed and interpreted by the law of England and English courts shall have jurisdiction over it regardless of the place of execution or performance.

11.2. This agreement shall come into full force and effect as and from the .... day of ....,

nineteen hundred and

AS WITNESS this Agreement has been signed on behalf of the parties hereto the day and year first written above

SIGNED
for and on behalf of
Duly authorized

SIGNED
for and on behalf of
Duly authorized

(i) DRAFT OPTION TO TAKE OUT A LICENCE AGREEMENT

This agreement is made the .... day of..........,
nineteen hundred and , between:
.... Hereinafter referred to as the Company,
and
.... Hereinafter referred to as the Client.

WHEREAS the Company has carried out research and development work in relation to .... and;

WHEREAS the Client intends to carry out in conjunction with the Company further development work on the said .... to establish the financial viability of the manufacture and marketing of the Technology and various discussions between the Company and the Client resulted in the signing of a non-binding letter of intent dated .... day of ....,
nineteen hundred and and;

WHEREAS the Company has now agreed to grant an option to take a licence for the exploitation of the Company's technology and to enter into a consultancy agreement on the terms and conditions hereinafter appearing;

NOW it is hereby agreed as follows.

1. Definitions

1.1. In this Agreement, the following expressions shall (unless the context otherwise admits) have the meanings set out against them.

Research Project	....
The Technology	....
Know-How	....
Field of Application	....
Territory	. . (e.g.) The Member States of the European Patent Convention 1973, United States of America and any other countries as may from time to time be mutually agreed.
Licence Agreement	The agreement set out in the Second Schedule hereto
Consultancy Agreement	The Agreement set out in the Third Schedule hereto.
Option Period	The period of one calendar year commencing on the date hereof

2. In consideration of the payment by the Client to the Company of a sum of .... receipt of which the Company hereby acknowledges, the Company hereby grants to the Client the option ('The Option') to require the company to enter into the Licence Agreement and the Consultancy Agreement with the Client at any time during the Option Period save that although the Client may exercise its option in respect of the Consultancy Agreement without exercise of the option in respect of the Licence Agreement the exercise of the option in respect of the Licence Agreement cannot take place prior to the exercise of the option in respect of the Consultancy Agreement.

3. In consideration of the Company making applications for patents in the Territory, the Client agrees to pay to the company a sum representing one half of the costs incurred by the Company of filing and prosecuting the said patent applica-

tions and of maintaining any patent application subsequently granted.

4. During the Option Period the Company agrees that it will not enter into any negotiations relating to the Technology in the Field of Application in the Territory with any other party.

5. The Client will seek the approval from the Company in respect of any arrangements with third parties for the development of the Technology. The Company will not unreasonably withhold consent in respect of such arrangements.

6. During the Option Period the parties hereto agree to keep each other fully informed as to developments and improvements to the Technology and such developments and improvements shall become embodied within the Technology.

7. Subject to Clause Two above the option shall be exercisable by serving written notice on the Company following which the Licence Agreement and/or Consultancy Agreement shall be signed by each of the parties within twenty one days of such notice being served by the Client.

8. The notice required to be served pursuant to Clause Seven of this Agreement shall be sent by recorded delivery post and shall be deemed to have served on the date on which delivery of it is recorded. The notice shall be addressed to the Company at:

....

....

....

or such other addresses as the Company may designate in writing from time to time

9. Law and Date

9.1. This agreement shall be governed and interpreted by the law of England and English courts shall have jurisdiction over it regardless of the place of execution or performance.

9.2. This agreement shall come into full force and effect as and from the .... day of .... nineteen hundred and

AS WITNESS this Agreement has been signed on behalf of the parties hereto the day and year first written above

SIGNED
for and on behalf of

Duly authorized

SIGNED
for and on behalf of

Duly authorized

(j) EMPLOYEE'S ACKNOWLEDGEMENT OF EMPLOYER'S TITLE

I........................., of ....

hereby declare that I am the sole / joint inventor of an invention, particulars of which are set out in Schedule I to this declaration, and that the said invention was made either:

(a) In the course of my normal duties or in the course of duties falling outside my normal duties but specifically assigned to me in circumstances in which an invention might reasonably have been expected to result from the carrying out of those duties; or

(b) In the course of my duties as an employee at a time when, because of the nature of those duties and / or because of the particular responsibilities arising from the nature of those duties, I had a special obligation to further the interests of my employer.

I acknowledge that, because of the circumstances set out above, title in the invention vests in my employer.

I hereby declare that the invention was not made while using any resources or equipment or facilities of any third party.

I hereby acknowledge that I will at no cost to my employer, save for the reimbursement of out of pocket expenses reasonably and properly incurred, promptly and at the request of the employer assist the employer in obtaining patents and any other registered rights in respect of the invention in any country in the world and will to that end sign or otherwise perfect any documents properly required to be signed or effected by me, provided that the costs, fees and other expenses of registration of such rights shall be borne by the employer.

I hereby acknowledge that all drawings or other materials relating to the said invention were drawn or otherwise created by myself and were drawn or created in the course of my employment and copyright in all such material vests in my employer and that I will at no cost to the company save for the reimbursement of out of pocket expenses reasonably and properly incurred promptly and at the request of the employer deliver to the employee any drawings, plans, phototypes or other materials relating to the invention in my possession, and ownership of which materials I acknowledge to vest in the employer.

I hereby agree to keep all information relating to the invention confidential unless and until the same is published by the employer, or appears on a patent specification or otherwise comes into public domain without my fault.

SCHEDULE I (brief description of invention)

SIGNED BY

(The inventor)

IN THE PRESENCE OF

(Witness)

Dated

(k) DRAFT ASSIGNMENT OF PATENT AND INVENTION RIGHTS

(It is necessary to effect an assignment of a patent or patent application to obtain an entry on the register of patents of the title of the assignee by supplying the Comptroller with Patents Form 21/77 and supplying him or her with a certified copy of a document transferring title in the patent drafted along the following lines)
THIS ASSIGNMENT is made the .... day of ...., 19 ....,
BETWEEN Joseph Bloggs, inventor, of 1 Acacia Avenue (hereinafter called 'the Assignor') of the first part AND Agrimechanicals Ltd of 13 Luckless Way Spoke on Trent (hereinafter called 'the Assignee') of the second part.

WHEREAS
(1) The Assignor is the registered proprietor of a patent [beneficial owner of a patent application] particulars of which are set out in the schedule to this assignment and which is hereinafter referred to as 'the Patent' ['the patent application'].
(2) The parties have by an agreement made between them the .... day of ...., 19 ...., agreed that the patent [patent application] should be transferred by the Assignor to the Assignee for the consideration hereinafter set out.

NOW THIS DEED WITNESSETH
1. In consideration of the sum of .... paid by the Assignee to the Assignor, receipt of which the Assignor hereby acknowledges, the Assignor as the sole beneficial owner of the patent [patent application] hereby assigns to the Assignee the patent [patent application] and all rights, powers, privileges and immunities arising or conferred thereby free from any incumbrance and including the right to sue for damages and other remedies in respect of infringement of the patent [patent application] which may have occurred prior to the date of this assignment [to the intent that the grant of a patent pursuant to the application shall vest in the Assignee].
[2. The Assignor shall promptly at the request and at the sole expense of the Assignee assist in the further prosecution of the patent application and shall execute all documents and do all acts which may be necessary or proper to obtain the grant of a patent in respect of the patent application.]
[3. In the event of any objection or opposition to the grant of a patent in respect of the patent application the Assignor will at the request of the Assignee and promptly furnish all such information and assistance to the Assignee as may lie within his power to satisfy the Comptroller that a grant of a patent in respect of the patent application should be made with as close a resemblance to the form of the patent application at the date of this assignment as possible.]
4. In the event of any objection or challenge to the validity of the patent the Assignor will promptly and at the request of the Assignee furnish all such advice information and assistance to the Assignee as may lie within his power to defend the validity of the patent as granted at no cost to the Assignee save for the full reimbursement of any out of pocket expenses reasonably and properly incurred in furnishing the said advice information and assistance.

5. The Assignor hereby warrants that he has neither done nor omitted to do any thing whereby the validity of the patent [patent application] may be impugned by any person or adversely affected.
6. (stamp duty clause)

IN WITNESS WHEREOF etc. etc. ....

SCHEDULE (listing the patent/patent application number)

(Signatures, Seals of corporate body etc. etc.) ....

[There are many forms and variations depending on the stage to which the application has been taken, overseas applications, warranties as to title, searches for any relevant prior art not disclosed, further assurances in respect of granted patents and defence of infringement claims etc. A similar form of assignment might be used in respect of a registered design or application with the appropriate modifications.]

(1) DRAFT ASSIGNMENT OF RIGHTS IN INVENTION NOT YET MADE THE SUBJECT OF A PATENT APPLICATION BY AN EMPLOYEE TO HIS OR HER EMPLOYER

WHEREAS the Employee has made an invention (hereinafter described as the Invention and more particularly set out in the Schedule to this agreement) and WHEREAS the Employee and the Employer have agreed that the Employee shall for a consideration of .... pounds, the receipt of which the Employee hereby acknowledges, assign all rights in the Invention and further assist the employer in developing and protecting the Invention throughout the world,

It is hereby agreed as follows:
1. The Employee hereby declares that the said Invention was devised by him or herself and that (s)he is the sole and beneficial owner of the said Invention and that the said Invention was not made while (s)he was employed by the Employer (or by any other person) in the course of his or her normal duties under his or her contract of employment or in the course of duties specifically assigned to him or her and in either event in circumstances where an invention might reasonably be expected to result from his or her performance of those duties.
2. The Employee hereby declares that the invention was not made while using any resources or equipment or facilities of the Employer or any other employer of the Employee, save those listed in the Schedule to this agreement.
3. The Employee hereby assigns all rights in the said Invention to the Employer and will at no cost to the Employer save for the reimbursement of out of pocket expenses reasonably and properly incurred promptly and at the request of the Employer assist the Employer in obtaining patents and any other registered rights in respect of the Invention in any country in the world and will to that end sign or

otherwise perfect any documents properly required to be signed or perfected by him or her, provided that the costs, fees and other expenses of registration of such rights shall be borne by the Employer.

4. The Employee agrees to keep all information relating to the invention confidential unless and until the same is published by the employer, or appears in a patent specification or otherwise comes into the public domain without the fault of the Employee.

5. [Consultancy and paid consultancy fees particulars for development of the invention may be included here]

(m) DRAFT ASSIGNMENT OF COPYRIGHT IN AN EXISTING WORK

In consideration of the sum of ...., receipt of which I hereby acknowledge, I Joseph Bloggs of 1 Acacia Avenue Muncaster hereby as beneficial owner assign to Agrimechanicals Ltd of 13 Luckless Way Spoke on Trent, the copyright in all drawings relating to a mechanical tilling implement devised by myself, complete copies whereof and numbered 1–6 have been signed by myself and delivered to the said Agrimechanicals Ltd.

Signed.

(n) DRAFT ASSIGNMENT OF COPYRIGHT IN A FUTURE WORK

In consideration of the sum of ...., receipt of which I hereby acknowledge, I Joseph Bloggs of 1 Acacia Avenue Muncaster hereby grant to Agrimechanicals Ltd of 13 Luckless Way, Spoke on Trent, the sole and exclusive copyright in all the drawings, relating to a mechanical tilling implement devised by myself, about to be drawn by myself, to the intent that the said copyright shall forthwith upon the completion of the said drawings vest in the said Agrimechanicals Ltd.

Signed.

[Note: a s36 assignment of an existing work, i.e. of a drawing, plan etc. requires writing but need not be under seal. It must be signed by or on behalf of the assignor. It will be effective in the United Kingdom and elsewhere. It must be stamped. A s37 assignment of copyright in material not yet produced will also be effective to vest title in the copyright without any further document being executed when that copyright material is completed provided the s37 assignment is signed by or on behalf of the assignor. There should also be consideration for it is probable that a full assignment of a future copyright even if under seal is ineffective without consideration. This consideration may be nominal i.e. one pound. Although a s37 assignment will be effective in the United Kingdom it will not be necessarily effective in other countries to which the *Copyright Act 1956* extends, and a further execution may be necessary in those other countries].

(o) DRAFT 'PACKAGE' INTELLECTUAL PROPERTY LICENCE

THIS AGREEMENT is made the day of

BETWEEN:

THE FOLLOWING PARTIES:

(1) *THE LICENSOR* full details of which are set out in Part I of Schedule I hereto

(2) *THE LICENSEE* full details of which are set out in Part II of Schedule I hereto

RECITALS:

(A) The Licensor is in possession of certain know-how relating to the manufacture of products full details of which are set out in Schedule II hereto

(B) The Licensor is the owner of the patents and is entitled to the benefit of the applications for patents full details of which are set out in Schedule III hereto

(C) The Licensor is the owner of the trade marks and is entitled to use the trade marks applications for the registration of which have been and full details of which are set out in Schedule IV hereto

(D) The Licensor is the owner of the registered design and is entitled to the benefit of applications for registered designs full details of which are set out in Schedule V hereto

(E) The Licensor is the owner of and is entitled to the benefit of the copyright in all the drawings and other written information full details of which are set out in Schedule VI hereto

(F) The Licensor has agreed with the Licensee to grant the licences hereinafter contained upon and subject to the terms hereinafter set out

NOW THIS AGREEMENT WITNESSETH AND IT IS HEREBY CONVENANTED AND *AGREED BEWEEN THE PARTIES HERETO* as follows:-

Article 1

(1) *Definitions*:

In this Agreement the following terms shall have the following meanings:-

Terms	*Meanings*
(a) the Know-how	The various techniques skills data and all technical information and other knowledge of a secret and confidential nature now in the possession of the Licensor and relating to the manufacture of the products as defined in Schedule II
(b) the Patents	The Patents listed in Schedule I and all or any patents that may be granted upon the said applications for patents
(c) the Inventions	All such Inventions as may from time to time be claimed by the Patents or by any one or more of them
(d) Improvements	All or any improvement to the Licensed Products which shall include any improvement upon or further invention relating to or an invention that may be expediently applied to or used in connection with any of the Inventions

(e) the Trade Marks	The Trade Marks listed in Schedule I and all other (if any) trade names or trade marks which are used by the Licensor in relation to the Licensed Products
(f) the Registered Designs	The Registered Designs listed in Schedule I and all or any registered designs that may be granted upon the said applications for registered design
(g) the Copyright Matter	All the drawings, designs, plans and other written information listed in Schedule VI
(h) the Licensed Products	The products specified in Schedule II hereto comprising any and all products made in accordance with any of the Inventions
(i) the Licensed Territory	

(2) *Headings and Commas*

Headings and Commas used in this Agreement are for the purpose of ease of reference or reading only and shall not affect its interpretation.

Article 2

Duration of Agreement

(1) This Agreement shall come into force on the day of 19 ('the Commencement date') and subject as herein provided it shall remain in force for a period of years from the Commencement date ('the first period') each such year being herein referred to as 'a year of this Agreement'

(2) This Agreement shall subject otherwise as herein provided continue in force for a further period of years from the expiry of the first period ('the second period') unless either party hereto shall give notice in writing to the other before the end of the year of this Agreement that this Agreement is not to continue for the second period

Article 3

Licence to Manufacture Assemble and Sell

(1) The Licensor hereby grants to the Licensee in the Licensed Territory the several exclusive/non-exclusive licences set out below

(2) The licences hereby granted shall comprise the following separate and independent licences:

(a) A licence under the Know-how to manufacture the Licensed Products

(b) A licence under the Patents to make use exercise and vend the Inventions

(c) A licence to use the Trade Marks in connection with the Licensed Products

(d) A licence under the Registered Designs to manufacture assemble and sell the Licensed Products

(e) A licence to use the Copyright Matter in connection with the Licensed Products

(3) The Licensee will if requested execute formal licences or agreements for registration purposes on being provided by the Licensor with an appropriate

document in standard form as required for such formal registration in the Licensed Territory in respect of the licences granted under sub-paragraph (2) above

Article 4

Communication of the Know-how and the Technical Assistance

(1) Immediately this Agreement shall come into force the Licensor shall communicate to the Licensee for its use within the Licensed Territory during the continuance in force of this Agreement the Know-how to enable the Licensee to manufacture the Licensed Products

(2) The Licensor shall if requested provide the Licensee from time to time with a person able to supply such technical advice and help as the Licensor considers necessary to assist with the manufacture of the Licensed Products

Article 5

Fees and Royalties

(1) On the coming into force of this Agreement the Licensee will pay to the Licensor a Design and Information fee of £

(2) During the currency of this Agreement the Licensee will pay to the Licensor royalties calculated at the rate of % of the selling price of each item of the Licensed Product

(3) The Licensee shall within 30 days of the end of each calendar quarter commencing with the calendar quarter ending on during the currency of this Agreement deliver to the Licensor in such form as shall be required by the Licensor a statement setting out full details of the quantity and nature of the Licensed Products manufactured and sold during such calendar quarter and showing the amount due in respect of royalties in respect thereof and the Licensee shall at the same time pay to the Licensor the amount shown by the statement to be due

(4) The Licensee shall keep at its usual place of business full and proper accounts containing such true entries (complete in every material particular) as may be necessary or appropriate for enabling the amount of the fees due hereunder to be conveniently ascertained and shall at all reasonable times produce the said accounts and all other relevant accounts vouchers documents and information to the Licensor or to a qualified accountant or other person duly authorized by the Licensor and shall permit him, her or them to inspect the same and take copies or extracts therefrom and shall give all such other information as may be necessary or appropriate to enable the amount of the royalties payable hereunder to be ascertained

Article 6

The Patents and the Registered Designs

(1) (a) In the event of any infringement or suspected infringement of the Patents or the Registered Designs coming to the notice of the Licensee the Licensee shall at once inform the Licensor and shall at the request of the Licensor or on the Licensee's own initiative commence all necessary legal

or other proceedings in respect of such infringement or suspected infringement of the Patents or any of them for effectually protecting and defending the same and shall take all such steps and do all such things as may be necessary or expedient for prosecuting any such proceedings to a successful conclusion and keep the Licensor informed of such steps and actions

(b) In the event of the Licensee so commencing or prosecuting any such proceedings then the Licensor shall execute and do all such documents and things and render all such assistance to the Licensee as may be necessary or expedient to enable the Licensee so to do

(c) The Licensee shall indemnify and keep indemnified the Licensor against all costs charges or expenses of whatsoever kind or description arising to the Licensor as a consequence of any such proceedings

(d) In the event of the Licensee being awarded any sum by way of costs and or damages as a result of commencing or prosecuting any such proceedings the Licensee shall first be entitled to recover all costs charges or expenses of whatsoever kind reasonably incurred in commencing or prosecuting the same and shall thereafter pay to the Licensor out of any sums remaining such sums as would have been payable to the Licensor hereunder had the infringing sales been sales made by the Licensee hereunder any balance remaining thereafter shall belong to the Licensee

(2) The Licensee shall mark the Licensed Products with a suitable indication that they are the subject matter of an application for Letters Patent and Registered Design in the Licensed Territory or are protected by Patents and Registered Designs as the case may be and shall set forth the numbers of the Patents and the Registered Designs as and when granted

(3) Nothing herein contained shall be construed as being or implying any representation or warranty of the validity of the Patents or the Registered Designs or any of them

Article 7

Improvements (if any)

(1) In the event that the Licensor shall make discover design or develop any Improvements in the manufacture of the Licensed Products it shall communicate the same to the Licensee

(2) In the event that the Licensee shall make discover design or develop any Improvements it shall communicate the same to the Licensor

(3) The Licensor shall alone be entitled to make application for and be granted Patents both in the Licensed Territory and elsewhere on any Improvements which are or may be of a patentable nature and the Licensee will give to the Licensor such assistance as the Licensor may reasonably require for this purpose but the Licensee shall be entitled to a royalty-free licence in respect of such Improvements

Article 8

Use of the Trade Marks

(1) The trade marks and no other marks shall be used by the Licensee in

connection with the sale in the Licensed Territory of the Licensed Products manufactured by the Licensee under the Provisions hereof

(2) The Licensee will act upon such instructions as the Licensor may from time to time give as to the manner in which the trade marks are used in connection with the sale of the Licensed Products

(3) The Licensee shall immediately it becomes aware thereof inform the Licensor of any improper or wrongful use in the Licensed Territory of the Know-how the Trade Marks or the Copyright Matter and shall render to the Licensor all necessary assistance in protecting its property in the Know-How the Trade Marks and the Copyright Matter in the Licensed Territory

Article 9

Secrecy of information

(1) The Licensee undertakes to treat the Know-how the Copyright Matter and all other information which it receives under the provisions hereof as strictly confidential and not to disclose the same to any other person at any time during the continuance in force of this Agreement or at any time after the expiry or termination of this Agreement and the Licensee further undertakes to use the Know-how and the Copyright Matter and such information only for the purpose of manufacturing and selling the Licensed Products in accordance with this Agreement and during the continuance in force thereof and the Trade Marks and all documents containing or disclosing the Know-how or the Copyright Matter shall be and at all times shall remain the property of the Licensor and shall at the expiry or termination of this Agreement be returned to the Licensor

(2) The Licensee shall ensure that its employees shall undertake in writing upon terms to be approved by the Licensor not to disclose to any person firm or company any of the information which such employees may receive in accordance with the provisions of this Agreement or relating to the Know-how

Article 10

Licensee's Covenants

(1) The Licensee hereby agrees with the Licensor that it shall be responsible for:

(a) establishing the maximum demand for the Licensed Products in the Licensed Territory and ensuring that such demand is met to the fullest possible extent

(b) identifying customers and initiating appropriate contacts and selling the Licensed Products

(c) manufacturing the Licensed Products in a proper and workmanlike manner in accordance with the methods and to the established standards of the Licensor

(2) Any claims brought by customers of the Licensed Products manufactured by the Licensee hereunder shall be the sole responsibility of the Licensee and the Licensor shall not be liable in respect of any loss or damage whatsoever suffered or incurred by any person in any circumstances and howsoever caused and whether as a consequence or arising out of or caused directly or indirectly by the Licensed Products or any use thereof and the Licensee convenants with the Licensor that it will indemnify and keep indemnified the Licensor against any such claim as aforesaid or any costs or expenses incurred as a result thereof

Article 11

Termination

(1) The Licensor shall on the happening of any of the following events be entitled by notice in writing to the Licensee to determine this Agreement with immediate effect namely

(a) If any payments hereunder shall be in arrears and shall remain unpaid for a period of 30 days after the same shall have been due; or

(b) If the Licensee shall have committed or knowingly permitted a breach of any of the other terms herein contained and on its part to be performed and observed; or

(c) If the Licensee shall become insolvent

(2) Any determination of this Agreement shall be without prejudice to any right of action vested in either party hereto or to any provisions herein relating to accounting or payment of fees or confidentiality or secrecy

Article 12

Amendments and language

(1) This Agreement has been drawn up on the understanding reached by both parties hereto and it overrides and supersedes any prior promises representations understandings or implications

(2) Any amendment to any of the provisions of this Agreement shall be in writing and shall be signed by duly authorized officials of each of the parties

Article 13

Notices

(1) Any notice document or communication given by either party to the other in relation to this Agreement or in any proceedings connected with it may (without prejudice to the use of any other method) be given by post and shall be sufficiently served if posted in a prepaid registered airmail cover addressed to the last known address of the party to be served and such service shall be deemed to have been effected 10 days after posting

(2) All documents given or served under or in connection with this Agreement shall be given in the English language

Article 14

Interpretation

(1) This Agreement shall be personal to the Licensee which shall not have the right to assign the benefit of this Agreement or grant any sublicence of the rights granted by this Agreement

(2) This Agreement shall be governed by the Law of

(3) In the event of any dispute or difference between the parties hereto arising out of this Agreement or upon or after its discharge or termination the parties will endeavour to reach an amicable settlement failing which such dispute or difference shall be submitted to arbitration in Paris to be finally settled in accordance with the Rules of Conciliation and Arbitration laid down by the International Chamber of Commerce by one or more arbitrators appointed in accordance with the Rules

(4) Judgement upon an arbitration award under the provisions of paragraph (3) may be entered in any court having jurisdiction or application may be made to such court for judicial acceptance of such arbitration award and an order for enforcement thereof as the case may be and the parties hereto hereby consent to such entry of award or judicial acceptance and enforcement thereof as the case may be

(5) In the event that any provision of this Agreement is declared or found to be prohibited by law then that provision shall be deemed to be a separate and distinct provision which shall not effect the validity or enforceability of the remaining provisions hereof and such provisions shall be deemed to have been inapplicable to this Agreement from the date hereof and the parties hereto shall negotiate an acceptable alternative to such provisions failing which this Agreement shall be deemed to have been terminated by the parties hereto in accordance with the provisions hereof but such termination shall be without prejudice to the provisions herein contained relating to secrecy or confidentiality

Article 15

Waiver of Defaults

In the event that the Licensee fails to perform any of its obligations under this Agreement no waiver of any breach of those obligations shall constitute a waiver of any further or continuing breach of the same or of a different kind nor shall any delay or omission of the Licensor to exercise any right arising from any default affect or prejudice the Licensor's rights as to the same or any future default

Article 16

Force Majeure

Both parties shall be released from their respective obligations in the event of a national emergency or war or prohibitive governmental regulations or if any other cause beyond the control of the parties shall render performance of the Agreement impossible whereupon

(a) All sums due to the Licensor shall be paid immediately and

(b) All rights conferred upon the Licensee under this Agreement shall terminate immediately

Provided that this article shall have effect only in the discretion of the Licensor except where the events above mentioned render performance of the Agreement impossible for a continuous period of twelve calendar months

Article 17

Approval of Authorities

(1) The Licensee shall at its own expense procure all necessary approvals and consents of those public authorities full details of which are set out in Schedule VII hereto applicable to the Agreement and its operation

(2) The rights of the Licensee conferred under this Agreement are conditional on the prior receipt in writing by the Licensor of all necesssary consents from those public authorities full details of which are set out in Schedule VII hereto either unconditionally or subject only to conditions which are in the opinion of the

Licensor acceptable

As witness the hands of the duly authorized representatives of the parties hereto on the dates and at the places written below

SCHEDULE I

PART I

The Licensor:

a company incorporated in accordance with the laws of

with registered number

and having its registered office at

and its corporate address at

PART II

The Licensee:

a company incorporated in accordance with the laws of

with registered number

and having its registered office at

and its corporate address at

SCHEDULE II

Details of products:

SCHEDULE III

The patents are:

The applications for patents are:

SCHEDULE IV

The trade marks are:

The applications for trade marks are:

SCHEDULE V

The registered designs are:

The applications for registered designs are:

SCHEDULE VI

The copyright matter is:

SCHEDULE VII

The public authorities are:

To be annexed to the Agreement:

(1) formal patent licence

(2) formal trade mark registered user agreement
(3) formal registered design licence

SIGNED by
duly authorized on behalf of

LIMITED

on the day of 19
at
in the presence of:-
Witness:
Address:

Occupation

(p) DRAFT LETTER DRAWING INFRINGEMENT TO ATTENTION OF INFRINGER OF PATENT RIGHTS

Recorded Delivery

Dear Sirs,
It has come to our attention that certain products advertised by yourselves would appear to bear a substantial resemblance to designs in respect of which we are registered proprietors. We enclose for your consideration copies of the designs as registered in the United Kingdom of which we are the registered proprietors, being Registered Designs Nos 123456 and 123457. We would invite you to compare these with your own advertised designs and to make any observations you may feel appropriate in writing to us within seven days of receiving this letter.
Thanking you for your attention.

Yours faithfully,

For J. Bloggs and Co. Ltd

Enc.

(q) DRAFT LETTER REQUESTING CORRECT USAGE OF TRADE MARK BY A TRADE JOURNAL

Bloggs and Co. Ltd

The Editor
HARDWARES Magazine

Dear Sirs,
An article which appeared in the November issue of your magazine 'HARDWARES' entitled 'Tool for Summer' used the trademark 'Stromper' of which we are the

registered proprietors as if it were the common name for all the brands of tool in question. In fact 'Stromper' is a registered trade mark applied only to our brand of that type of tool.

We hope that you will understand that frequent use of our registered trade mark as if it were the generic or common name for this type of product could lead to our loss of the right to the trade mark registration. We would therefore appreciate it if in future issues of your magazines care is taken to ensure proper use of our trade mark, making it clear that 'Stromper' is a trade mark and not the name of a product, for instance by using the phrase 'Stromper' garden tool.

Yours faithfully

For Bloggs and Co. Ltd

Appendix 4
Licensing negotiation checklists

(a) TECHNOLOGY EVALUATION CHECKLIST FOR LICENSEE / LICENSOR

(1) What is the technology?
(2) What is its status in the market place?
(3) Test reports on the technology if possible by independent expert bodies and on alternative solutions, performance specifications, information about raw materials samples, etc.
(4) Method of presenting the technology – technical slides show, film, video. Can such method be used on location?
(5) Details of the nature of the property which comprises the technology
  (i) Invention
  (ii) Design
  (iii) Information / know-how / show-how
  (iv) Documents and drawings
(6) How is the property protected?
  (i) Patents – application or registered and where?
  (ii) Designs – application or registered and where?
  (iii) Secrecy or non-disclosure agreement / black box?
  (iv) Copyright – scope and duration?
(7) Who owns each item of the intellectual property package which makes up the technology package? Do they own them as:
  (i) Original owner?
  (ii) Assignee?
  (iii) Licensee?
(8) What internal records exist in relation to the licensed technology?
(9) Is documentation containing all the necessary information available now to hand over to the licensee if the agreement is concluded? This documentation should normally comprise:
  (i) Shop drawings ready for copying
  (ii) Assembly drawings and list of items
  (iii) Installation and operation manuals
  (iv) Specification of raw materials

(v) Specification of the finished product
(vi) Calculation basis step by step: labour and machine time
(vii) Manufacturing instructions/process description
(viii) Instructions for use of the product
(ix) Safety instructions, advice numbers, etc.
(x) Packing and stock and transport instructions

(10) What arrangements will be necessary for training of personnel to operate the technology?

(b) COMMERCIAL EVALUATION CHECKLIST FOR LICENSEE/LICENSOR

(1) What is the commercial significance of the technology?
(2) What information can be provided relating to the return from the sale of the technology?
(3) Market information including gross figures
(4) Details of products for which the new products could be substitutes
(5) List of areas for which your technology can be used
(6) Potential life of the product and its intellectual property protection
(7) Marketing ideas – publicity, customers, etc.
(8) Flow of information during the life of the agreement and new marketing factors, ideas, common marketing concepts etc.
(9) A reasoned and reasonable expectation of the return expected on the licence; downpayments, royalties, minimum royalty clauses, territories
(10) What are the taxation consequences of the different elements in the financial package in item 9?
(11) Is the licence to be exclusive or non-exclusive and in what territories? Prepare separate calculation for each option so that they can be raised separately

(c) LEGAL EVALUATION CHECKLIST FOR LICENSEE/LICENSOR

(1) Full details and status of the parties
(i) Full name
(ii) Full legal status i.e. incorporated where etc.
(iii) Full corporate address

(2) Nature of the agreement and its legal effect
(3) Full details of all the technology to be licensed: work through the details in Appendix 4(a), especially items 5–7.
(4) Full details of the commercial return expected. How far can performance criteria be written into the contract as obligations and undertakings or warranties?
(5) Patent aspects

(i) Is the technology patentable?

(ii) Are there any third party rights which might hinder the free use of the technology: check results of patent search in the licensed territories

(iii) Should a licence be taken from a third party to avoid infringement or can the process be modified to avoid this?

(6) Trade mark aspects
- (i) Is the product sold under a trade mark?
- (ii) If not can a trade mark be used for the product?
- (iii) Is the trade mark suitable?
- (iv) What is the status of the trade mark in the market place: check results of a trade mark search in the licensed territories

(7) Other rights
- (i) Design registration
- (ii) Copyright

Check items 5, 6 again for these rights

(8) Secrecy and legal conditions
- (i) Who will have access to the secret information?
- (ii) Is the technology protected against disclosure by employees and others by legal and or physical means?

(9) Legal restrictions, i.e. government consents to technology transfer, customs, fiscal obstacles, currency exchange, etc.

(10) Have available a suitable secrecy or non-disclosure agreement for submission to interested party at the appropriate stage.

(11) Have available a suitable draft agreement or draft heads of agreement for submission to interested party at the appropriate stage

(d) ADMINISTRATION CHECKLIST FOR LICENSEE/LICENSOR

(1) Details of the other party
- (i) Address, telephone number, telex etc. of head office
- (ii) Same for subsidiary offices and operation plants
- (iii) Names of contacts and positions
- (iv) Other parties involved, controlling parties, controlled parties, location and background
- (v) Details of directors and authority
- (vi) Details of staff and work force

(2) Financial information about other party
- (i) Accounts for say three years
- (ii) Banker's reference
- (iii) Business references or reports e.g. brokers, market service reports, etc.

(3) General reputation
- (i) With suppliers
- (ii) With customers
- (iii) With employees

(4) Reputation of products for
- (i) Quality of design
- (ii) Price
- (iii) Availability

(5) Local factors affecting the parties
- (i) Regional grants
- (ii) Government restrictions
- (iii) Language

Index

Index

Index